The Patient Factor

Volume I

The Patient Factor

Theories and Methods for Patient Ergonomics

Volume I

Edited by
Richard J. Holden and Rupa S. Valdez

CRC Press
Taylor & Francis Group
Boca Raton London New York

CRC Press is an imprint of the
Taylor & Francis Group, an **Informa** business

Cover illustration by Michael Morgenstern

First edition published 2021
by CRC Press
6000 Broken Sound Parkway NW, Suite 300, Boca Raton, FL 33487-2742

and by CRC Press
2 Park Square, Milton Park, Abingdon, Oxon, OX14 4RN

CRC Press is an imprint of Taylor & Francis Group, LLC

ISBN: 978-0-367-24558-0 (hbk)
ISBN: 978-0-367-72095-7 (pbk)
ISBN: 978-0-429-29299-6 (ebk)

Typeset in Times
by codeMantra

We created this book for the patients and their families
We lovingly dedicate it to our families, for their patience

To Carly and Roman

To my first circle:

"people to whom [I] feel so close, it is hard to imagine life without them."[1]

[1] Antonucci, T. C., & Akiyama, H. (1987). Social networks in adult life and a preliminary examination of the convoy model. *Journal of Gerontology*, 42(5), 519–527.

Contents

SECTION V Conclusion

Foreword

Gloria, a pseudonym for a real patient, was a 72-year-old grandmother admitted to a hospital for a flareup of her chronic obstructive pulmonary disease (COPD). Over the years, she learned how to control her symptoms to prevent flareups, but this time she just could not control them with her inhalers at home. After four days in the hospital, she was discharged home with antibiotics and tapering corticosteroids. She had to incorporate the new medicines in addition to those prescribed for her other health issues, which included insomnia and heart conditions. As she settled back home, she had more questions about the new medicines, such as whether she could take antibiotics with other medicines. She wanted to succeed and feared that if she made mistakes, she may have to go to the hospital again and interrupt the lives of her whole family.

Gloria's daughter, Tracey, participated in discussions with the doctors and nurses at the hospital and learned of the additional goal of reducing medications for insomnia at the hospital and at home upon discharge. After discharge, Tracey visited her mother a couple of times per day, helping Gloria to stay organized with pillboxes, medicine schedules, doctors' appointments, hospital paperwork, test results, and prescription inserts. Tracey used a large file folder provided by a primary care doctor to organize the drug brochures and visit documents from hospitals and doctors' offices. She also started logging her mother's medicines in a notebook, a strategy she learned from one of the home health nurses. Tracey always used the same pharmacy in the community and knew the pharmacists there. On the first day after discharge, Tracey woke her mother early to take the morning medications and realized that her mother was sleepy most of the day from a new allergy medicine. Tracey stopped all sleeping aids until her mother's next doctor's appointment a few days later.

Around the world, variations of Gloria's activities at home play out daily and contribute to health outcomes. People living with diverse chronic conditions spend more time on self-management than with their healthcare professionals. The increasing burden and costs of chronic diseases, coupled with incentive-driven innovative care models, have accelerated a shift away from expensive specialized and episodic care toward self-care that is facilitated by information technology, social support, and clinical expertise from healthcare professionals (Nelson et al., 2014). This evolving paradigm carries with it risks and uncertainties, but will likely prevail if society continues to recognize healthcare services as a coproduction (Nelson et al., 2014) and questions the traditional view of healthcare services as something given to patients as passive recipients of care. Patient roles must be recognized beyond passively complying with professionals' orders and instead should range from assessing and accepting risks versus benefits from therapies, to seeking information on their health goals, to devising systems at home, to performing health-related tasks successfully.

The health-related work performed by non-professionals is relatively understudied compared to work performed by professionals. Our vocabularies about health-related activities of patients are often limited by simplistic terms such as patient errors, support systems, adherence, and health literacy. For example, one model of medication self-management at home delineates tasks in medication use as fill, understand, organize, take, monitor, and sustain (Bailey et al., 2013). This model makes no references to tools used or constraints expected at home to accomplish medication management tasks. Another example involving medication instructions (Davis et al., 2006) demonstrated that although low literacy limits a patient's abilities to understand hospital communications and to read medication instructions, how well a patient manages medications depends on multiple barriers and facilitators, including tools, social support, utilization of community resources, and user-friendly discharge summaries and medication labels (Murray et al., 2004). The amount of medications tends to increase as one ages and the nature of medication management tasks changes, usually with increased safety-related activities, such as managing drug–drug interactions and inconvenient medication schedules (Kongkaew et al., 2013). At home, the support "infrastructure" is entirely different from that in a hospital setting. One is not expected to have a safety-net system to check on drug–drug interactions, control inventory, activate electronic information systems to log and remind medication taking, or manage accounting systems.

Innovative concepts and tools are needed to support patient activities to achieve health outcomes. A good starting point is to examine patient health-related activities the way other types of professional care activities have been studied. Patient ergonomics, as envisioned by the authors in this and the following volume, starts to outline the challenges, opportunities, and approaches to develop patient-centered tools that can potentially redefine the roles of the healthcare industry to support individuals' pursuit of health goals. Human factors engineering provides a wealth of tools to understand and optimally support patient activities. We present two broad goals of patient work research that may serve to improve health, either through self-care or through working with healthcare professionals.

UNDERSTANDING AND SUPPORTING PATIENT WORK

As in the case of ergonomic studies of professional work, we may look at various aspects of patient and family caregiver activities in different settings, including home and acute care hospitals. As illustrated in our opening vignette, patients engage in cognitive activities such as decision-making, planning, and information seeking. They engage in teamwork in a broad sense, by working with caregivers and health professionals, and are subject to variables of teamwork similar to those in professional work such as trust and communication barriers. The physical demands of patient work may overwhelm the patient, and as he or she ages, diminishing physical functions will lead to the inability to open blister packs and cut small pills. Digital and non-digital tools are often designed to support patient work and require careful examination of interactions between users and tools. The chapters in this volume provide human factors engineering methods and conceptual tools to better understand the socio-technical environment in which patient work is performed, along with physical and cognitive aspects of patient work. As an example, the chapter on the application of cognitive theories to patient work by Morrow reviews the similarities and differences between lessons learned in professional domains and some of the early work extending these concepts to non-professional environments (see Chapter 2). Human factors engineering approaches to work systems design for professional work can help us understand patient ergonomics. Methods such as usability assessment and participatory design can help designers to truly engage end users of tools.

The amount and importance of health-related activities performed by patients and their caregivers are reflected in measures of the burden of care or treatment to patients and caregivers. We use existing measures of patient burden to highlight the value of the human factors engineering approach in contrast with the overall assessment of burden and patient work (Shih et al., 2013). Treatment burden or illness work refers to the workload imposed by health-related tasks on patients (Tran et al., 2014). For example, for medication-related tasks, treatment burden may be measured by the amount of medications taken, efforts involved in obtaining and taking medications, required monitoring, frequency of clinical appointments, and the amount of administrative burden involved. One systematic review published in 2014 found that a chronic condition may impose two hours of work per day on patients and their caregivers (Jowsey et al., 2012). Other studies, such as one by Presley et al. (2017), found indirect indications of substantial workload placed on the patients. These indications include the number of days of encounters with healthcare professionals, the number of physicians interacted with, and the number of prescribed medicines received.

Each of these indications reflects patient work associated with managing logistics with healthcare professionals, documents, insurance, and information seeking. In the example of non-small-cell lung cancer patients, Presley et al. (2017) found that patients spent one in three days interacting with healthcare systems during the initial sixty days of treatment and called for future research to understand and ameliorate patient burden in health care.

Gallacher et al. (2011) proposed a framework for treatment burden to outline important aspects of patient work, including work involved in developing and understanding treatment and illness, in working with professionals, in attending clinical encounters, in managing medications, in making changes to lifestyles, and in appraising treatment. This framework identified workload impact due to multiple medications, appointments with multiple healthcare professionals, financial and logistical barriers to access services, and areas for bridging fragmented health care and inadequate communication between professionals. Boyd et al. (2014) proposed a similar framework, through a patient self-assessment instrument, to score difficulties in accomplishing self-care tasks including obtaining medications, planning medication schedules, administering medications, deciding to change medications, managing medical bills, scheduling appointments, arranging transportation, and getting information. This framework captured tasks performed by older adults living with multiple chronic conditions, many of whom spend enormous mental and physical efforts accomplishing medication-related tasks. Neither framework includes the burden on caregivers.

Frameworks such as these illustrate the need to fully understand patient work and to understand from the patient's perspective their work in actively participating in their own health care. Human factors engineering concepts and tools, such as those described in this volume, are designed to study patient work beyond overall burden assessment: what constitutes patient work; how such work is performed; what support tools may be useful; how best the patient and family members may be educated and trained; and so on. In other words, we need to study patient ergonomics in a similar fashion as we study the ergonomics of professional workers.

Many concepts used to study professional work may apply to patient work, especially by those patients and family members managing multiple chronic conditions and those patients whose very illnesses limit their abilities to perform health-related activities. Additionally, patient work is performed in the context of physical and social environments, as illustrated by the frameworks proposed by Holden et al. (2013) and Carayon et al. (2020). Inter-personal interactions within the home environment and with healthcare professionals may be defined by concepts often used in human factors studies of professional work such as common ground, mutual trust, relationships, and psychological safety. Patient strategies may also be defined as part of the expertise in addressing complexity in patient work and in achieving resilient performance with expected and unexpected disturbances and setbacks, such as recurrence of cancer and gaps in access to medications (Lippa et al., 2008). For example, one application of human factors methods to understand patient work at home is the study by Keller et al. (2018), which looked at the physical environment, tools, required tasks, and patient strategies in maintaining antimicrobial therapy at home through an infusion port.

One important principle in human factors is to fit the task to the person who is to perform the task. This may be achieved through providing support tools and modifying the tasks in consideration of the human limitations and abilities expected. One example of such supporting strategies is to make medication instructions easier to follow (Klein & Isaacson, 2003; Morrow et al., 2005). Instead of framing issues of not understanding instructions as that of poor health literacy, human factors approaches attempt to support the person performing the task by redesigning other aspects of their work system.

PATIENT WORK SYSTEM INFORMED HEALTHCARE DECISIONS

Understanding patient work is not only important to improve systems to support patient tasks at home, but it is also important to healthcare professionals when they make therapeutic decisions that may impose requirements beyond patient and family capabilities. For example, using insulin requires the patient to closely monitor and control his or her caloric intake in addition to using correct insulin products. This requirement is challenging for some older adult patients. Consequently, decisions on how to manage diabetes should be made not only on clinical risk–benefit analysis collaboratively with the patient, but also on a joint assessment of the patient's abilities at home to use insulin products safely. In one analysis of emergency department visits due to insulin-related hypoglycemia (Geller et al., 2014), patients eighty years or older had the highest risks of harms from insulin therapies due to the inability to coordinate food intake with insulin regimens and to ensure correct dosing and products.

There have been efforts to incorporate patient burdens in treatment decisions, such as through shared decision-making, with the goal of alleviating burden of treatment and tailoring treatment regimens to the realities of people's daily lives (Sav et al., 2015). Healthcare professionals may be encouraged, or—even better—processes may be designed to hardwire patient preferences and goals in decision processes through collaborative discussions (Tinetti et al., 2019). Advances in data science and artificial intelligence may also reduce patient burdens in care coordination (Randhawa et al., 2019).

Similarly, tools to assess patient capacity in self-care are needed to see if self-care requirements exceed the abilities of the patient's work system at home (Irvine-Meek & Gould, 2011). When a patient is discharged from the hospital to home, the patient's role changes from a mostly passive recipient of care (with meals and medications provided) to mostly self-management (which includes reconciling new medications with those taken before going to the hospital). A well-structured assessment of the patient work system at home, based on human factors methodologies, can inform healthcare professionals so that they can better partner with the patient to bridge the transition period safely (Xiao et al., 2019).

One important distinction between professional work and patient work is that patients sometimes have conditions that limit their ability to perform required tasks, which include daily living dependency, emotional distress, social isolation, and symptom burdens such as pain, fatigue, breathing difficulty, sleeping difficulty, depressed mood, and anxiety (Patel et al., 2019).

Consequently, health-related tasks performed by patients need to be assessed against expected performance limiting factors. Healthcare professionals may use their understanding of patient work to better tailor treatment regimens according to family circumstances, comorbidity, high use of medications, characteristics of treatment, and trust in healthcare providers (Sav et al., 2015). Living in poverty itself limits the patient's ability in self-care and should require additional efforts to simplify patient health-related tasks (Nwadiuko & Sander, 2017).

BACK TO GLORIA

Gloria's daughter performed a large number of health-related activities that often went unrecognized or underappreciated. For example, she used the hospitalization as an occasion to learn more about managing her mother's conditions. She learned medication management practices from home nurses and utilized tools to manage a large amount of paper documents collected from encounters in different settings. Gloria's work system also included the neighborhood pharmacists, as she recognized the pharmacists as important resources that were easier to access than hospital- or clinic-based clinicians.

The chapters in this book contribute to our understanding of such patient work, ways to support patient work, and necessary methodologies. However, this should be considered as only initial steps because it is difficult to overstate the challenges in developing new scientific tools that place the patient at the center, understanding patient needs, designing for patients, and supporting patients to achieve health-related goals. Each patient's work system has a set of unique characteristics, ranging from managing chemotherapy, to post-surgery recovery, to managing chronic health conditions. Patient ergonomics, as envisioned by the authors in this volume, starts to outline the challenges as well as present opportunities to develop patient-centered tools that can potentially redefine the roles of the healthcare industry to better support individuals' pursuits of health-related goals. It will take creativity and true passion to study patient work performed in an environment organically emerging and consistently changing.

Yan Xiao, PhD
University of Texas at Arlington, Arlington, Texas
Richard Young, MD
John Peter Smith Health Network, Fort Worth, Texas

REFERENCES

Bailey, S. C., Oramasionwu, C. U., & Wolf, M. S. (2013). Rethinking adherence: A health literacy-informed model of medication self-management. *Journal of Health Communication, 18*(Suppl 1), 20–30.

Boyd, C. M., Wolff, J. L., Giovannetti, E., Reider, L., Weiss, C., Xue, Q. L., . . . Rand, C. (2014). Healthcare task difficulty among older adults with multimorbidity. *Medical Care,* 52(Suppl 3), S118–125.

Carayon, P., Wooldridge, A., Hoonakker, P., Hundt, A. S., & Kelly, M. M. (2020). SEIPS 3.0: Human-centered design of the patient journey for patient safety. *Applied Ergonomics, 84*, 103033.

Davis, T. C., Wolf, M. S., Bass, P. F., 3rd, Middlebrooks, M., Kennen, E., Baker, D. W., . . . Parker, R. M. (2006). Low literacy impairs comprehension of prescription drug warning labels. *Journal of General Internal Medicine, 21*(8), 847–851.

Gallacher, K., May, C. R., Montori, V. M., & Mair, F. S. (2011). Understanding patients' experiences of treatment burden in chronic heart failure using normalization process theory. *Annals of Family Medicine, 9*(3), 235–243.

Geller, A. I., Shehab, N., Lovegrove, M. C., Kegler, S. R., Weidenbach, K. N., Ryan, G. J., & Budnitz, D. S. (2014). National estimates of insulin-related hypoglycemia and errors leading to emergency department visits and hospitalizations. *Journal of the American Medical Association Internal Medicine, 174*(5), 678–686.

Holden, R. J., Carayon, P., Gurses, A. P., Hoonakker, P., Hundt, A. S., Ozok, A. A., & Rivera-Rodriguez, A. J. (2013). SEIPS 2.0: A human factors framework for studying and improving the work of healthcare professionals and patients. *Ergonomics, 56*(11), 1669–1686.

Irvine-Meek, J. M., & Gould, O. N. (2011). Psychometric evaluation of a self-medication assessment tool in an elderly population. *The Canadian Journal of Hospital Pharmacy, 64*(1), 16–24.

Jowsey, T., Yen, L., & Mathews W, P. (2012). Time spent on health related activities associated with chronic illness: A scoping literature review. *BioMed Central Public Health, 12*, 1044.

Keller, S. C., Cosgrove, S. E., Kohut, M., Krosche, A., Chang, H. E., Williams, D., & Gurses, A. P. (2018). Hazards from physical attributes of the home environment among patients on outpatient parenteral antimicrobial therapy. *American Journal of Infection Control, 47*(4), 425–430.

Klein, H. A., & Isaacson, J. J. (2003). Making medication instructions usable. *Ergonomics in Design, 11*, 7–11.

Kongkaew, C., Hann, M., Mandal, J., Williams, S. D., Metcalfe, D., Noyce, P. R., & Ashcroft, D. M. (2013). Risk factors for hospital admissions associated with adverse drug events. *Pharmacotherapy, 33*(8), 827–837.

Lippa, K. D., Klein, H. A., & Shalin, V. L. (2008). Everyday expertise: Cognitive demands in diabetes self-management. *Human Factors, 50*(1), 112–120.

Morrow, D. G., Weiner, M., Young, J., Steinley, D., Deer, M., & Murray, M. D. (2005). Improving medication knowledge among older adults with heart failure: A patient-centered approach to instruction design. *Gerontologist, 45*(4), 545–552.

Murray, M. D., Morrow, D. G., Weiner, M., Clark, D. O., Tu, W., Deer, M. M., . . . Weinberger, M. (2004). A conceptual framework to study medication adherence in older adults. *The American Journal of Geriatric Pharmacotherapy, 2*(1), 36–43.

Nelson, E. C., Meyer, G., & Bohmer, R. (2014). Self-care: The new principal care. *The Journal of Ambulatory Care Management, 37*(3), 219–225.

Nwadiuko, J., & Sander, L. D. (2017). Simplifying care: When is the treatment burden too much for patients living in poverty? *BMJ Quality & Safety, 27*(6), 484–488.

Patel, K. V., Guralnik, J. M., Phelan, E. A., Gell, N. M., Wallace, R. B., Sullivan, M. D., & Turk, D. C. (2019). Symptom burden among community-dwelling older adults in the United States. *Journal of the American Geriatric Society, 67*(2), 223–231.

Presley, C. J., Soulos, P. R., Tinetti, M., Montori, V. M., Yu, J. B., & Gross, C. P. (2017). Treatment burden of Medicare beneficiaries with stage I non-small-cell lung cancer. *Journal of Oncology Practice, 13*(2), e98-e107.

Randhawa, G. S., Xiao, Y., & Gorman, P. N. (2019). Designing a "thinking system" to reduce the human burden of care delivery. *Generating Evidence & Methods to Improve Patient Outcomes (Wash DC), 7*(1), 18.

Sav, A., King, M. A., Whitty, J. A., Kendall, E., McMillan, S. S., Kelly, F., . . . Wheeler, A. J. (2015). Burden of treatment for chronic illness: A concept analysis and review of the literature. *Health Expectations, 18*(3), 312–324.

Shih, Y. C., Ganz, P. A., Aberle, D., Abernethy, A., Bekelman, J., Brawley, O., . . . Schnipper, L. (2013). Delivering high-quality and affordable care throughout the cancer care continuum. *Journal of Clinical Oncology, 31*(32), 4151–4157.

Tinetti, M. E., Naik, A. D., Dindo, L., Costello, D. M., Esterson, J., Geda, M., . . . Blaum, C. (2019). Association of patient priorities-aligned decision-making with patient outcomes and ambulatory health care burden among older adults with multiple chronic conditions: A nonrandomized clinical trial. *JAMA Internal Medicine, 179*(12), 1688–1697.

Tran, V. T., Harrington, M., Montori, V. M., Barnes, C., Wicks, P., & Ravaud, P. (2014). Adaptation and validation of the Treatment Burden Questionnaire (TBQ) in English using an internet platform. *BioMed Central Medicine, 12*, 109.

Xiao, Y., Abebe, E., & Gurses, A. P. (2019). Engineering a foundation for partnership to improve medication safety during care transitions. *Journal of Patient Safety and Risk Management, 24*(1), 30–36.

Preface

Better to hunt in fields for health unbought
Than fee the doctor for a nauseous draught.
The wise for cure on exercise depend;
God never made his work for man to mend.

-John Dryden, 1700, "Epistle 15, To my Honour'd Kinsman, John Driden, of Chesterton"

Just don't let the human factor fail to be a factor at all.

-Andrew Bird, 2005, "Tables and Chairs"

Health is like a democracy, largely governed by the people it affects. The work of health spans decades for most. Much of that work is done on an everyday basis by the patient, family member, or other non-professional. This is called *patient work* and it often occurs in homes and communities, although patients and other non-professionals contribute to care in professional care settings, as well. The work of health is also performed in the form of clinical care and treatment carried out by healthcare professionals. This is called *professional work* and it occurs in hospitals, primary care clinics, and similar settings, although increasingly in home and community settings. Rarely is it one or the other. There is no merit to pitting patient and professional forms of health work against one another, as a resentful Dryden did in the epistle to his cousin John. We would even say the two forms of work are co-dependent and equally necessary: "the patient" drinks the "doctor's draught"; clinicians consult with us about our "exercise" and what food from the "fields" we put in our bodies.

However, our research, funding, policymaking, and attention are disproportionately allocated to formal healthcare delivery mechanisms. While irreplaceable, professional care is estimated to contribute only 10% or less to a person's total health. Therefore, it is time to expand our vision of health care beyond stethoscopes, lab coats, and X-ray machines. This book is our attempt to bring more research, funding, policymaking, and attention to patient work, the effortful health-related activities patients, families, and other non-professionals perform for themselves and one another. We strive to augment the mental images of health and health care by adding "the patient" and those around them.

As researchers and engineers (and other things—including patients, caregivers, advocates, and leaders of non-profit organizations), our ultimate goal is to study and improve patient work. The discipline of human factors can aid in this goal. Readers of this handbook will learn how, from chapters contributed by members of the *patient ergonomics* community who have pioneered the application of human factors to study and improve patient work.

We thank these contributors and other members of the patient ergonomics community. You are our friends, our collaborators, and often our inspiration. Thank you to our mentors and mentees, for helping us be who we are and where we are. We appreciate the support of our professional societies, particularly the Human Factors and Ergonomics Society (HFES) and American Medical Informatics Association (AMIA), who have allowed us to convene sessions, workshops, and town halls on patient ergonomics since 2014. Lasting gratitude is owed, with interest for the occasional late payment, to our families and friends as well.

As we write this Preface, something is happening around us that is unprecedented in current times. The COVID-19 pandemic is changing the world and our daily lives. We are sheltering in place for the sake of our health and the public's. We are as grateful as ever for the heroic healthcare professionals delivering essential care, laboring at great risk to themselves. At the same time, we acknowledge that billions of people must take their physical and mental health into their own hands. As patients we are thinking, almost uncontrollably, about hand washing, getting exercise, keeping our distance, keeping our minds occupied, taking our medicine, taking our temperature, watching for flu-like symptoms, watching what we eat, calming our children, calming ourselves… More than ever, we recognize the vast amount of health work we do for ourselves and for others. This current reality highlights the essential need for patient ergonomics to help understand and advance how we do this work.

We will therefore end with this message to you, whether you are a human factors expert considering applying your skills to an area of patient ergonomics, a health sciences professional seeking to learn from a different discipline, a student, a patient, a family member, or Hollywood moviemaker: *just don't let the patient factor fail to be a factor at all.*

Rupa S. Valdez,
Charlottesville, Virginia
Richard J. Holden,
Indianapolis, Indiana

Editors

Richard J. Holden, PhD, is an Associate Professor of Medicine at the Indiana University (IU) School of Medicine and the Chief Healthcare Engineer at the IU Center for Health Innovation and Implementation Science. He earned a joint PhD in Industrial Engineering and Psychology from the University of Wisconsin. He founded and directs the Health Innovation Lab and co-directs the Brain Safety Lab. Dr. Holden's research applies human-centered design and evaluation methods to improve health outcomes, especially for older adults. He specializes in research on technology for patients with chronic diseases, such as dementia and heart failure, and their family caregivers. He is a scientist in the Regenstrief Institute and in 2020 he received the 2019 Outstanding Investigator Award and the Regenstrief Institute Venture Fellowship. Dr. Holden has led or played key roles in more than 20 federally funded research and demonstration projects, totaling in excess of $75 million. He has authored more than 150 peer-reviewed works in the fields of human factors engineering, patient safety and quality, health informatics, and research methods. He is most proud of being an innovator, mentor, and connector of dots.

Rupa S. Valdez, PhD, is an Associate Professor at the University of Virginia jointly appointed in the Schools of Medicine and Engineering and Applied Sciences. She is also affiliated with Global Studies and the Disability Studies Initiative. Dr. Valdez merges human factors engineering, health informatics, and cultural anthropology to understand and support the ways in which people manage health at home and in the community. Her research and teaching focus on underserved populations, including populations that are racial/ethnic minorities, of low socioeconomic status, and/or living with disabilities. Her work draws heavily on community engagement and has been supported by the National Institutes of Health (NIH), Agency for Healthcare Research and Quality (AHRQ), and the National Science Foundation (NSF), among others. She serves as Division Chair of Internal Affairs for the Human Factors and Ergonomics Society (HFES) and as Associate Editor for *Journal of American Medical Informatics Association (JAMIA) Open*. She is the founder and president of Blue Trunk Foundation, a non-profit organization dedicated to making it easier for people with chronic health conditions, disabilities, and age-related conditions to travel. Dr. Valdez lives with multiple chronic health conditions and disabilities, which have and continue to influence her work and advocacy.

Contributors

Armagan Albayrak
Faculty of Industrial Design Engineering
Delft University of Technology
Delft, Netherlands

Onur Asan
School of Systems and Enterprises
Stevens Institute of Technology
Hoboken, New Jersey

Malaz A. Boustani
Center for Health Innovation and Implementation Science
Indiana University School of Medicine
Indianapolis, Indiana

Barrett S. Caldwell
School of Industrial Engineering
Purdue University
West Lafayette, Indiana

Pascale Carayon
Department of Industrial and Systems Engineering
University of Wisconsin-Madison
Madison, Wisconsin

Lora Cavuoto
Department of Industrial and Systems Engineering
University at Buffalo, The State University of New York
Buffalo, New York

Yong K. Choi
School of Medicine
University of California
DavisUniversity of Washington
Sacramento, California

Avishek Choudhury
School of Systems and Enterprises
Stevens Institute of Technology
Hoboken, New Jersey

Bradley H. Crotty
Department of Medicine
Medical College of Wisconsin
Wauwatosa, Wisconsin

Colleen Ewart
NIHR Applied Research Collaborations East Midlands
Leicester, United Kingdom

Dominic Furniss
Human Reliability Associates Ltd.
Lancashire, United Kingdom

Richard Goossens
Faculty of Industrial Design Engineering
Delft University of Technology
Delft, Netherlands

Joel S. Greenstein
Department of Industrial Engineering
Clemson University
Clemson, South Carolina

Siobhan M. Heiden
Puyallup, WA

Michelle Jahn Holbrook
Bold Insight
Chicago, Illinois

Richard J. Holden
Department of Medicine
Indiana University School of Medicine
Indianapolis, Indiana

Peter Hoonakker
Wisconsin Institute for Healthcare Systems Engineering
University of Wisconsin-Madison
Madison, Wisconsin

Bat-Zion Hose
Department of Industrial and Systems Engineering
University of Wisconsin-Madison
Madison, Wisconsin

Michelle M. Kelly
Department of Pediatrics
University of Wisconsin School of Medicine and Public Health
Madison, Wisconsin

Rohit Ashok Khot
School of Design
Royal Melbourne Institute of Technology
Melbourne, Australia

Barbara J. King
School of Nursing
University of Wisconsin-Madison
Madison, Wisconsin

Alexandra R. Lang
Trent Simulation and Clinical Skills Centre
Nottingham University Hospitals NHS Trust & Human Factors Research Group
University of Nottingham
Nottingham, United Kingdom

Annie Y. S. Lau
Centre for Health Informatics
Australian Institute of Health Innovation, Macquarie University
Sydney, Australia

Kapil Chalil Madathil
Departments of Civil and Industrial Engineering
Clemson University
Clemson, South Carolina

Marijke Melles
Faculty of Industrial Design Engineering
Delft University of Technology
Delft, Netherlands

Dan Morrow
Department of Educational Psychology
University of Illinois at Urbana-Champaign
Champaign, Illinois

Laurie Lovett Novak
Department of Biomedical Informatics
Vanderbilt University Medical Center
Nashville, Tennessee

Mustafa Ozkaynak
College of Nursing
University of Colorado Anschutz Medical Campus
Aurora, Colorado

Elizabeth Lerner Papautsky
Department of Biomedical & Health Information Sciences
University of Illinois at Chicago
Chicago, Illinois

Megan E. Salwei
Department of Biomedical Informatics
Vanderbilt University Medical Center
Nashville, Tennessee

Linsey M. Steege
School of Nursing
University of Wisconsin-Madison
Madison, Wisconsin

Rupa S. Valdez
Public Health Sciences & Engineering Systems and Environment
University of Virginia
Charlottesville, Virginia

P. Jon White
Veterans Affairs Salt Lake City Health Care System
Salt Lake City, Utah

Yan Xiao
College of Nursing and Health Innovations
University of Texas at Arlington
Arlington, Texas

Kathleen Yin
Centre for Health Informatics
Australian Institute of Health Innovation
Macquarie University
Sydney, Australia

Richard Young
John Peter Smith Health Network
Fort Worth, Texas

Teresa Zayas-Cabán
Office of the National Coordinator for Health Information Technology
Washington, District of Columbia

Section I

Introduction to Patient Ergonomics

1 Patient Ergonomics
The Science (and Engineering) of Patient Work

Richard J. Holden
Indiana University School of Medicine

Rupa S. Valdez
Department of Public Health Sciences
Department of Engineering Systems and Environment
University of Virginia

CONTENTS

Patient ergonomics is the science (and engineering) of patient work. More formally, patient ergonomics is the application of human factors and ergonomics (HFE) or related disciplines (e.g., human–computer interaction, usability engineering) to study or improve patients' and other nonprofessionals' performance of effortful work activities in pursuit of health goals (Holden & Valdez, 2018). These work activities can be performed independently of or in concert with healthcare professionals. Embedded in the definition are three core assumptions of patient ergonomics:

1. Individuals who have no professional health-related training—including, for example, patients, families, and community members—nevertheless perform goal-driven, effortful, health-related, and consequential activities called *patient work*;

2. The theories and methods of HFE are applicable and useful for studying and improving patient work; and
3. Studying and improving patient work requires adapting existing and developing new HFE approaches to suit the specific characteristics of patient work and the various contexts in which it is performed.

1.1 DO PATIENTS DO WORK?

Instead of offering a simplistic answer (which would be "yes"), let us examine a less controversial premise that it takes work to manage a person's health. Work is involved in preventing disease, treating health conditions, and keeping the conditions from becoming worse or causing undesirable outcomes, such as life disruption, morbidity, and mortality. Managing health is justifiably "work" because it is: (1) effortful, (2) goal-driven, and (3) consequential, meaning it results in something important. This conceptualization of work coincides with Hal Hendrick's (2002, p. 1) definition of work as "any form of human effort or activity, including recreation and leisure pursuits." This broad scope is predated by Polish philosopher Jastrzebowski (1857), who coined the term ergonomics and defined it as "encompassing all aspects of human activity, including labour, entertainment, reasoning, and dedication" (Karwowski, 2005, p. 437).

Among the people who perform health-related work are healthcare professionals, individuals who are trained to deliver health services to others and are typically formally employed and paid. Family, friends, and other members of a community (e.g., church, village, or online platform) constitute another group of people caring for others. These individuals (sometimes called "informal caregivers," "family caregivers," "support persons," or "carers") perform or assist with health-related work and may or may not be paid for it. Last, but not least, people perform health-related work for themselves. People who find recipes, shop for groceries, and prepare meals for themselves are doing health-related work. Such individuals have no formal designation in the health system because they are not yet care recipients or patients, but one could call them "health seekers," "health participants," members of the "public," or just "people." As a different example, a patient with a lacerated hand—an acute condition—performs temporary work when they seek and pay for urgent care, attempt to avoid post-treatment infection, and alter their routines (e.g., around dressing, typing, and exercising). A chronically ill patient whose second heart attack resulted in a diagnosis of chronic heart failure and wheelchair reliance does daily work to learn to live with their new disease and disability identities, monitor and respond to symptoms, interact with clinicians, and manage new medication activities (purchasing, preparing, administering, monitoring, and so on). In all three examples, we discern an *exertion of effort*, although the type, amount, difficulty, timing, distribution, and burden of the effort may vary. In each case, there are also *health-related goals* motivating the effort. Finally, in each example, what one does, how, and how well produce *outcomes of consequence*, ranging from quality of life to the need for subsequent costly or life-altering treatment or death. These outcomes can be proximal or distal, desirable or undesirable (Holden et al., 2013).

1.1.1 Defining Patient Work

The term patient work was popularized by sociologist Anselm Strauss, who defined it as "exertion of effort and investment of time on the part of patients or family members to produce or accomplish something" (Strauss, 1993, pp. 64–65). Strauss and colleagues studied patient work in hospitals and in home and community settings and identified several types or "lines" of work (Corbin & Strauss, 1985). For example, they distinguished between medical tasks comprising "illness work," personal or household activities comprising "everyday life work," and illness-related adjustments to identity and self-concept comprising "biographical work." These and subsequent researchers also wrote about the "articulation" or logistic work people—both patients and their informal caregivers (Timmermans & Freidin, 2007)—do to coordinate, plan, and facilitate the other forms of work, for example, paying bills, driving to appointments, obtaining supplies in the service of medical tasks, or cultivating a network of people to assist life and rehabilitation after an acute event (Corbin & Strauss, 1988).

1.1.2 Patient Work—Visible or Not?

An important finding in social science research on patient work is that all or part of patient work may be invisible: not recognized to occur, not acknowledged as work, or undervalued. Often, the concept of visibility and invisibility is framed in relation to healthcare professionals. Ancker et al. (2015) have described how some healthcare professionals are unaware of the time their patients spend on dealing with insurance companies, errors in their medical information, and conveying medical information between clinicians, a role one colleague calls "being the Human Thumbdrive" (personal communication, April Savoy). The thousands of recommended self-care activities a chronically ill person must perform annually (Steiner, 2012), often consuming several hours per day (Jowsey et al., 2012), suggest the inadequacy of terms such as adherence and compliance to convey the true effort of patient work. Moreover, Brennan's concept of "the care between the care" illustrates that what patients and families do between formal healthcare encounters is actually far greater in volume, frequency, and consequence than the encounters themselves (Brennan & Casper, 2015). To illustrate, in a person's roughly 5,840 annual waking hours, even monthly half-hour medical appointments account for only 0.1%. Indeed, the time and effort required to arrange for appointments, acquire and use transportation to attend them, and complete related paperwork might surpass the annual value-adding time spent with clinicians.

When patient work activities are unseen or underappreciated, patients may be denied formal assistance and expected to bear the burden alone (Ancker et al., 2015) or may not receive help in advance, resulting in "reactive and bursty" activity (Unruh & Pratt, 2008, p. 45). Clinicians and patients may also end up with discordant ideas about what a patient is able to do, needs, or wants (Gorman et al., 2018; Werner et al., 2019). When designers fail to see patient work, they fail to design systems to accommodate patients and other nonprofessionals' roles in their own health. A colleague once said patient work "never shows up in anyone's flowchart" (personal communication, Laurie L. Novak); one of the goals of this handbook is to make sure it does.

1.1.3 A New Image of Health-Related Work

Recently, we performed an informal search using the Google Images search engine in an exercise to explore how the public envisions health and health care. The results were surprising (Holden & Valdez, 2019b). The imagery of both "health" and "health care" was dominated by professionals wearing medical garb, typically a lab-coated physician with their arms crossed. Patients and other nonprofessionals did not show up often, although images produced by the "health" keyword search occasionally included slim, young people jogging or in yoga poses. The typical setting was a hospital room or hallway for "health care" and sunny, outdoor settings for "health." Both keywords produced a preponderance of medical iconography, such as the caduceus, red crosses, hearts, and electrocardiograph (EKG) waveforms. Amazingly, the primary image of "health care" was the stethoscope—an instrument for healthcare professionals—and the keyword search for "health" demonstrated *a surprising persistence of stethoscopes* (Holden & Valdez, 2019b, p. 62).

Although patients were largely absent in the above exercise, Strauss et al. 1982, p. 978) famously wrote,

> the classic picture of the patient—whether painted by a discerning Dutch realist or more recently described by Parson[s]...is of an acutely sick person, hence temporarily passive and acquiescent, being treated by an active physician and helped by equally vigorous caretakers.

The Dutch painting in question may be the one at the top of Figure 1.1, seventeenth-century Dutch painter Jan Steen's oil painting, *The Doctor's Visit* (c.1660). Characteristic of other works of art of the era (Dixon, 1995), Steen's painting depicts the physician as the central figure, calm and confident, and the patient as a passive, disengaged sufferer positioned off-center. (Incidentally, a Google Images search for "patient" produces an overabundance of passive, though content, individuals lying supine on a hospital bed or examination table, with a compassionate healthcare professional standing over them.) The tasks being performed in the painting—if one ignores the flirtatious exchange between the two central characters—are medical: the taking of the pulse by the physician and the collection of urine specimen (in the glass container) by the assistant. We have observed a notable connection between this image and the application of HFE in health care (Holden & Valdez, 2019a, p. 726):

> It serves as the cover art of Donchin and Gopher's [2014] superb edited volume, *Around the Patient Bed: Human Factors and Safety in Health Care*. The book's title implies the centrality of patient, but its content largely focuses on the clinical team surrounding the patient. Its 24 chapters cover important HFE applications to improving the "professional work" (Holden et al., 2013) of clinical teams, often in settings such as the adult intensive care unit (ICU), neonatal ICU, emergency department, or operating room, where the patient's passivity may be due to incapacity or sedation.

The work of healthcare professionals is incredibly important and deserving of the most attentive efforts of the HFE community. However, it is only part of the work

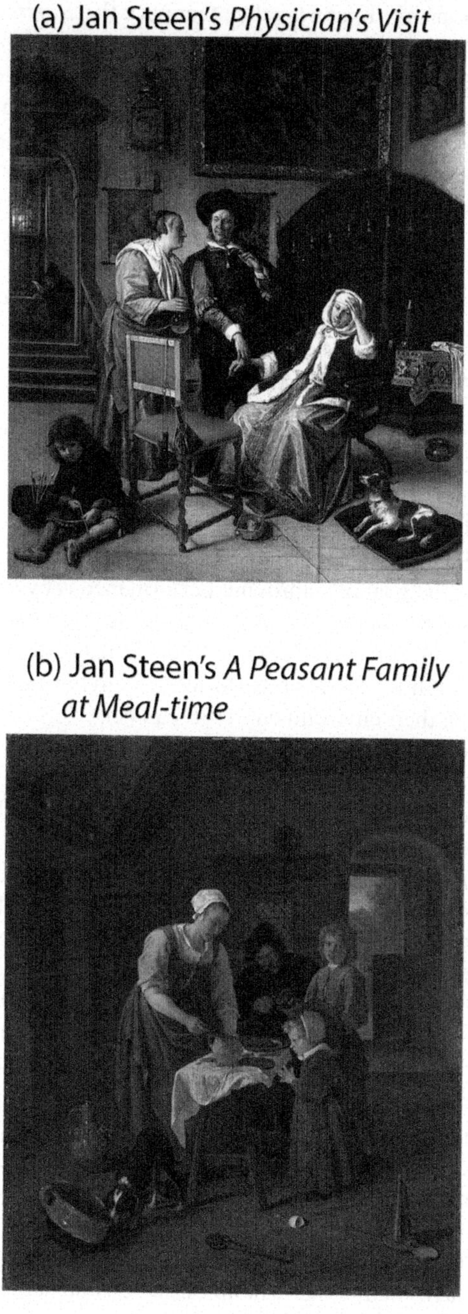

(a) Jan Steen's *Physician's Visit*
(b) Jan Steen's *A Peasant Family at Meal-time*

FIGURE 1.1 (a) Jan Steen's *The Doctor's Visit* (c.1660) showcases a physician caring for a passive, suffering patient. (b) Steen's *A Peasant Family at Meal-time ('Grace before Meat')* (c.1665) represents an alternate view of health as everyday activities performed by people in home and community settings. Images were obtained from Wikimedia Commons and are faithful photographic reproductions of two-dimensional, public domain works of art.

HFE can study and improve. To convey the other part, the *patient work*, we proposed the painting at the bottom of Figure 1.1, Steen's *A Peasant Family at Meal-time ('Grace before Meat')* (c.1665). The painting of a family gathered around a meal in their home, we argued, could serve as "an updated image of health and health care [depicting] people in everyday life activities that have a bearing on their health, from taking medications at home to mealtime and exercise" (Holden & Valdez, 2019a, pp. 725–726). Of course, both professional and nonprofessional work are important, and they sometimes occur together or in interrelated cycles (Ozkaynak et al., 2018). It is our contention that HFE approaches can be applied to both, and at their intersection, which we term "collaborative work" (Holden et al., 2013; Valdez et al., 2015).

1.2 CAN HUMAN FACTORS APPROACHES HELP?

The end-goal of HFE is "enhancing human performance" (Dempsey et al., 2000, p. 6). Initially, applications of HFE in health care interpreted this goal as supporting healthcare professionals (Donchin & Gopher, 2014; Karsh et al., 2006). Over time, it became evident that HFE can improve performance for a broader set of actors, including professionals, patients, and families (Holden et al., 2013). The way HFE professionals achieve the goal of improving performance is by studying and designing interactions between humans and other elements (e.g., technologies, tasks) in the context of sociotechnical systems (Carayon, 2006; Wilson, 2014). Thus, if a patient uses a technology, a family member performs caregiving tasks, or a community member interacts with their environment, HFE can help.

HFE, along with related sociotechnical disciplines—human–computer interaction, usability engineering, and applied psychology, to name a few—are remarkable for their many methods to study and improve work performance. Lists of these methods are usually at the scale of hundreds (Stanton et al., 2013), and there are categories of HFE methods such as workflow analysis (Carayon et al., 2012), cognitive task analysis (Cooke, 1994), and design (Kumar, 2012) that themselves have hundreds of variants. To illustrate the diversity of HFE methods, Figure 1.2 shows a word cloud analysis informally performed in 2014 on the tables of contents of several HFE methods textbooks. The figure makes it evident that many methods are available, with concentrations in areas such as tasks and processes, simulation, visualization, design and usability, teams and communication, cognition and decisions, errors and resilience, and workload.

These methods and their related theories and tools can be applied in three broad ways in patient ergonomics, corresponding to the three general phases of engineering: studying or analyzing the problem, designing and developing a solution, and testing solutions. Table 1.1 illustrates how a series of studies with older adults with chronic heart failure used HFE theories and methods across those three phases. As this handbook will show through many other examples, there is no shortage of problems for patient ergonomics to study and attempt to solve, including ones related to patient–professional communication (Chapter 6), self-care (Chapter 7), and patient safety (Chapter 8). There is also a clear need to design patient- and family-centered solutions and to test them for usability, acceptability, and other outcomes (Chapters 5 and 10).

FIGURE 1.2 The many methods in human factors and ergonomics that can be applied to study and improve patient work.

1.3 IS PATIENT ERGONOMICS "ITS OWN THING"?

Patient ergonomics is not entirely new, as we will show in the next section. Rather, it is a boundary drawn around a community of practice comprising scientists, practitioners, and others whose work fits the definition given above. Members of this community of practice may have diverse interests apart from patient ergonomics. For example, most contributors to this textbook have also applied HFE to study or improve clinical work, a considerable number have applied HFE outside health-related domains, and some have addressed patient work by applying disciplines outside of HFE (e.g., nursing, geriatrics, public health). Such diversity results in a welcome level of cross-fertilization and multidisciplinarity in patient ergonomics.

Applying the label patient ergonomics to a community of practice renders it a distinct area and arguably its own discipline or subdiscipline. Doing so promotes patient ergonomics and can attract outside scholars and practitioners. For those who apply HFE to study and improve healthcare delivery, patient ergonomics offers a way to broaden their focus and recognize the role of patient work in instances where professionals and nonprofessionals work together. For example, designers of electronic health records systems used by clinicians will progressively benefit from additionally considering tools such as patient portals, secure patient–clinician messaging, and patient-facing apps, as these technologies become integrated to facilitate patient–clinician collaboration (Valdez et al., 2015). As another example, scholars who study clinical decision-making cannot avoid the movement toward shared decision-making and consumer decision aids (Stacey et al., 2017; Stiggelbout et al., 2012).

TABLE 1.1
Examples from Research with Older Adults with Chronic Heart Failure of the Application of Human Factors and Ergonomics Theories and Methods across Three Broad Phases of Engineering

Example	Phase	Theory or Framework	Method
Study of work systems barriers to and facilitators for self-care (Holden, Schubert, Eiland, et al., 2015; Holden, Schubert, & Mickelson, 2015)	Study	Sociotechnical systems	Work system analysis
Study of cognitive artifacts that patients, caregivers, and clinicians use for medication management (Mickelson et al., 2015)	Study	Distributed cognition	Artifacts analysis
Study of the macrocognitive workflow of medication management (Mickelson et al., 2016)	Study	Macrocognition	Workflow analysis
Study of intentional and unintentional medication nonadherence events (Mickelson & Holden, 2018a)	Study	Human error taxonomy/resilience engineering	Incident analysis
Study of everyday health-related decision-making (Holden, Daley, et al., 2020; Mickelson & Holden, 2018b)	Study	Naturalistic decision-making	Cognitive task analysis
Creation of personas representing discernible user groups (Holden, Daley, et al., 2020; Holden et al., 2017)	Design	Interaction design	Personas
Iterative prototyping of mobile apps to support patient self-care management (Cornet et al., 2020; Srinivas et al., 2017)	Design	User-centered design	Rapid prototyping
Participatory design sessions with patients to develop novel displays of health information (Ahmed et al., 2019)	Design	Participatory design	Participatory design
Laboratory-based testing of prototypes of apps for patients, assessing usability, acceptability, and mental workload (Cornet et al., 2017, 2019)	Evaluation	Usability engineering	Task-based usability testing
Expert inspection of apps developed for patients (Cornet et al., 2020) (see also Chapter 10 in this volume)	Evaluation	Usability engineering	Heuristic evaluation

1.3.1 The Need for Patient Ergonomics-Specific Adaptations

Patient ergonomics is related to various disciplines by borrowing from and contributing to them, as well as through the overlapping membership of its members with other communities of practice, as described above. However, an important theme in the patient ergonomics literature is the extent to which theories, methods, and tools developed in other fields or applied to the work of clinical professionals can be applied in their original form or need to be modified to suit patient work. The argument for adapting methods rests on the unique characteristics of patient work and the settings where it takes place. Some of the unique characteristics mentioned by scholars have included:

- Nonprofessionals who perform patient work may have little or no training for this work;
- There appears to be great variation between individuals performing the same kind of patient work, for example, in terms of motivation, knowledge, or prior experience;
- The people who do patient work may be volunteers and may not take on the work willingly;
- Patient work may not be standardized; it often occurs around the clock and is entangled with everyday life, creating privacy, ethical, and other implications;
- Tools and other resources used for patient work are usually self-acquired rather than provided to workers;
- Patient work occurs across many settings, including formal care settings, home and community settings, and in transit;
- The outcomes of patient work are often personal and affect the person doing it;
- Patient work may not be overtly governed by organizational policies, payment structures, or laws (although social norms and other rules may have a great impact); and
- One's patient work may greatly evolve over the course of a lifetime and during transitions (such as new disease onset or childbirth).

These and other differences help make the case for not only applying theories from HFE but also adapting them, as argued by several chapters in this volume (e.g., distributed cognition in Chapter 2, work systems in Chapter 4, situation awareness in Chapter 7, and teamwork in Chapter 8). Traditional HFE methods can also be applied but in many cases need to be adapted, as illustrated in this volume as well in Chapters 9–13; in particular, Chapter 13 and several others argue for adapting approaches to better involve patients, families, and other nonprofessional stakeholders as partners in, not subjects of, research and design (see also Carayon et al., 2020).

1.4 WHY PATIENT ERGONOMICS AND WHY NOW?

Two paradigm shifts support the present need for and growth of patient ergonomics (for further discussion, see Valdez & Holden, 2016). The first is the evolution of HFE from micro to macro or systems levels of analysis (Carayon, 2006; Waterson, 2009). This shift makes it no longer acceptable to investigate individual tools or tasks in controlled laboratory settings (Wilson, 2014). Instead, HFE professionals must consider work from a broader perspective, observing multiple physical, cognitive, and organizational interactions between multiple agents or teams in field settings, across time, settings, and levels of analysis (Hendrick & Kleiner, 2002). Such a lens applied to health care or other health-related activity will unavoidably include patients and families as agents in this work and will call for HFE interventions that address patient work alongside professional work. The second paradigm shift is in the public opinions, regulations, practices, and scientific and technical advances in the fields of health and health care. An obvious example is the attention being paid this century to patient engagement, autonomy, and co-production (Batalden et al., 2016; Coulter, 2002; Coulter & Ellins, 2007; Dentzer, 2013). Other changes in this area include the proliferation of "consumer"-facing technologies, the democratization of data, and payment reforms promoting health-oriented and holistic rather than disease-oriented and episodic models of care (for more, see e.g., Holden, Cornet, et al., 2020). We have argued before that not only is the time ripe for HFE to study and improve health-related work performed by both professionals *and* nonprofessionals, but that doing so is critical to HFE's ongoing value proposition in the domains of health and health care (Holden et al., 2013). In short, if HFE fails to account for "the patient factor" in the new paradigm of health, it risks obsolescence.

Fortunately, the two coinciding paradigm shifts have been met with enthusiasm by many in HFE and related disciplines. In the United States, this can be seen annually in the various panels and town halls organized at the Human Factors and Ergonomics Society (HFES) annual meeting (Holden & Valdez, 2018, 2019a; Holden, Valdez, et al., 2020; Holden, Valdez, et al., 2015; Valdez et al., 2016; Valdez et al., 2014; Valdez et al., 2017; Valdez et al., 2019) and international healthcare symposium (Papautsky et al., 2018, 2019, 2020). In Europe, the question "What about the patient?" has been raised by HFE experts since early in the global patient safety movement at the turn of the century (e.g., Vincent & Coulter, 2002).

The growth in patient ergonomics is owed in part to seminal efforts in the history of HFE, from early writing on hospital design to improve patient experience (Rappaport, 1970) to the popular series of HFE for aging (Czaja et al., 2019) to national reports calling for HFE approaches to promote health at home (National Research Council, 2011). These and other works have contributed to the state of patient ergonomics today, reviewed next.

1.5 STATE OF THE SCIENCE AND PRACTICE OF PATIENT ERGONOMICS

A recent 10-year mapping review examined published proceedings papers on patient ergonomics in two HFES conferences, 2007–2017 (Holden, Cornet, et al., 2020). The review's objectives were to describe the scope and content of patient ergonomics and identify areas needing further attention. Box 10.1 summarizes several conclusions from the review. Several well-studied areas of patient ergonomics such as technology (Chapter 5 in this volume) and older adults (Chapter 9 in Volume II) are covered in this handbook. Importantly, several topics that were found in need of more attention are also covered, for example, care transitions and vulnerable populations (Chapters 3 and 10 in Volume II). The variety of methods in the literature depicted in the mapping review is also well represented in the methods section of this volume (Chapters 9–12).

BOX 1.1 CONCLUSIONS FROM THE 10-YEAR MAPPING REVIEW OF PATIENT ERGONOMICS

- A growth in the patient ergonomics community of practice is evident in the past decade.
- The most frequent target of patient ergonomics papers was the patient, with relatively few studies of the general public and fewer of informal caregivers and other collectives (e.g., households).
- Among patients, older adults and chronically ill individuals made up the vast majority. Few papers examined children or vulnerable or marginalized patient populations.
- Most papers (90%) examined some form of care process, divided across studies of communication, medication use, learning or knowledge, safety or error management, disease self-care, patient experience, and health-promoting behavior. A smaller literature was devoted to other processes such as decision-making and care transitions from the patient or family perspective.
- About one-third of papers (34%) examined some form of technology, often information technology (e.g., apps, websites), but also medical or wearable devices and robots. Most technology studies could also be classified as related to a care process, for example, work on software to support self-care, medication use, or patient–clinician communication.
- The built environment, especially in the home, was a less commonly studied topic. Physical ergonomic factors of patient work were also less commonly studied than cognitive or organizational ones.
- Most papers ($n=85$) had the goal of understanding a phenomenon, such as self-care of a chronic illness, and these studies used mostly interviews, observations, surveys, and focus group methods.
- The next most frequent goal was to test or evaluate an intervention ($n=49$), often using performance measurement in the context of usability tests.
- A smaller but sizeable set of papers ($n=22$) aimed to generate or report on intervention design.
- There were few papers in these publication venues reporting experimental or longitudinal research on the effectiveness or safety of interventions.

1.6 ABOUT THE HANDBOOK

This handbook arose from a collective desire among its contributors to accelerate and expand the patient ergonomics community of practice. The handbook builds on the growing parallel movements in health and health care and HFE toward studying and improving patients and other nonprofessionals' health-related activities. It also capitalizes on a growing corpus of prior work, much of which is reviewed in individual chapters and some of which has been presented as part of an ongoing series of HFES panels on patient ergonomics and the patient in patient safety since 2014.

One of this handbook's objectives is to inform. The handbook, for the first time, provides a single reference for patient ergonomics, collecting in one place theory, research, methods, and applications that heretofore have been distributed across many venues and disciplines (e.g., HFE, gerontology, public health, nursing, medical informatics, human–computer interaction, and so on). As a result, much of the content of individual chapters consists of reviews and syntheses of prior work. Each chapter also presents new findings and case studies.

Another objective is to further develop the field of patient ergonomics. We have included a breadth of topics that together define a comprehensive, inclusive community of practice. Each contribution includes thoughtful commentaries on the current state of the science and expert recommendations for future work. In keeping with the themes of inclusiveness and multidisciplinarity, we are proud of the diversity of contributors to the handbook, representing different nations and regions, racial and ethnic identities, disciplines, and perspectives. Some of the contributors identify as scientists or researchers; others as clinicians, practitioners, HFE professionals, or government officials; some as patients or caregivers; and most as a combination of some of these. This diversity is a core strength of the handbook.

The handbook's final objective is to inspire and encourage others to join the patient ergonomics community—or minimally to learn from it meaningful lessons for their own work. If you are a student or professional in HFE or related field, this handbook can be of value in future applications of HFE to patients' and other nonprofessionals' work. On the other hand, perhaps you will find the patient ergonomics perspective helpful and complementary as you study or improve the work of healthcare professionals. If you belong to another discipline or community of practice, we are just as delighted. Your expertise can help improve patient ergonomics and we hope you obtain value from the HFE approach. Patient ergonomics is, in many ways, a multidisciplinary effort with overlap and connections with other fields, from patient- and family-centered health sciences to the social sciences to systems engineering and design, and more. If you are a patient, family member, or just a "person" reading this because you perform health-related activities and identify with the content of this handbook, then we hope we have done justice to your experiences and needs. You are, after all, the reason that patient ergonomics and *The Patient Factor* exist in the first place.

1.7 ABOUT VOLUME I

This volume is organized in three substantive sections, Theories (Chapters 2–4), Domains (Chapters 5–8), and Methods (Chapters 9–13) of patient ergonomics. The theories are presented into the three categories typically dividing the HFE discipline, namely, cognitive, physical, and organizational (or macro-) ergonomics. The domains represent the four areas of patient ergonomics most represented in the literature: consumer health information technology; patient–professional communication; self-care; and patient safety. The bulk of patient work is in some way related to these domains, although they are not the only ones. The chapters in the methods section in this volume are diverse, covering both quantitative and qualitative approaches, both field and laboratory settings, both research and design, and both traditional and emerging methods. Each chapter in this volume contains one or more case examples helping to connect the content to the "real world."

1.8 PARTING WORDS

We hope you are eager to read on, so we will keep it brief: Patient and other nonprofessionals do work; HFE and related disciplines can help. Read on to find out how and, to paraphrase singer-songwriter Andrew Bird, *just don't let the patient factor fail to be a factor at all.*

REFERENCES

Ahmed, R., Toscos, T., Ghahari, R. R., Holden, R. J., Martin, E., Wagner, S., . . . Mirro, M. (2019). Visualization of cardiac implantable electronic device data for older adults using participatory design. *Applied Clinical Informatics, 10*(4), 707.

Ancker, J. S., Witteman, H. O., Hafeez, B., Provencher, T., Van de Graaf, M., & Wei, E. (2015). The invisible work of personal health information management among people with multiple chronic conditions: Qualitative interview study among patients and providers. *Journal of Medical Internet Research, 17*(6), e137.

Batalden, M., Batalden, P., Margolis, P., Seid, M., Armstrong, G., Opipari-Arrigan, L., & Hartung, H. (2016). Coproduction of healthcare service. *The BMJ Quality & Safety, 25*(7), 509–517.

Brennan, P. F., & Casper, G. (2015). Observing health in everyday living: ODLs and the care-between-the-care. *Personal and Ubiquitous Computing, 19*(1), 3–8.

Carayon, P. (2006). Human factors of complex sociotechnical systems. *Applied Ergonomics, 37*, 525–535.

Carayon, P., Cartmill, R., Hoonakker, P., Schoofs Hundt, A., Karsh, B., Krueger, D., . . . Wetterneck, T. B. (2012). Human factors analysis of workflow in health information technology implementation. In P. Carayon (Ed.), *Handbook of Human Factors and Ergonomics in Patient Safety* (2nd ed., pp. 507–521). Mahwah, NJ: Lawrence Erlbaum.

Carayon, P., Wooldridge, A., Hoonakker, P., Hundt, A. S., & Kelly, M. M. (2020). SEIPS 3.0: Human-centered design of the patient journey for patient safety. *Applied Ergonomics, 84*, 103033.

Cooke, N. J. (1994). Varieties of knowledge elicitation techniques. *International Journal of Human-Computer Studies, 41*(6), 801–849.

Corbin, J., & Strauss, A. (1985). Managing chronic illness at home: Three lines of work. *Qualitative Sociology, 8*, 224–247.

Corbin, J., & Strauss, A. (1988). *Unending Work and Care: Managing Chronic Illness at Home*. San Francisco, CA: Jossey-Bass.

Cornet, V. P., Daley, C., Bolchini, D., Toscos, T., Mirro, M. J., & Holden, R. J. (2019). Patient-centered design grounded in user and clinical realities: Towards valid digital health. *Proceedings of the International Symposium on Human Factors and Ergonomics in Health Care, 8*(1), 100–104.

Cornet, V. P., Daley, C. N., Srinivas, P., & Holden, R. J. (2017). User-centered evaluations with older adults: Testing the usability of a mobile health system for heart failure self-management. *Proceedings of the Human Factors and Ergonomics Society Annual Meeting, 61*(1), 6–10.

Cornet, V. P., Toscos, T., Bolchini, D., Ghahari, R. R., Ahmed, R., Daley, C., . . . Holden, R. J. (2020). Untold stories in user-centered design of mobile health: Practical challenges and strategies learned From the design and evaluation of an app for older adults with heart failure. *JMIR mHealth and uHealth, 8*(7), e17703.

Coulter, A. (2002). *The autonomous patient.* Oxford: Radcliffe Medical Press.

Coulter, A., & Ellins, J. (2007). Effectiveness of strategies for informing, educating, and involving patients. *The BMJ, 335*, 24–27.

Czaja, S. J., Boot, W. R., Charness, N., & Rogers, W. A. (2019). *Designing for older adults: principles and creative human factors approaches* (3rd ed.). Boca Raton, FL: CRC Press.

Dempsey, P. G., Wogalter, M. S., & Hancock, P. A. (2000). What's in a name? Using terms from definitions to examine the fundamental foundation of human factors and ergonomics science. *Theoretical Issues in Ergonomics Science, 1*, 3–10.

Dentzer, S. (2013). Rx for the 'blockbuster drug' of patient engagement. *Health Affairs, 32*, 202.

Dixon, L. S. (1995). *Perilous chastity: Women and illness in pre-enlightenment art and medicine.* Ithaca, NY: Cornell University Press.

Donchin, Y., & Gopher, D. (Eds.). (2014). *Around the patient bed: Human factors and safety in health care.* Boca Raton, FL: CRC Press.

Gorman, R. K., Wellbeloved-Stone, C. A., & Valdez, R. S. (2018). Uncovering the invisible patient work system through a case study of breast cancer self-management. *Ergonomics, 61*(12), 1575–1590.

Hendrick, H. W. (2002). An overview of macroergonomics. In H. W. Hendrick & B. M. Kleiner (Eds.), *Macroergonomics: Theory, methods and applications* (pp. 1–23). Mahwah, NJ: Lawrence Erlbaum Associates.

Hendrick, H. W., & Kleiner, B. M. (2002). *Macroergonomics: Theory, methods and applications.* Mahwah, NJ: Lawrence Erlbaum Associates.

Holden, R. J., Carayon, P., Gurses, A. P., Hoonakker, P., Hundt, A. S., Ozok, A. A., & Rivera-Rodriguez, A. J. (2013). SEIPS 2.0: A human factors framework for studying and improving the work of healthcare professionals and patients. *Ergonomics, 56*(11), 1669–1686.

Holden, R. J., Cornet, V. P., & Valdez, R. S. (2020). Patient ergonomics: 10-year mapping review of patient-centered human factors. *Applied Ergonomics, 82*, 102972.

Holden, R. J., Daley, C. N., Mickelson, R. S., Bolchini, D., Toscos, T., Cornet, V. P., . . . Mirro, M. J. (2020). Patient decision-making personas: An application of a patient-centered cognitive task analysis (P-CTA). *Applied Ergonomics, 87*, 103107.

Holden, R. J., Kulanthaivel, A., Purkayastha, S., Goggins, K. M., & Kripalani, S. (2017). Know thy eHealth user: Development of biopsychosocial personas from a study of older adults with heart failure. *International Journal of Medical Informatics, 108*, 158–167.

Holden, R. J., Schubert, C. C., Eiland, E. C., Storrow, A. B., Miller, K. F., & Collins, S. P. (2015). Self-care barriers reported by emergency department patients with acute heart failure: A sociotechnical systems-based approach. *Annals of Emergency Medicine, 66*, 1–12.

Holden, R. J., Schubert, C. C., & Mickelson, R. S. (2015). The patient work system: An analysis of self-care performance barriers among elderly heart failure patients and their informal caregivers. *Applied Ergonomics, 47*, 133–150.

Holden, R. J., & Valdez, R. S. (2018). Town hall on patient-centered human factors and ergonomics. *Proceedings of the Human Factors and Ergonomics Society Annual Meeting, 62*(1), 465–468.

Holden, R. J., & Valdez, R. S. (2019a). 2019 town hall on human factors and ergonomics for patient work. *Proceedings of the Human Factors and Ergonomics Society Annual Meeting, 63*(1), 725–728.

Holden, R. J., & Valdez, R. S. (2019b). Beyond disease: Technologies for health promotion. *Proceedings of the International Symposium on Human Factors and Ergonomics in Health Care, 8*(1), 62–66.

Holden, R. J., Valdez, R. S., Anders, S., Ewart, C., Lang, A., Montague, E., & Zachary, W. (2020). The patient factor: Involving patient and family stakeholders as advisors, co-designers, citizen scientists, and peers. *Proceedings of the Human Factors and Ergonomics Society Annual Meeting, 64*(1), in press.

Holden, R. J., Valdez, R. S., Hundt, A. S., Marquard, J., Montague, E., Nathan-Roberts, D., . . . Zayas-Cabán, T. (2015). Field-based human factors in home and community settings: Challenges and strategies. *Proceedings of the Human Factors and Ergonomics Society Annual Meeting, 59*(1), 562–566.

Jastrzebowski, W. B. (1857). An outline of ergonomics, or the science of work based upon the truths drawn from the science of nature, Part I. *Nature and Industry, 29*, 227–231.

Jowsey, T., Yen, L., & Mathews, P. (2012). Time spent on health related activities associated with chronic illness: a scoping literature review. *BioMed Central Public Health, 12*(1), 1044.

Karsh, B., Holden, R. J., Alper, S., & Or, C. (2006). A human factors engineering paradigm for patient safety: designing to support the performance of the healthcare professional. *Quality and Safety in Health Care, 15*(suppl 1), i59–i65.

Karwowski, W. (2005). Ergonomics and human factors: The paradigms for science, engineering, design, technology and management of human-compatible systems. *Ergonomics, 48*(5), 436–463.

Kumar, V. (2012). *101 design methods: A structured approach for driving innovation in your organization*. Hoboken, NJ: John Wiley & Sons.

Mickelson, R. S., & Holden, R. J. (2018a). Medication adherence: Staying within the boundaries of safety. *Ergonomics, 61*, 82–103.

Mickelson, R. S., & Holden, R. J. (2018b). Medication management strategies used by older adults with heart failure: A systems-based analysis. *European Journal of Cardiovascular Nursing, 17*, 418–428.

Mickelson, R. S., Unertl, K. M., & Holden, R. J. (2016). Medication management: The macrocognitive workflow of older adults with heart failure. *Journal of Medical Internet Research Human Factors, 3*, e27.

Mickelson, R. S., Willis, M., & Holden, R. J. (2015). Medication-related cognitive artifacts used by older adults with heart failure. *Health Policy & Technology, 4*, 387–398.

National Research Council. (2011). *Health care comes home: The human factors*, National Academies Press (Washington, DC).

Ozkaynak, M., Valdez, R., Holden, R. J., & Weiss, J. (2018). Infinicare framework for integrated understanding of health-related activities in clinical and daily-living contexts. *Health Systems, 7*, 66–78.

Papautsky, E. L., Holden, R. J., Ernst, K., & Kushniruk, A. (2020). The patient in patient safety: Unique perspectives of researchers who are also patients. *Proceedings of the International Symposium on Human Factors and Ergonomics in Health Care, 9*(1), 292–296.

Papautsky, E. L., Holden, R. J., Valdez, R. S., Belden, J., Karavite, D., Marquard, J., . . . Muthu, N. (2018). The patient in patient safety: Starting the conversation. *Proceedings of the International Symposium on Human Factors and Ergonomics in Health Care, 7*(1), 173–177.

Papautsky, E. L., Holden, R. J., Valdez, R. S., Gruss, V., Panzer, J., & Perry, S. J. (2019). The patient in patient safety: Clinicians' experiences engaging patients as partners in safety. *Proceedings of the International Symposium on Human Factors and Ergonomics in Health Care, 8*(1), 265–269.

Rappaport, M. (1970). Human factors applications in medicine. *Human Factors, 12*(1), 25–35.

Srinivas, P., Cornet, V., & Holden, R. J. (2017). Human factors analysis, design, and testing of Engage, a consumer health IT application for geriatric heart failure self-care. *International Journal of Human-Computer Interaction, 33*(4), 298–312.

Stacey, D., Légaré, F., Lewis, K., Barry, M. J., Bennett, C. L., Eden, K. B., . . . Wu, J. H. C. (2017). Decision aids for people facing health treatment or screening decisions. *Cochrane Database of Systematic Reviews, 4*(4), CD001431.

Stanton, N. A., Salmon, P. M., Rafferty, L. A., Walker, G. H., Baber, C., & Jenkins, D. P. (2013). *Human factors methods: A practical guide for engineering and design.* Surrey, UK: Ashgate.

Steiner, J. F. (2012). Rethinking adherence. *Annals of Internal Medicine, 157*(8), 580–585.

Stiggelbout, A. M., Van der Weijden, T., De Wit, M. P., Frosch, D., Légaré, F., Montori, V. M., . . . Elwyn, G. (2012). Shared decision making: Really putting patients at the centre of healthcare. *The BMJ, 344*(7842), e256.

Strauss, A. (1993). *Continual permutations of action.* New York: Aldine de Gruyter.

Strauss, A., Fagerhaugh, S., Suczek, B., & Wiener, C. (1982). The work of hospitalized patients. *Social Science & Medicine, 16*, 977–986.

Timmermans, S., & Freidin, B. (2007). Caretaking as articulation work: The effects of taking up responsibility for a child with asthma on labor force participation. *Social Science & Medicine, 65*(7), 1351–1363.

Unruh, K. T., & Pratt, W. (2008). The invisible work of being a patient and implications for health care: "[the doctor is] my business partner in the most important business in my life, staying alive". *Proceedings of the Ethnographic Praxis in Industry Conference (EPIC '08), 1*, 34–44.

Valdez, R. S., & Holden, R. J. (2016). Health care human factors/ergonomics fieldwork in home and community settings. *Ergonomics in Design, 24*, 44–49.

Valdez, R. S., Holden, R. J., Caine, K., Madathil, K., Mickelson, R., Lovett Novak, L., & Werner, N. (2016). Patient work as a maturing approach within HF/E: Moving beyond traditional self-management applications. *Proceedings of the Human Factors and Ergonomics Society Annual Meeting, 60*(1), 657–661.

Valdez, R. S., Holden, R. J., Hundt, A. S., Marquard, J. L., Montague, E., Nathan-Roberts, D., & Or, C. K. (2014). The work and work systems of patients: A new frontier for macroergonomics in health care. *Proceedings of the Human Factors and Ergonomics Society Annual Meeting, 58*(1), 708–712.

Valdez, R. S., Holden, R. J., Khunlerkit, N., Marquard, J., McGuire, K., Nathan-Roberts, D., . . . Ramly, E. (2017). Patient work methods: Current methods of engaging patients in systems sesign in clinical, community and Extraterrestrial settings. *Proceedings of the Human Factors and Ergonomics Society Annual Meeting, 61*, 625–629.

Valdez, R. S., Holden, R. J., Madathil, K., Benda, N., Holden, R. J., Montague, E., . . . Werner, N. (2019). An exploration of patient ergonomics in historically marginalized communities. *Proceedings of the Human Factors and Ergonomics Society Annual Meeting, 63*(1), 914–918.

Valdez, R. S., Holden, R. J., Novak, L. L., & Veinot, T. C. (2015). Technical infrastructure implications of the patient work framework. *Journal of the American Medical Informatics Association, 22*(el), e213–e215.

Vincent, C., & Coulter, A. (2002). Patient safety: What about the patient? *Quality & Safety in Health Care, 11*, 76–80.

Waterson, P. E. (2009). A critical review of the systems approach within patient safety research. *Ergonomics, 52*, 1185–1195.

Werner, N. E., Tong, M., Borkenhagen, A., & Holden, R. J. (2019). Performance-shaping factors affecting older adults' hospital-to-home transition success: A systems approach. *The Gerontologist, 59*, 303–314.

Wilson, J. R. (2014). Fundamentals of systems ergonomics/human factors. *Applied Ergonomics, 45*, 5–13.

Section II

Patient Ergonomics Theories

2 Cognitive Patient Ergonomics

Application of Cognitive Theories to Patient Work

Dan Morrow
University of Illinois at Urbana-Champaign

CONTENTS

Patient work, the behaviors that patients and their informal caregivers perform to manage health (see Chapter 1 in this volume), involves many tasks such as taking medication that depend on cognitive processes, including making sense of and acting on information. Therefore, theories about these processes are crucial for improving patient work by designing tools and environments that support this work or by training patients and their informal caregivers. This chapter summarizes cognitive theories related to patient work with a focus on distributed cognition approaches because patient work, like much complex work, is accomplished by individuals interacting with each other and the environment to accomplish shared goals. Distributed cognition approaches have traditionally focused on clinicians' work in healthcare organizations, although they often include some dimensions of patient work. The present chapter describes frameworks and example studies that extend this general approach to analyzing patient work in the home. It focuses on how older adults self-manage chronic illness because older adults are typical patients in healthcare systems.

2.1 COGNITIVE THEORIES OF PATIENT WORK

In general, theory explains how patient work is accomplished in sufficiently general terms to guide research that can ultimately improve patient outcomes. Since theory also guides the development of measures and other aspects of methodology, theory and research methods are often tightly connected.

Cognitive theories in particular are important for explaining the patient work involved in self-care, defined as the behaviors involved in managing symptoms and progressions of chronic illness (see Chapter 7 in this volume). Although self-care depends on myriad factors related to patients (or their informal caregivers), tasks, tools, and contexts, cognitive factors in particular are essential. For example, taking medication requires understanding why the medication is important and how to take it in the context of daily life. Moreover, frameworks that conceptualize patients as empowered agents rather than passive recipients of care (e.g., Bodenheimer et al., 2002) often assume that patients or their informal caregivers have the cognitive capacity to manage the patients' own illnesses. Therefore, it is important to characterize patients' cognitive processes and resources as patients interact with informal caregivers, clinicians, and other people, as well as with tools and other aspects of the environment, to accomplish self-care goals. Such theories can parse patient work into components, explain how these components interact, identify causal processes involved (e.g., how patients accomplish work goals), and suggest the mechanisms by which interventions (e.g., technology design, training) may improve outcomes.

Whereas cognitive theories relevant to patient work have traditionally focused on the internal or mental cognition of individuals, more recent theories analyze how cognition is distributed or extended across individuals (e.g., team members) and the external context. Distributed cognition theories in turn vary along many dimensions (Baber, 2020). After considering internal cognition theories, this chapter will first describe distributed cognition theories that focus on how individuals extend their cognition to task partners and external representations and then describe theories that take a broader, more systems-based view of distributed cognition (Figure 2.1).

2.1.1 Theories Focused on Individual Cognition in Health Care

Theories focused on individuals (clinicians, patients) have long been used to explain how individuals' cognitive representations and processes are involved in accomplishing healthcare tasks (for review, see Morrow et al., 2006). The information-processing

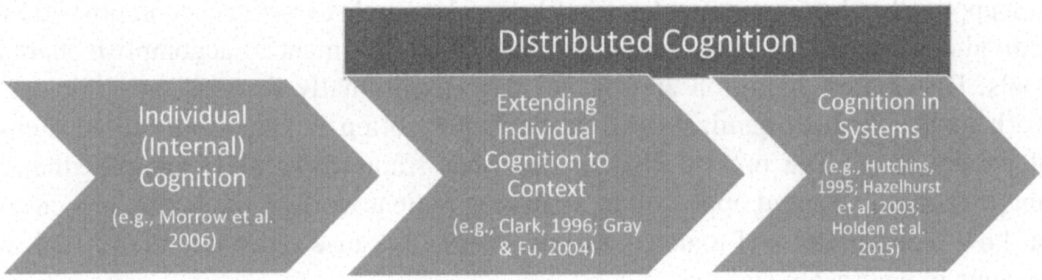

FIGURE 2.1 Cognitive theories of patient work.

framework underpinning theories of individual cognition is generally useful for describing cognitive abilities and resources involved in patient work and how they are influenced by task and situational factors (see Morrow et al., 2006). Cognitive task performance is analyzed in terms of stages of processing: sensing (e.g., vision, audition); interpreting information; deciding what to do; and responding. General cognitive resources are viewed as constraints on or as facilitators of performance by influencing these processing stages. For example, limited working memory capacity may contribute to error by decreasing the quality of decision-making or by increasing the likelihood of forgetting to respond. Knowledge, on the other hand, may reduce errors by increasing processing efficiency, which improves decision-making (Morrow et al., 2006). Thus, this general approach can identify processing bottlenecks that interact with task and situational factors to contribute to errors such as misdiagnosing an illness or forgetting to take medication. A key concept within the information-processing framework is cognitive workload, which articulates relationships between task demands and individual resources such as attention and working memory (e.g., cognitive load theory; Sweller, 2011). A patient may struggle to manage complex multi-medication regimens because of the workload imposed by integrating multiple requirements and constraints such as which medications can be taken together or which must be taken with food. Methodological contributions of the information-processing framework include measurement of abilities and performance (e.g., response time, accuracy), computational modeling of cognitive processes, and cognitive task analysis (how task requirements influence cognitive workload).

Cognitive theories often guide experimental studies of processes and outcomes involved in patient self-care. For example, Rogers et al. (2001) conducted a task analysis of blood glucose monitors and identified aspects of the interface (e.g., confusing feedback) that contributed to patient error (e.g., miscalibration) by increasing demands on sensory (visual acuity) or central (working memory) resources that typically decline with age. This led to an experiment in which the investigators found that enhancing audio instruction with video reduced device-use errors, presumably by reducing working memory demands involved in understanding how to use the device (McLaughlin et al., 2002).

Theories focused on individual cognition have also supported extensive work on how individual differences in cognition influence different stages of processing in self-care tasks. Park and colleagues (1999) analyzed the role of individual differences in cognitive resources (e.g., speed of processing and working memory capacity) and illness beliefs on medication adherence. Age-related differences in adherence were partly explained by differences in cognitive ability, which they argued impaired older adults' ability to understand and manage complex regimens (Park et al., 1999). Another study focused on the role of prospective memory (remembering to do a task at the right time) in adherence. The cognitive processes involved in remembering to take medication at the right time were identified, which included encoding into memory the need to take the medication, retrieving and executing this intention at the right time, and monitoring performance to ensure completion (Insel et al., 2016). This theoretical analysis guided the development of an intervention in which older adults were trained to use strategies that support these processes (e.g., integrating medication taking into daily routines that cue retrieval of the intention at appropriate times). The intervention primarily improved adherence for participants with lower levels of working memory capacity, perhaps because those with higher capacity did not need such strategies or developed them on their own (Insel et al. 2016). Finally, the impact of health literacy on older adults' memory for information about self-care of hypertension is explained in part by differences in processing capacity (e.g., working memory), especially for older adults with lower levels of general and health knowledge (Chin et al., 2015). Older adults with more knowledge may have understood and remembered the information more efficiently, reducing demands on processing capacity. Such findings underline the value of cognitive theory for developing targeted interventions by identifying which processes to support and who is more likely to benefit. For example, older adults who are newly diagnosed with hypertension (and therefore lack knowledge about the illness) may benefit from explanations of the illness and self-care that reduce demands on processing capacity. This might be accomplished by using simple, direct language, or leveraging general knowledge by using metaphors (e.g., high blood pressure is like the build-up of water in a hose).

2.1.2 From Individual to Distributed Cognition Theories in Health Care

It has long been recognized that theories focused on individual cognition are insufficient to explain complex tasks. Instead, performance of tasks, such as controlling aircraft speed during flight, is explained in terms of interactions between pilot cognition and external representations (e.g., flight displays; Hutchins, 1995). This point is captured by Herbert Simon's famous metaphor of the mind and world meshing like the blades of scissors during cognitive work (Simon, 1979).

Similarly, the fluent performance of complex daily tasks is difficult to explain in light of the severe limits on working memory capacity that increase with age (Hartshorne & Germine, 2015). Whereas age differences often occur in laboratory-based prospective memory tasks (e.g., remembering to press a specific key when a target word appears in a word list memorization task), such differences can diminish in everyday prospective memory tasks such as remembering to attend appointments (Henry et al., 2004). One reason for this discrepancy is that older adults rely on external supports or external memory cues such as reminders, as well as motivation and other factors. The availability of external supports such as note-taking can eliminate age differences in memory-based performance in complex tasks such as aircraft pilot communication (Morrow et al., 2003). A similar conclusion comes from evidence that interventions based primarily on individual cognition theories (and tested in the lab) may fail to improve performance in daily life, in part because the interventions are not designed to address contextual factors beyond individual cognition, and thus are not robust to situational factors (Holden et al., 2015).

2.2 DISTRIBUTED COGNITION APPROACHES: EXTENDING INDIVIDUAL COGNITION

This section describes work focused on how individuals extend their cognitive resources to external contexts, including collaborators and tasks (Figure 2.1). This focus has long been central to theories of everyday conversation, which is essential to self-care and other complex task domains. Conversation is more than a sequence of individual cognitive processes such as producing messages by speakers and understanding these messages by listeners. Rather, speakers and listeners co-create meaning: speakers may modulate word choice moment to moment in response to listeners' nonverbal signals; their utterances may be completed by listeners; and so on (Morrow & Fischer, 2013).

According to common ground and other communication theories (Clark, 1996), explaining conversation requires extending the concept of individual cognitive resources, or effort, to collaborative resources. Conversational partners coordinate the communication process (e.g., speakers ensure that addressees are attentive and addressees signal that they are paying attention) as well as the content (e.g., what is presupposed as known, what should be stated) to ensure information is grounded, or mutually understood and accepted as relevant. Communication partners use a variety of resources to ground information. Speakers provide information through linguistic and nonlinguistic signals (e.g., words, gaze, gesture), and listeners signal that they understand either implicitly by providing a relevant turn or more explicitly by "back-channel" responses (e.g., "uh-huh") or verbal acknowledgment. Listeners may signal a problem by requesting clarification or expressing confusion, or may unwittingly indicate a problem by making an unexpected response. Partners try to efficiently use these resources by coordinating individual effort to minimize collaborative effort (Clark & Wilkes-Gibbs, 1986). More generally, they adhere to a principle of minimizing joint effort, or co-efficiency in accomplishing joint goals where two partners coordinate actions (Török et al., 2019). So, a home health worker, in explaining to a patient how to use a glucometer, may point to the display after inserting a calibration strip rather than verbally describing where the display is located on the device. Noticing that the patient is looking at the display, the worker can go on to explain what the displayed numbers mean without first asking the patient to confirm they understand, unless the patient indicates nonverbally (e.g., puzzled look) or verbally (e.g., question) that they do not understand something.

The concept of common ground has influenced the field of human–computer interaction (HCI) because the idea of a collaborative effort has been extended to computer-mediated communication (Clark & Brennan, 1991; Monk, 2008). Telephone or other forms of voice communication, with partners temporally but not physically co-located, allow rapid turn-taking that supports grounding (e.g., immediately indicating and repairing misunderstanding), but eliminate the visual nonverbal cues afforded by face-to-face communication, and thereby increasing the effort needed to produce and accept messages (Clark & Brennan, 1991). In the example of the health worker explaining the use of a glucometer, the explanation would take more effort if the worker and patient were talking over the phone because the worker would need to verbally reference the device display rather than simply pointing to it before explaining what the displayed information means. The telephone also eliminates shared gaze or similar nonverbal evidence of shared understanding. Grounding becomes even more challenging for asynchronous communication such as email, which lacks synchronicity and sometimes even sequentiality in turn-taking. More effort may be involved in producing messages (typing vs. speaking), which influences individual and collaborative effort involved in grounding. On the other hand, email can support message revisability and reviewability, reducing demands on working memory (Clark & Brennan, 1991).

Distributed cognition approaches have also influenced the HCI field by investigating how individual cognition is extended by technology that enhances or replaces internal capacities, with performance adaptively emerging from the interplay of internal and external resources (e.g., Gray & Fu, 2004; Zhang & Norman, 1994). People can more efficiently accomplish tasks by relying on external information sources when their memory is less reliable or when more effort is required to access information from memory than from the environment. As external information look-up becomes more effortful or memory retrieval requires less effort (e.g., with practice; Gray & Fu, 2004; Touron, 2015), though, they switch to memory. Thus, since a home health worker is very familiar with the specific glucometer, she can rely on her memory when telling the patient how to use the device, but would need to consult instructions while explaining how to use a less familiar device.

Distributed, as well as individual, cognition approaches address individual differences (e.g., age-related) in cognition and performance. The concept of environmental support was developed within a general person–environment framework to address the consequences of mismatches between individual resources (e.g., working memory) and demands and affordances related to tasks and environments (Craik, 1994). Morrow and Rogers (2008) conceptualized environmental support in terms of distributed cognition. Age-related cognitive declines can impair internal determinants of performance, making them less reliable than external determinants, so that greater reliance on external resources may be needed to maintain high levels of performance. Unsurprisingly, older adults are sometimes more likely than younger adults to rely on external resources (Rogers et al., 2000).

Similar to traditional individual cognition theories, distributed cognition research focuses on individual–context interaction and uses cognitive task analysis and experiments to evaluate theory-based predictions about performance. However, the focus is on how performance emerges from interactions of cognitive processes and external representations, often through task analyses informed by observation and interview. For example, because older adults' self-care increasingly relies on web-based tools such as patient portals to health records (see Chapter 5 in this volume), experiments have investigated sources of age-related use problems and evaluated potential design solutions that support impaired processes. Taha et al. (2013) analyzed typical portal-based tasks, from the relatively simple (e.g., identifying appointment date and time) to the more complex (e.g., identifying whether cholesterol test scores are out of critical range). More complex tasks of using the portal were assumed to increase demands on numeracy and general reasoning abilities. Investigation of performance by middle-aged (40 to 59-year-olds) and older adults (60 to 85-year-olds) in a simulated portal environment revealed age-related impairment for the more complex tasks, which was explained in part by differences in numeracy and reasoning abilities. Demands on numeracy and reasoning included interpreting test scores and comparing them to cut-off values to determine whether the scores were out of range. Performance on such tasks can be improved by redesigning the presentation of portal-based test results to clarify how the scores map onto regions of risk on the scale and highlighting the score's status (e.g., in/out of range, high/low risk) in graphic displays to reduce the need for reasoning about the information and thus the need for limited cognitive resources (Morrow et al., 2019; Zikmund-Fisher et al., 2017). Making portal-based information more cognitively accessible should increase older adults' use of portals as a tool for self-care, so that they do not need to rely on their memory for what their doctors told them about their results or other health information. External representations also support collaborative planning for self-care. The "medtable," designed as an external workspace to reduce internal working memory demands during collaborative medication review, was found to increase patients' satisfaction with communication and improve knowledge of medication purpose in an intervention in which nurses used the aid to review and optimize complex medication regimens with older adults who had diabetes (Graumlich et al., 2016).

Although making external information easy to access, understand, and use encourages older adults to rely more on external rather than internal cognitive resources, this strategy may not always be adaptive. Older adults tend to over-rely on external strategies (e.g., looking up answers rather than relying on memory even when they have memorized the answers), which reduces efficiency in real-world as well as laboratory tasks (Mayr & Lindenberger, 2014). Providing feedback about performance or giving instructions to increase confidence in memory-based strategies can increase older adults' use of memory strategies and improve performance, suggesting that experience can shape their judgments about trade-offs between the use of environment and internal memory to accomplish tasks (Touron, 2015; also see Morrow et al., 2003). Encouraging older adults' use of memory when possible may also help protect against age-related impairment of memory abilities, so they can support performance when external support is unavailable.

2.2.1 Distributed Cognition Approaches: System Perspectives on Patient Work

Distributed cognition approaches must be expanded and elaborated to account for patient work in actual contexts such as self-care of chronic illness, where patients collaborate with family members and healthcare professionals in multiple environments (e.g., home, healthcare settings, and community organizations). Analyzing such complex activity involves explaining how cognitive representations are coordinated across people, space, and time (e.g., analysis of information flow across agents, tools, and physical space). Such approaches are similar to macroergonomic theories such as the Systems Engineering Initiative for Patient Safety (SEIPS) model (e.g., Carayon et al., 2006), which analyze how patient safety and other organizational outcomes emerge from interacting agents, tasks, environments, and organizations. These approaches were developed to analyze healthcare organizations and the impact of technology on work processes and outcomes, but they are increasingly applied to patient work as well (e.g., Holden et al., 2013).

Hazlehurst and colleagues (2003) built on the seminal distributed cognition work of Hutchins (1995). The unit of analysis in their approach is an activity system, composed of agents, tools, and environment, organized around shared goals. Like individual cognition approaches, behavior is viewed as emerging from an internal organization that responds to a task or other demands. However, rather than driven only by individuals' mental processes, behavior depends on the propagation of representations through physical media (internal to people or external such as digital tools) across the activity system as influenced by physical and social structures (Hazlehurst et al., 2003). This activity system analysis also broadens the scope of inquiry from how an individual extends their own internal resources to a task or to collaborative partners, to a system of interacting agents and external resources in which actions by multiple actors are coordinated through external representations over time, with the actions shaped by constraints imposed by tools, environment, tasks, and multiple goals (e.g., accomplishing specific tasks, training novices, catching errors). The systems approach provides a rich context for considering how to design systems to increase efficiency and safety. It has been used to analyze how medication orders are accomplished in intensive care units (ICUs) (Hazlehurst et al., 2003), how clinicians coordinate during surgery (Hazlehurst et al., 2007), and other aspects of clinical work. Dovetailing with this research is work focused on how external artifacts serve multiple functions for clinicians in coordinating their work, such as representing shared information and helping to manage the joint attention required for collaboration. Examples include analysis of whiteboards, sign-out sheets, and other artifacts to support coordination and planning in operating rooms and ICUs (Nemeth et al., 2004; Xiao, 2005).

This systems-based distributed cognition approach is increasingly applied to patient work such as self-care in the home because, like clinical work, patient work is a situated set of complex activities involving interactions among people, external resources (often including technology), and the environment, even though it is typically not organized by explicitly defined roles, procedures, and other characteristics of formal work (Holden et al., 2013; Monk, 2000; National Research Council, 2011). The distributed cognition approach has the potential to provide a more comprehensive analysis of cognitive processes and representations involved in self-care by building on and expanding previous work that focused on patients' use of self-care technology, such as how use is influenced by the perceived usefulness and usability of the technology (Or et al., 2011).

Blandford and colleagues (e.g., Rajkomar et al., 2015) used a distributed cognition approach to analyze how patients with kidney failure use hemodialysis systems at home. The general distributed cognition approach (Hazlehurst et al., 2003) was instantiated as a framework (Distributed Cognition for Teamwork) with information flow, physical layout, social structure, and artifact models describing how patients use the technology. The framework guided the design of methods for collecting and interpreting observation and interview data to understand how system components interacted to constrain and facilitate self-care performance, identify sources of use problems, and suggest causes of incidents that compromised safety. According to their analysis, home hemodialysis is an activity system involving several agents (the patient, informal caregivers, providers) and artifacts in addition to the technology system itself (the hemodialysis machine). It is embedded in and interacts with other systems such as the home system (physical and social contexts) and the dialysis unit system (part of the healthcare system). The framework situates a traditional HCI task analysis (how the patient uses the dialysis machine) into a more comprehensive analysis of the interacting dimensions of the home hemodialysis activity system. For example, central to the dialysis activity at home are the steps that patients or their informal caregivers must do to operate the machine, including preparation (e.g., starting auto-disinfection of the machine, inserting needle into the patient), treatment (e.g., programming parameters, monitoring machine readings), and termination (e.g., remove the needle from the patient, treat wound).

Analysis of this human–machine interaction includes considering how it depends on dimensions of the activity system as described by the different models derived from the distributed cognition framework. For example, the information flow model analyzes how the interaction depends on information flowing from machine to patient (e.g., through displays), between patient and caregiver (e.g., face-to-face communication), or between patient and dialysis technicians (e.g., typically by telephone). Important issues addressed within this flow model include how actors respond to machine alarms and how decisions depend on intersecting information flows (e.g., programming parameters based on physician prescription and current patient state). The artifact model analyzes how patient-created artifacts (e.g., notes attached to the machine) support performance. Artifacts vary in the extent to which they explicitly represent the status of goal accomplishment, such as the gap between the current task step and the goal ("representation-goal parity;" Hutchins, 1995). More explicit artifacts (e.g., more informative machine feedback about current step in dialysis) are more likely to provide external support that reduces the need for internal cognitive resources (e.g., attention). The physical layout model articulates how the location of artifacts, machine, actors, and so on constrains self-care performance and influences the likelihood of error. This encompasses the analysis of how the physical home layout influences tailoring of the dialysis workspace to facilitate activities, such as co-locating the dialysis machine, supplies, manuals, and artifacts such as phone contacts and records. This spatial arrangement can transparently indicate goals, action sequences, and so on, thus influencing the need for agents' internal cognitive resources.

This distributed cognition framework helps integrate into a more comprehensive analysis of self-care activity systems the insights from previous observational studies about artifact arrangement in the home (e.g., use of a "command center" with medication co-located with calendars, instructions, and other resources for managing medication; Palen & Aaløkke, 2006), and the findings from a focused analysis of patient self-care activities such as understanding portal-based self-care information (Morrow et al., 2019) and using medical devices (Rogers et al., 2001).

Holden and colleagues used a similar distributed cognition approach to analyze patient work related to self-care among patients with heart failure (e.g., Holden et al., 2015; Mickelson et al., 2015). Their model, the Patient Work System, integrates a macroergonomic approach to clinical work with cognitive aging models. Similar to the approach developed by Blandford and colleagues, this model views patient work as encompassing person (patient, caregiver, clinician), task (self-care activities), and tool factors that interact through multiple contexts. It also guides a multi-method approach (interview, observation, record review) to collecting and interpreting patient work data (see Chapter 9 in this volume). Whereas Blandford and colleagues analyzed a specific self-care activity (home-based dialysis) to explain how performance emerged from the interaction of situational resources and constraints, Holden et al. take a broader view of person and task factors that influence a range of self-care tasks among patients with heart failure and how these factors interact through multiple context domains (physical, social, organizational) and "context spaces" (home, community, health facility). For example, their model is more inclusive of patient characteristics involved in patient work—physical, affective, and illness and self-care beliefs, as well as cognitive factors such as knowledge and working memory (Holden et al., 2017).

Like other systems approaches, there is a potential for integrating insights from previous research on isolated facets of self-care into a more comprehensive analysis that may more successfully guide interventions. For example, echoing the previous self-care literature, Holden et al. (2015) found that medication management is central to self-care among patients with heart failure. Task factors such as regimen complexity (e.g., number and type of medications and doses) were identified as an important barrier to adherence that may interact with person-related factors such as perceptual-cognitive limitations and inadequate knowledge of illness and self-care. These findings converge with previous evidence that regimen complexity is associated with patient adherence and health outcomes (e.g., Pollack et al., 2010), perhaps reflecting age-related health literacy declines due to limited health knowledge and processing capacity (Chin et al., 2015). However, the Patient Work System model (Holden et al., 2015) enables more comprehensive analysis of adherence. Adherence is analyzed as more than a function of medication task demands and patient resources (similar to the workload concept from individual cognition approaches), and as emerging from a set of interacting system constraints. For example, the findings from Holden et al. suggest that complex medication regimens requiring more frequent daily doses may be more challenging to integrate with daily routines and lead to more disruptive side effects (e.g., diuretics increasing the frequency of night-time visits to the bathroom), which in turn impose greater workload on patients with fewer internal resources (e.g., physical, sensory, cognitive) and external resources (e.g., tools such as pill organizers and reminders; social support). This, in turn, can exacerbate illness progression, further diminishing patient resources and increasing regimen complexity as new medications are prescribed for worsening symptoms. Such multi-factorial analyses that trace interacting situational constraints as they evolve over time are challenging to accomplish, yet may ultimately provide the theoretical lens that supports comprehensive interventions that improve self-care outcomes for patients (also see Mickelson et al., 2016).

2.3 RECOMMENDATIONS AND IMPLICATIONS FOR FUTURE RESEARCH AND PRACTICE

This chapter has sketched an evolution of the literature related to cognitive theories relevant to patient work from individual cognition, to distributed cognition theories focused on how individuals extend cognition to the physical or social environment, to distributed cognition theories that take a broader systems view of patient work (see Figure 2.1).

2.3.1 Distributed Cognition Theories of Patient Work: Advantages

Distributed cognition theories are well suited for guiding research on patient work for several reasons. First and most obviously, broader theoretical views are needed to begin to account for the complex interacting resources and constraints that characterize the work by clinicians in healthcare organizations or by patients at home. Moreover, the focus on distributed cognitive representations and processes supports analysis of how patient and clinician resources that support and constrain these processes contribute to more or less successful outcomes. In short, distributed cognition views are more descriptively adequate than those that focus on individual cognition, capturing more dimensions of the phenomena of interest.

Second, the review of systems approaches, such as the Distributed Cognition for Teamwork (Rajkomar et al., 2015) and the Patient Work System (Holden et al., 2015), revealed the potential to integrate findings from a more focused (and narrower) investigation of facets of patient work, such as medication adherence (Park et al., 1999) or using medical devices at home (Rogers et al., 2001), to help bring coherence to a sprawling literature.

Third, these more comprehensive frameworks have the potential to enrich our understanding of theoretical constructs related to self-care, such as environmental support. Simply adding information or tools to the environment as an external resource does not in itself constitute environmental support, as shown by the fact that environmental support can increase as well as decrease age differences in task performance. Morrow and Rogers (2008) addressed these disparate findings in the literature by analyzing the relationship between individual cognition and environmental support design (e.g., pre-requisites for using environmental support). For example, using a pill organizer may actually decrease adherence if it is difficult to place pills in the appropriate compartments. A more comprehensive distributed cognition account may provide further insight that guides the design of effective environmental support, such as the relation of a tool to other external information representations (e.g., user-generated notes) or how well the tool meshes with multiple agents' goals (e.g., representation-goal parity; Hutchins 1995). Such a view places a premium on environmental support that promotes efficient performance by integrating external as well as internal resources.

2.3.2 Distributed Cognition Theories of Patient Work: Challenges

The present chapter also identified challenges for distributed cognition theories of patient work. First, these theories more successfully address descriptive rather than explanatory adequacy. The processes and mechanisms underlying patient work may need to be analyzed in greater detail in order to explain how they contribute to error, and therefore can guide interventions. Integrating insights about cognitive processes from more focused analyses of how patients interact with specific self-care tasks and tools may help distributed cognition frameworks "drill-down" to finer-grain analyses of the cognitive processes that underlie successful performance. For example, previous work has analyzed at a micro-level individual strategy for accessing information to accomplish task goals. This work has examined how these strategies adapt to changes in the cognitive cost of internal versus external access (Gray & Fu, 2004), as well as the impact on the strategies of practice and age-related differences in cognition (Touron, 2015). Similarly, insights from the common ground theory applied to technology-mediated communication (e.g., Clark & Brennan, 1991) help explain barriers to information flow across tools and agents working together to accomplish joint goals in complex patient work. Examples include how easily patients or informal caregivers can reach dialysis technicians or nurses during home hemodialysis and the impact of communication medium (e.g., telephone vs. video) on troubleshooting a self-care device problem. An important challenge is how to incorporate such fine-grained analyses with broader theories that explain how patients and providers interact to accomplish complex joint self-care goals, such as controlling hemoglobin A1c levels among patients with diabetes or maintaining kidney function among patients with renal failure.

A second related challenge is more methodological. The integration of different levels of analysis within a theory to explain patient work requires methodological innovation. System frameworks have made strides in developing mixed methods approaches to research (integrating insights from qualitative and quantitative data; e.g., Carayon et al., 2015). In addition, integrating different levels of analysis depends on integrating different research paradigms. Experiments that manipulate key variables in controlled conditions are well suited for identifying the processes involved in individual and collaborative cognition that are part of patient work (see Chapter 11 in this volume). However, to be directly relevant to actual patient work, the design of such studies needs to be informed by observation and other techniques better suited to analyzing larger scale phenomena such as patient work in situ, where experimental control is not possible (see Chapters 9 and 12 in this volume). It may be possible to combine paradigms, given rapid progress in simulation technology (embedding experimental manipulations in simulations of complex work activities; Drews & Bakdash, 2013).

A third challenge relates to the need to make distributed cognition theories of patient work even more comprehensive, for example by taking a broader view of patient resources, including affect, motivation, and other noncognitive factors. More micro-level theories relevant to patient work have integrated affective and cognitive processes related to self-care (see Chapter 4 in this volume). For example, affective responses to information are central to shaping risk perception (Slovic & Peters, 2006), gist-based memory for health information (Reyna, 2011), and more generally, patients' illness representations (Meyer et al., 1985; Mishel, 1999). Interactions between affect and cognition have motivated design approaches to supporting older adults' understanding and use of self-care information such as clinical test results (Morrow et al. 2019; Zikmund-Fisher et al., 2017). Although noncognitive factors have played a role in technology acceptance frameworks (e.g., influencing perceived usefulness of technology; Or et al., 2011), this is rarely the case for distributed cognition frameworks, which focus on memory as the key internal resource that may trade off with external resources (off-loading memory to environment; Hazlehurst et al., 2003; Rajkomar et al., 2015). However, affective design can promote users' affective and emotional responses to tools and other external resources (Helander & Khalid, 2006), which in turn may reduce demands on limited cognitive resources in health-related tasks (Mikels et al., 2010).

A final theoretical challenge relates to how well (or how much) distributed cognition approaches to clinicians' work apply to patient work. This is an important issue given that theories of patient work often build on theories of clinician work. As pointed out in this chapter and elsewhere (Holden et al., 2013), both clinical and patient work can be analyzed as a system with activities distributed across people, tools, and the environment that are organized around shared goals. However, some aspects of clinician work do not translate to patient work, and patient work has unique characteristics. For example, it is unlikely that patients, informal caregivers, and clinicians in patient work systems need to develop the kind of shared dynamic mental models that support team situation awareness involved in orchestrating coordinated work during surgery. Patient work, on the other hand, is more likely to be supported by informal, experiential knowledge rather than more structured and formally learned knowledge about biomedical domains and work processes. Patient work is rarely as formally organized as clinical work, lacking the formally defined rules, roles, prescribed tasks, and incentive or evaluation systems that are accountable to regulatory organizations. However, to the extent that system analyses of patient work include organizational and cultural factors (e.g., Holden et al., 2013), some formal aspects of clinician work may also apply to patient work, such as norms, expectations, division of labor, and hierarchies. Given these differences between clinician and patient work, it is important to understand how the more formally defined clinical work systems and the more informal patient work systems do or do not mesh at points of intersection (e.g., provider/patient collaboration). These "cultural differences" may create tensions between actors across the systems. Common examples include difficulties in clinician–patient communication (Haidet, 2007) (see Chapter 6 in this volume) or medication adherence (Brundisini et al., 2015), reflecting differences in perspective and agenda associated with different roles in medical systems versus the everyday lives of patients disrupted by illness.

Although meeting these challenges is daunting, doing so will help refine theories of patient work to specify target mechanisms precisely enough to be addressed by interventions, while also identifying important contextual factors in broad terms that enable generalization across self-care contexts.

ACKNOWLEDGMENTS

Preparation of this chapter was supported by the Jump Applied Research for Community Health through Engineering and Simulation (ARCHES) program, UIUC/OSF Hospital, Peoria IL. Any opinions, findings, and conclusions or recommendations expressed in this publication are those of the authors and do not necessarily reflect the views of the funding agencies.

REFERENCES

Baber, C. (2020). Is expertise all in the mind? How embodied, embedded, enacted, extended, and situated theories of cognition account for expert performance. In P. Ward, J. M. Schraagen, J. Gore, & E. Roth (Eds.), *The Oxford Handbook of Expertise: Research & Application* (pp. 243–261). Oxford University Press, Oxford, UK.

Bodenheimer, T., Lorig, K., Holman, H., & Grumbach, K. (2002). Patient self-management of chronic disease in primary care. *Journal of the American Medical Association*, 288, 2469–2475.

Brundisini, F., Vanstone, M., Hulan, D., DeJean, D., & Giacomini, M. (2015). Type 2 diabetes patients' and providers' differing perspectives on medication nonadherence: A qualitative meta-synthesis. *The BMJ Health Services Research, 15*(1), 516.

Carayon, P., Hundt, A. S., Karsh, B. T., Gurses, A. P., Alvarado, C. J., Smith, M., & Brennan, P. F. (2006). Work system design for patient safety: The SEIPS model. *The BMJ Quality & Safety, 15*(suppl 1), i50–i58.

Carayon, P., Kianfar, S., Li, Y., Xie, A., Alyousef, B., & Wooldridge, A. (2015). A systematic review of mixed methods research on human factors and ergonomics in health care. *Applied Ergonomics, 51*, 291–321.

Chin, J., Madison, A., Stine-Morrow, E. A. L., Gao, X., Graumlich, J. F., Murray, M. D., Conner-Garcia, T., & Morrow, D. G. (2015). Cognition and health literacy in older adults' recall of self-care information. *The Gerontologist, 57*, 261–268.

Clark, H. H. (1996). *Using Language*. Cambridge: CUP.

Clark, H. H., & Brennan, S. E. (1991). Grounding in communication. In L. B. Resnick, J. Levine, & S. D. Teasley (Eds.), *Perspectives on Socially Shared Cognition* (pp. 127–149). Washington, DC: American Psychological Association.

Clark, H. H., & Wilkes-Gibbs, D. (1986). Referring as a collaborative process. *Cognition, 22,* 1–39.

Craik, F. I. (1994). Memory changes in normal aging. *Current directions in Psychological Science, 3*(5), 155–158.

Drews, F. A., & Bakdash, J. Z. (2013). Simulation training in health care. *Reviews of Human Factors and Ergonomics, 8*(1), 191–234.

Graumlich, J. F., Wang, H., Madison, A., Wolf, M. S., Kaiser, D., Dahal, K., & Morrow, D. G. (2016). Effects of a patient-provider, collaborative, medication-planning tool: A randomized, controlled trial. *Journal of Diabetes Research, https://doi.org/10.1155/2016/2129838.*

Gray, W. D., & Fu, W. T. (2004). Soft constraints in interactive behavior: The case of ignoring perfect knowledge in-the-world for imperfect knowledge in-the-head. *Cognitive Science, 28*(3), 359–382.

Haidet, P. (2007). Jazz and the 'art' of medicine: Improvisation in the medical encounter. *The Annals of Family Medicine, 5*(2), 164–169.

Hartshorne, J. K., & Germine, L. T. (2015). When does cognitive functioning peak? The asynchronous rise and fall of different cognitive abilities across the life span. *Psychological Science, 26*(4), 433–443.

Hazlehurst, B., McMullen, C. K., & Gorman, P. N. (2007). Distributed cognition in the heart room: How situation awareness arises from coordinated communications during cardiac surgery. *Journal of Biomedical Informatics, 40*(5), 539–551.

Hazlehurst, B., McMullen, C., Gorman, P., & Sittig, D. (2003). How the ICU follows orders: Care delivery as a complex activity system. In *AMIA Annual Symposium Proceedings* (Vol. 2003, p. 284). American Medical Informatics Association, Bethesda, MD.

Helander, M. G., & Khalid, H. M. (2006). Affective and pleasurable design. In G. Salvendy (Ed.), *Handbook of Human Factors*, Third Ed (pp. 543–572). New York: Wiley.

Henry, J. D., MacLeod, M. S., Phillips, L. H., & Crawford, J. R. (2004). A meta-analytic review of prospective memory and aging. *Psychology and Aging, 19*(1), 27–39.

Holden, R. J., Carayon, P., Gurses, A. P., Hoonakker, P., Hundt, A. S., Ozok, A. A., & Rivera-Rodriguez, A. J. (2013). SEIPS 2.0: A human factors framework for studying and improving the work of healthcare professionals and patients. *Ergonomics, 56*(11), 1669–1686.

Holden, R. J., Schubert, C. C., & Mickelson, R. S. (2015). The patient work system: An analysis of self-care performance barriers among elderly heart failure patients and their informal caregivers. *Applied Ergonomics, 47*, 133–150.

Holden, R. J., Valdez, R. S., Schubert, C. C., Thompson, M. J., & Hundt, A. S. (2017). Macroergonomic factors in the patient work system: Examining the context of patients with chronic illness. *Ergonomics, 60*(1), 26–43.

Hutchins, E. (1995). How a cockpit remembers its speeds. *Cognitive Science, 19*(3), 265–288.

Insel, K. C, Einstein, G. O., Morrow, D. G., & Hepworth, J. T. (2016). Multifaceted prospective memory intervention to improve medication adherence. *Journal of the American Geriatrics Society*, 64, 561–568.

Lindenberger, U., & Mayr, U. (2014). Cognitive aging: is there a dark side to environmental support?. *Trends in cognitive sciences*, 18(1), 7-15.

McLaughlin, A. C., Rogers, W. A., & Fisk, A. D. (2002, September). Effectiveness of audio and visual training presentation modes for glucometer calibration. In *Proceedings of the Human Factors and Ergonomics Society Annual Meeting* (Vol. 46, No. 25, pp. 2059–2063). Sage, CA: Los Angeles, CA: SAGE Publications.

Meyer, D., Leventhal, H., & Gutmann, M. (1985) Common-sense models of illness: The example of hypertension. *Health Psychology*, *4*(2), 115–135.

Mickelson, R. S., Unertl, K. M., & Holden, R. J. (2016). Medication management: The macrocognitive workflow of older adults with heart failure. *Journal of Medical Internet Research Human Factors, 3*, e27: https://humanfactors.jmir.org/2016/2012/e2027/.

Mickelson, R. S., Willis, M., & Holden, R. J. (2015). Medication-related cognitive artifacts used by older adults with heart failure. *Health Policy & Technology, 4*, 387–398.

Mikels, J. A., Löckenhoff, C. E., Maglio, S. J., Carstensen, L. L., Goldstein, M. K., & Garber, A. (2010). Following your heart or your head: Focusing on emotions versus information differentially influences the decisions of younger and older adults. *Journal of Experimental Psychology: Applied, 16*, 87.

Mishel, M. H. (1999). Uncertainty in chronic illness. *Annual Review of Nursing Research, 17*(1), 269–294.

Monk, A. (2000). User-centered design. In *International Conference on Home-Oriented Informatics and Telematics* (pp. 181–190). Springer, Boston, MA.

Monk, A. F. (2008). Common ground in electronically mediated conversation. In John M. Carroll (Ed.), *Synthesis Lectures on Human-Centered Informatics #1* (pp. 1–50). Morgan & Claypool, London, UK.

Morrow, D. G. Azevedo, R. F. L., Garcia-Retamero, R., Hasegawa-Johnson, M., Huang, T., Schuh, W., Gu, K., & Zhang, Y. (2019). Contextualizing numeric clinical test results for gist comprehension: Implications for EHR patient portals. *Journal of Experimental Psychology: Applied, 25*(1), 41–61.

Morrow, D. G., & Fischer, U. M. (2013). Communication in socio-technical systems. In John D. Lee & A. Kirlik (Eds.), *Oxford Handbook of Cognitive Engineering* (pp. 178–199). New York: Oxford University Press.

Morrow, D. G., North, R., & Wickens, C. D. (2006). Reducing and mitigating human error in medicine. In R. Nickerson (Ed.), *Reviews of Human Factors and Ergonomics* (Vol. 1, pp. 254–296). Santa Monica, CA: Human Factors & Ergonomics Society.

Morrow, D. G., Ridolfo, H. E., Menard, W. E., Sanborn, A., Stine-Morrow, E. A. L., Magnor, C., Herman, L., Teller, T., & Bryant, D. (2003). Environmental support promotes expertise-based mitigation of age differences in pilot communication tasks. *Psychology and Aging, 18*, 268–284.

Morrow, D. G., & Rogers, W. A. (2008). Environmental support: An integrative framework. *Human Factors, 50*, 589–613.

National Research Council. (2011). *Health Care Comes Home: The Human Factors*. National Academies Press, Washington DC.

Nemeth, C. P., Cook, R. I., O'Connor, M. F., & Klock, P. A. (2004). Using cognitive artifacts to understand distributed cognition. *Institute of Electrical and Electronics Engineers Transactions on Systems, Man and Cybernetics - Part A: Systems and Humans, 34*, 726–735.

Or, C. K., Karsh, B. T., Severtson, D. J., Burke, L. J., Brown, R. L., & Brennan, P. F. (2011). Factors affecting home care patients' acceptance of a web-based interactive self-management technology. *Journal of the American Medical Informatics Association, 18*(1), 51–59.

Palen, L., & Aaløkke, S. (2006, November). Of pill boxes and piano benches: Home-made methods for managing medication. In *Proceedings of the 2006 20th Anniversary Conference on Computer Supported Cooperative Work* (pp. 79–88). ACM, New York, NY.

Park, D. C., Hertzog, C., Leventhal, H., Morrell, R. W., Leventhal, E., Birchmore, D., Martin, M., & Bennett, J. (1999). Medication adherence in rheumatoid arthritis patients: Older is wiser. *Journal of the American Geriatrics Society, 47*(2), 172–183.

Pollack, M., Chastek, B., Williams, S. A., & Moran, J. (2010). Impact of treatment complexity on adherence and glycemic control: An analysis of oral antidiabetic agents. *Journal of Clinical Outcomes Management, 17*(6), 257–265.

Rajkomar, A., Mayer, A., & Blandford, A. (2015). Understanding safety–critical interactions with a home medical device through distributed cognition. *Journal of Biomedical Informatics, 56*, 179–194.

Reyna, V. (2011). Across the life span. *Communicating Risks and Benefits: An Evidence-based User's Guide* (pp. 111–120). Silver Springs, MD: Federal Drug Administration.

Rogers, W. A., Hertzog, C., & Fisk, A. D. (2000). An individual differences analysis of ability and strategy influences: Age-related differences in associative learning. *Journal of Experimental Psychology: Learning, Memory, and Cognition, 26*(2), 359.

Rogers, W. A., Mykityshyn, A. L., Campbell, R. H., & Fisk, A. D. (2001). Analysis of a "simple" medical device. *Ergonomics in Design, 9*(1), 6–14.

Simon, H. A. (1979). *Models of Thought* (Vol. 352). Yale University Press, London, UK.

Slovic, P., & Peters, E. (2006). Risk perception and affect. *Current Directions in Psychological Science, 15*(6), 322–325.

Sweller, J. (2011). Cognitive load theory. In J. P. Mestre & B. H. Ross (Eds.) *Psychology of Learning and Motivation* (Vol. 55, pp. 37–76). Academic Press, Cambridge, MA.

Taha, J., Czaja, S., Sharit, J., & Morrow, D. (2013). Factors affecting the usage of a personal health record (PHR) to manage health. *Psychology and Aging, 28*, 1124–1139.

Török, G., Pomiechowska, B., Csibra, G., & Sebanz, N. (2019). Rationality in joint action: Maximizing coefficiency in coordination. *Psychological Science.* doi: 10.1177/0956797619842550

Touron, D. R. (2015). Memory avoidance by older adults: When "old dogs" won't perform their "new tricks". *Current Directions in Psychological Science, 24*(3), 170–176.

Xiao, Y. (2005). Artifacts and collaborative work in healthcare: Methodological, theoretical, and technological implications of the tangible. *Journal of Biomedical Informatics, 38*(1), 26–33.

Zhang, J., & Norman, D. A. (1994). Representations in distributed cognitive tasks. *Cognitive Science, 18*(1), 87–122.

Zikmund-Fisher, B. J., Scherer, A., Witteman, H. O., Solomon, J. B., Exe, N. L., Tarini, B. A., & Fagerlin, A. (2017). Graphics help patients distinguish between urgent and non-urgent deviations in laboratory test results. *Journal of the American Medical Informatics Association, 24*(3), 520–528.

3 Physical Patient Ergonomics

Understanding and Supporting Physical Aspects of Patient Work

Linsey M. Steege
University of Wisconsin-Madison

Lora Cavuoto
University at Buffalo, The State University of New York

Barbara J. King
University of Wisconsin-Madison

CONTENTS

> Mr. A is a 67-year-old male who recently retired and lives at home with his wife. He and his wife enjoy several outdoor activities, such as hiking and kayaking, and have plans to travel. Mr. A is concerned about recent weight gain and the onset of joint pain to his knees and hands. He talked with his healthcare team about weight loss, his increasing joint pain and reduced strength and dexterity in his hand and wrist—perhaps as a result of his work in a laboratory during much of his career. The pain and changes in hand function have contributed to some difficulty in opening packages, including medications, and he has concerns that at some point it may impact his ability to eat and groom.

In the anecdote above, Mr. A has multiple health goals that can be supported by ergonomics research and applications to better design the tools, tasks, and environments within Mr. A's "daily life" to ensure they are compatible with his goals, needs, and abilities. Within the discipline of human factors or ergonomics, the International Ergonomics Association defines physical ergonomics as the domain of specialization "concerned with human anatomical, anthropometric, physiological, and biomechanical characteristics as they relate to physical activity" (*Definition and Domains of Ergonomics*, 2019). Physical ergonomics aims to understand interactions among humans and other elements of a system and then guide design to improve human physical health and safety, performance, and satisfaction or physical comfort (*Definition and Domains of Ergonomics*, 2019). One of the primary foci within physical ergonomics research and practice has been industrial and occupational ergonomics, which emphasizes the understanding of physical demands within work settings and designing tools, tasks, environments, and processes to ensure that they are compatible with worker capacity and to minimize safety risks. For example, work demands and safety risks associated with non-neutral postures at work; lifting, pushing, and pulling tasks; repetitive movements; strength and dexterity requirements for interacting with physical tools and devices; and the layout of physical environments to "fit" workers and ensure sufficient reach have all been well studied and interventions and guidelines exist to minimize risks for worker injury and promote safety, performance, and satisfaction/comfort. This includes research specifically targeting the physical demands on healthcare workers and the design of healthcare physical environments; for example, guidelines for safe movement of patients and design of hospital beds to reduce low back strain and pushing forces have been studied with nurses.

However, the "work" of the general public including persons such as Mr. A has not been as well studied, outside of the occupational setting. The US Department of Health and Human Services' Healthy People program identifies national health objectives and improvement priorities, including attaining high-quality, longer lives free of preventable disease, disability, injury, and premature death; achieving health equity, eliminating disparities, and improving the health of all groups; creating social and physical environments that promote good health for all; and promoting quality of life, healthy development, and healthy behaviors across all life stages (*Healthy People 2030 Framework*, 2014). Achieving these goals will require human factors and ergonomics, including the domain of physical ergonomics, to expand its focus to consider the work required to pursue health in day-to-day life beyond the occupational setting. Particularly, the work of patients and their non-professional caregivers requires attention to the unique activities, conditions, and environments experienced by this subset of the general public.

3.1 DEFINING PHYSICAL PATIENT ERGONOMICS

Physical patient ergonomics can be defined as a focused area within patient ergonomics, like the sub-discipline of physical ergonomics within the broader discipline of human factors and ergonomics. Physical patient ergonomics involves applying physical ergonomics methods, theories, and tools to a phenomenon where:

- A patient (or non-professional caregiver) is involved as a primary actor;
- The primary actor performs effortful physical work activities or work activities that require physical perception or function or interactions with physical tools or environments; and
- The activity is performed in pursuit of health goals.

The criteria listed, although helpful for beginning to establish potential boundaries and opportunities for the focus of physical patient ergonomics, are open to some interpretation and blurriness in the lines between physical patient ergonomics and other related areas of work and study. For example, as definitions of health evolve from a historical biomedical focus on the presence or absence of disease or infirmity to a broader view of multidimensional well-being (World Health Organization (WHO), 2014), it may be difficult to discern what is a health goal as opposed to a function of daily life. For example, a patient with impaired vision may experience challenges navigating physical environments or driving. Maintaining independence and being able to drive or navigate community environments to attend social events may be goals that are important to an overall state of well-being and high quality of life for these individuals. However, would navigating environments be considered pursuit of a health goal in line with the broader work on patient ergonomics? Another important question to address may be who is considered a patient?

The criterion that a patient or non-professional caregiver be involved in the work as the primary actor may also be open to interpretation. Physical ergonomics methods and tools may be valuable for activities that include joint work or collaboration between healthcare professionals and patients. For example, research can consider the physical demands on both nursing staff and patients as they work together to engage in patient transfer or patient ambulation activities to guide tools that might improve the safety of both actors. In this scenario, the patient is one of the primary actors and therefore this could be considered physical patient ergonomics. If the focus of the research is solely on reducing the biomechanical load (e.g., to reduce risk of low back injuries) for the nurse, or if the patient is passive and not engaged in the work of completing the transfer or mobility task, then this would fall outside the bounds of physical patient ergonomics and within the realm of industrial or occupational ergonomics.

There is ample work in other disciplines that accounts for physical needs, abilities or limitations, and implications for the design of interventions, tools, or environments to improve human performance, safety, or comfort. For example, many architects and industrial designers have promoted the concept of "universal design," where products and environments are created so that all people can easily use them (Steinfeld & Mullick, 1990). One application is the universal design of housing that includes zero-level entrance into homes and having all rooms accessible on one level, allowing adults to age in place. A second example is rehabilitation, where gait analysis for patients in a clinical environment is often used to measure disease or rehabilitation progression. However, current research in these other disciplines often does not include the application of formal physical ergonomics tools such as anthropometric assessment for equipment used or formal observation-based risk assessment techniques. Physical ergonomics can be used to answer questions such as: which strategies for completing a physical task are easier to do; which devices are easier to use; can the task be safely accomplished by a range of people for the required duration using the prescribed tools and methods; and can a specific individual perform a task safely?

3.2 PHYSICAL PATIENT ERGONOMICS IN THE LITERATURE

To better understand the breadth of physical patient ergonomics and identify gaps in knowledge and opportunities for future research, a scoping review was conducted within human factors and ergonomics literature. The review focused primarily on research published in the human factors and ergonomics literature in the last 20 years as the healthcare system and patients' roles in care delivery and managing their own health have evolved.

Existing literature on physical patient ergonomics can be organized based on the types of patient work activities included or addressed; the specific characteristics of the patient population included or targeted; and the physical ergonomics approach utilized in the research (Table 3.1).

TABLE 3.1
Themes for Organizing Existing Research on Patient Physical Work

Patient Physical Work Activities	Patient Characteristics	Physical Ergonomics Approaches
• Monitoring health and activities • Navigating the physical environment • Activities of daily living • Medication management	• Age • Chronic health condition • Physical function impairment • Fitness level • Medication use • Prior injuries • Experience • Sensory ability (vision, hearing, touch sensitivity)	• Force/strength requirements • Range of motion assessment • Gait/balance analysis • Task and device redesign • Design for anthropometric variability • Heat stress

3.2.1 Patient Physical Work Activities

Patient physical work comprises various tasks requiring physical function (e.g., movement, strength, dexterity), physical fit or alignment with tools or the environment (e.g., postural demands, anthropometry), and physical perception (e.g., vision, hearing, tactile sensation). Considerations for understanding and ultimately designing to support patient work include the capabilities or capacity of patients to carry out the physical demands of a task, maintaining health and safety and ensuring minimal risk of physical harm or injury, and ensuring comfort.

3.2.1.1 Monitoring Health and Physical Activities

There is a growing body of literature around health monitoring systems, and particularly the use of technologies and modeling tools to monitor patients' health and physical activities outside of a formal healthcare setting; cue patient behaviors related to health maintenance and management; and guide delivery of services by the healthcare system (Mshali et al., 2018). An overarching aim of this research is to enable patient independence and reduce the burden on the health system by minimizing interactions with healthcare institutions (Li et al., 2019; Mshali et al., 2018). These technologies can also help to reduce demands on patients or caregivers by automating work related to self-management, monitoring health behaviors, activity levels, safety, and decision-making (see Chapters 5 and 7 in this volume).

Smart monitoring systems are context-aware and composed of sensor, communication, and processing components (Mshali et al., 2018). Existing literature has largely focused on technical design issues and modeling approaches for these systems; however, there is a recognition of the need to consider human factors to ensure usability, acceptability, and safety (Hossain, 2014; Lemlouma & Chalouf, 2012; Mshali et al., 2018). Human factors research has addressed user acceptance of home-based and wearable smart monitoring and activity tracking systems (Ehmen et al., 2012; Fausset et al., 2013; Li et al., 2019; Mshali et al., 2018; Preusse et al., 2017). Key factors associated with user acceptance of monitoring systems include perceived usefulness, compatibility, comfort, and reported health status. Other studies have considered design requirements for patients and caregivers (Gonzalez et al., 2014; Zulas et al., 2014) and evaluated the effectiveness of monitoring technologies for supporting patient self-management, tracking physical activity, and detecting changes in health and safety (Wilkinson et al., 2017).

Physical ergonomics research can help improve the understanding of patient function and activities to guide sensor selection and design for monitoring systems. Sensing systems include environmental sensors that capture information from the patient's environment such as noise, light, or temperature data, as well as interactions between patients and their environments (Mshali et al., 2018). Such interactions might include movement between rooms using motion sensors or use of household devices (microwave oven, sink) or objects (medication bottle) to carry out a task (e.g., preparing a meal, taking a medication) (Mshali et al., 2018; Suryadevara et al., 2013). Body or wearable sensors can measure physiologic data (e.g., heart rate, blood pressure, glucose, respiration, temperature), and both wearable and non-wearable sensors can capture body movement or posture (Li et al., 2019; Mshali et al., 2018). Physical ergonomics research has generated new knowledge about which parameters (e.g., balance) are most predictive of critical health changes or safety events (e.g., falls) (Dueñas et al., 2016). This is an example of how laboratory-based physical ergonomics research can help in determining what physical function variables need to be captured to guide design of monitoring sensors.

3.2.1.2 Navigating the Physical Environment and Interactions with Physical Devices

In pursuit of health, patients must interact with the physical environment, including navigation of physical spaces. Health-related physical environments for patients might include the patients' home, community settings, and healthcare delivery settings. A large portion of human factors research related to this patient work activity is on addressing the navigation needs of patients with impaired perception or physical function. Multiple studies describe the design of tools to support improved wayfinding and safe navigation of spaces for patients with visual impairment (Lee, 2019; Lee et al., 2014; Lewis et al., 2015; Rousek & Hallbeck, 2011). One of the challenges in evaluating this area of work is whether wayfinding in public or community spaces is an activity performed in pursuit of health goals, therefore meeting the definition of physical patient ergonomics. Knowledge gained from these studies may contribute to improved design of physical environments to support wayfinding within healthcare settings. Studies report design guidelines and evaluation of the effectiveness of different wayfinding interfaces within hospital environments (Harper et al., 2017; Rangel & Mont'Alvão, 2011; Rousek & Hallbeck, 2011). This work addresses both physical ergonomics considerations for physical perception and physical access to displays or interfaces and cognitive ergonomics considerations for comprehension and attention.

In addition to research on wayfinding, interactions with physical elements of the built environment have been studied in both home and healthcare environments. Bonenberg et al. (2019) present an overview of opportunities for inclusive design in home kitchen environments to better support the physical needs of persons with disabilities and older adults. Being able to physically access and safely manipulate kitchen appliances and storage spaces is important for health-related tasks such as preparing meals or managing and taking medications. Considerations related to physically accessing spaces in the home may also be important for exercise or rehabilitation activities. In healthcare settings, researchers have evaluated the accessibility of medical equipment for patients with disabilities or impaired physical function (Story et al., 2010). Identified physical ergonomics access and safety barriers related to medical equipment include orienting and positioning the body, body support, manipulation or operation of controls, sensory barriers, lack of compatibility with the use of assistive technology, and inability to access the medical equipment due to other elements in the physical environment (e.g., furniture blocking access, insufficient space to maneuver) (Story et al., 2010). Other work has also considered comfort and body forces associated with patient use of medical equipment such as imaging devices (Boute et al., 2018). A well-designed built environment that considers the physical capabilities and needs of the patient can support patient independence, allowing persons to navigate without reliance on a caregiver or to remain at home rather than a specialized living environment, such as Assisted Living, Group Home, or Skilled Nursing Facility.

Interactions with assistive devices, such as wheelchairs or crutches to enable mobility, have also been considered. Research has considered both the impact of wheelchair or crutch use on patient comfort and potential risk of hand or joint injury by evaluating hand pressures, shoulder forces, and joint movement during wheelchair propelling or crutch walking (Kabra et al., 2015; Kloosterman et al., 2016; Sherif et al., 2016). Findings from this area of research can help address injury risks, reduce physical demands, and improve comfort for patients who use these devices and guide design of gloves, hand grips, or other tools. The particular physical functional abilities of patients with specific health conditions or diagnoses have also been studied to support improved design of assistive devices. For example, research has identified new wheelchair designs to address the loss of mobility for patients with hemiplegia (Tsai et al., 2008) and design recommendations for accessibility of diabetes management devices to accommodate patients with vision impairments (Story et al., 2009).

3.2.1.3 Performing Activities of Daily Living

Activities of Daily Living (ADLs) are tasks associated with basic self-care, independence, and maintaining the quality of life such as feeding, bathing, dressing, grooming, work, toileting, social participation, and leisure. A patient's ability to independently complete ADLs is often used by health professionals as a measure of physical function (Krapp, 2007). In addition, the completion of ADLs is likely associated with the pursuit of health goals. As defined by the WHO, health is "a state of complete physical, mental, and social well-being and not merely the absence of disease or infirmity" (WHO, 2014, p. 1).

Physical ergonomics research has also used the performance of ADLs as a measure to evaluate the effectiveness of different rehabilitation devices. White et al. (2017) considered ADL performance in conjunction with ratings of perceived exertion and muscle load to evaluate the impacts of exoskeletal orthotic devices on patients' upper extremity rehabilitation work following a stroke. Findings from this work suggested that some orthotics may cause undesirable loads on patients' arms due to the weight of the device, which may limit patients' use of devices or increase muscle fatigue and risk of additional injury.

Other research has explored how the design of consumer products impacts posture and physical forces required to complete ADLs. For example, a study by Hensler et al. (2015) evaluated a new food packaging design for ease of opening by patients with hand osteoarthritis. Patients reported that food package design differentially influenced the force required, tear tab size, and palpability, which may reduce physical demands for patients with decreased hand strength and dexterity completing feeding tasks. In other work, Roda-Sales et al. (2019) investigated the effects of commercially available assistive devices on hand and arm posture, precision grasps, and contact forces during meal preparation (opening cans or bottles, pouring), eating (utensil use and drinking), or grooming (brushing teeth and hair, and dressing) ADLs. Physical ergonomics can support selection of appropriate assistive devices, for example, which type of fork design will result in minimal postural deviation and allow for the best reach during eating, and inform the design of new products (e.g., a new utensil design) to support this type of patient work.

Outside of self-care ADLs, physical ergonomics research also focuses on supporting patient work related to engagement in social and leisure activities. Mobility disabilities and impairments can impact how patients are able to navigate physical environments, potentially limiting opportunities for participating in activities outside the home (C. Brown et al., 2012; Carlson & Myklebust, 2002; Kayes et al., 2011). The design of clothing can also act as a barrier to patient participation in activities. Physical ergonomics approaches can help evaluate the impact of clothing on patients' thermal regulation; anthropometric requirements; compatibility of clothing design with patient medical and assistive devices, such as colostomy bags, catheters, and wheelchairs; and the physical function requirements (strength, flexibility, dexterity) required to dress in different types of clothing (Kabel et al., 2017; Wang et al., 2014). An increased understanding of the use requirements for patient clothing can inform ergonomic design of functional clothing to support patient work (Kabel et al., 2017).

3.2.1.4 Medication Management

Safe medication administration is an important work activity for many patients. Much of the human factors and ergonomics research related to medication management and administration in the home environment has focused on understanding and supporting patient cognitive work related to remembering to take medication accurately as prescribed. Medication administration may also impose physical demands that would benefit from physical ergonomics approaches to guide improved design. For example, differences in visual perception that can occur as the result of aging or a health condition such as diabetic retinopathy may negatively impact a patient's ability to accurately read a medication label or discern the color of a pill (Ward et al., 2010). Opening medication containers also requires gripping, squeezing, and turning actions (often performed at the same time). Many patients and their caregivers lack sufficient strength and dexterity to open pill bottles, thus influencing their ability to take medications as directed (Ward et al., 2010). The physical design of medication packaging to account for differences in functional ability may help reduce physical demands and increase medication adherence and safety.

In addition to medication packaging, physical ergonomics approaches can also be used to improve syringe design for injecting medications. A study by Sheikhzadeh et al. (2012) compared a conventionally versus ergonomically designed syringe for injection of biological medications by patients with rheumatoid arthritis. They found that patients were able to exert higher isometric forces with the ergonomic syringe design and rated the new design as easier to use and control during the injection process.

3.2.2 Patient Characteristics

Much attention has focused on changes in physical function and perception associated with normal aging and identifying design requirements to support the physical work of older adults. Normal changes in aging include decreased ability to focus on near objects (presbyopia) and decreases in high-frequency hearing (presbycusis). Furthermore, older adults often experience a loss of muscle mass and strength and a decrease in vibratory sensation. These age-related changes should be considered and addressed when designing monitoring sensors or measures of physical function, work tools, environments, and tasks (Bonenberg et al., 2019; Dueñas et al., 2016; Li et al., 2019; Mshali et al., 2018).

There is also literature addressing physical patient ergonomics in younger children and adolescents. Kenward (1971) considered the anthropometric measures of young users to guide improved wheelchair design and Lang et al. (2013) and Howard et al. (2017) examined adolescent user perceptions of design characteristics for medical devices used to treat cystic fibrosis and asthma, respectively. Overall, children and young adults are included less frequently in the literature on physical patient ergonomics. Yet, these patients may be important to consider, particularly as they progress toward independence from parent caregivers in managing their health. For example, adolescent patients with chronic diseases such as diabetes will take on increasing responsibilities for monitoring their blood sugar and administering medications as they age. Physical size and functional abilities, along with cognitive abilities, need to be accounted for when creating monitoring products for younger populations.

The physical functional abilities of patients are also central to much of the physical patient ergonomics literature. Changes in perception due to impaired vision or hearing are addressed in studies related to wayfinding or interactions with the physical environment (Lee, 2019; Lee et al., 2014; Lewis et al., 2015). Impaired mobility either in patients with disabilities or as a result of an injury or medical condition (e.g., stroke) is addressed in studies on physical impact and use of assistive devices (Kabra et al., 2015; Sherif et al., 2016; Tsai et al., 2008), access to medical devices (Story et al., 2010), and completion of ADLs (Kabel et al., 2017; White et al., 2017). Patients also experience changes in strength or dexterity, and these aspects of physical function, particularly in the hand and arms, are also addressed (Hensler et al., 2015; Sheikhzadeh et al., 2012). Again, these changes are often tied to a specific medical condition, such as rheumatoid arthritis or hand osteoarthritis, but not always. Whereas deviations from "normal" physical function are more frequently studied, less is known about the potential physical demands or risks that might be associated with physical work related to the pursuit of health goals for "healthy" patients or patients with typical physical perceptual and functional abilities. For example, are there risks that may emerge due to repetitive actions for patients that administer injection medications? Physical patient ergonomics could also add value to evaluating the physical toll that a caregiver without any limitations in perception or physical function may experience when helping to support their loved one in performing ADLs, predicting potential injury risk, and designing new tools that can support patient transfer and lifting activities within home versus healthcare settings.

3.2.3 Physical Ergonomics Approaches

A diverse set of physical ergonomics concepts are studied within the physical patient ergonomics literature using various research methods. Physical ergonomics topics that are included as part of understanding patient functional abilities and guiding design to support patient work in this area include visual perception; tactile perception; contact stress; heat stress; biomechanics; mobility; posture; anthropometry; comfort; muscle activity and strength; and balance. A summary of physical ergonomics approaches used in physical patient ergonomics literature is provided in Table 3.2. Notably, some concepts within the field of physical ergonomics are not as prevalent, including energy expenditure and physical fatigue. These topics may warrant exploration in future research to understand these aspects of patient physical work.

Researchers have used quantitative and qualitative study designs and data collection approaches to contribute knowledge on patient physical function as it relates to patient work and implications for design (see Chapters 9, 11, and 12 in this volume). Quantitative studies have utilized all four main types of research design—descriptive, correlational, quasi-experimental, and experimental. Data have been collected using surveys (e.g., Lee et al., 2014), interviews (e.g., Holden et al., 2015), direct observation (e.g., Rousek & Hallbeck, 2011), and direct measurement of patients' physical function and physiological responses (e.g., Tsai et al., 2008). Physical ergonomics measures and tools such as electromyography (e.g., White et al., 2017), postural demand assessment tools (e.g., Rapid Upper Limb Assessment—RULA) (e.g., Roda-Sales et al., 2019), posturography (e.g., Dueñas et al., 2016), grip and finger dynamometers (e.g., Hensler et al., 2015; Sheikhzadeh et al., 2012), heart rate monitors, pressure mapping (e.g., Kabra et al., 2015), force plates, and kinematics marker tracking and camera systems were all used. Commonly used subjective measures assessed patients' perceptions of exertion, pain, accessibility, and comfort (e.g., Hensler et al., 2015;

TABLE 3.2
Examples of Physical Ergonomics Approaches Used in Physical Patient Ergonomics Literature

Category	Description	Example Measurement Parameters	Example Application(s) to Design
Visual perception	Measurement and design of the visual environment, including lighting and displays	Accommodation, color distinction, visual acuity, lighting standards for luminance, contrast, glare	Color contrast such as black on white Bright-colored tape on stair risers Bright task lighting for reading medication labels Cooler color temperature lighting is better for older adults
Auditory perception	Assessing the ability to detect and interpret auditory displays and to minimize discomfort or annoyance	Frequency/pitch, sound level, speech intelligibility, masking	Reduced background noise for older adults to improve hearing Assisted listening devices for phone, tablet "apps" or closed-circuit systems

(Continued)

TABLE 3.2 (*Continued*)
Examples of Physical Ergonomics Approaches Used in Physical Patient Ergonomics Literature

Category	Description	Example Measurement Parameters	Example Application(s) to Design
Heat stress	Thermal comfort in indoor and outdoor environments based on the air temperature, humidity, airflow, amount of clothing, and physical activity level	Objective measurements of wet bulb globe temperature or expected comfort based on ISO 7730 (International Organization for Standardization (ISO), 2005), subjective ratings of comfort	Air conditioning in homes Outdoor activities in the morning or evening when it is cooler Ensure proper hydration Clothing design for persons with impaired thermoregulation
Biomechanics and strength	Compatibility of task force requirements compared to muscle/ joint strength	Grip and pinch strength, lifting capability	Handgrip design for tools for completing ADLs Design of packaging or open/close fasteners for medications Handling and use of mobile devices (including phones and tablets)
Mobility	Navigation of the built environment	Gait analysis, balance testing, clinical assessments such as Timed Up and Go (TUG) test, Tinetti Balance Test, or POMA Test	Assistive walking devices Sensor systems to detect movement and physical activity Declutter hallways and walking spaces Remove cords from walking pathways Remove throw rugs
Tactile perception and dexterity	Ability to distinguish controls and haptic alerts and to properly activate the necessary controls	Esthesiometer for tactile sensitivity	Design of medication packing and administration tools
Posture	Assessment of postural requirements to identify potentially awkward postures	Rapid assessment methods such as RULA or REBA	Supporting caregiver posture for assisting in ADLs Design of tools for completing ADLs
Anthropometry	Design for a population of users based on defined body size measurements	Fit assessment	Size of assistive technologies or mobility aids (crutches, wheelchair)
Balance	Stability in maintaining either a standing or seated position	Postural sway using posturography, clinical assessments such as TUG or 2-min walk	Placement of grab bars Use of chairs without wheels and with stable armrests

Kabel et al., 2017). Performance measures were also included in multiple studies. Data were collected in simulated environments, laboratory settings, and in the field, which included patient homes, healthcare settings, and the community.

In addition to the wide array of general physical ergonomics methods, tools, and theories described above, several studies identified correctable gaps in existing approaches. The importance of user-centered design processes is increasingly recognized across industries, including health care (see Chapter 10 in this volume). However, there can be challenges to effectively and efficiently engaging users in the design process and these challenges may be exacerbated when working with certain patient populations. For example, two studies discuss the challenges of representing diverse patients in design processes (Gyi et al., 2004; Vincent & Blandford, 2014). Collecting data on physical function or the impact of design on physical work may not be feasible with all patient populations or may add an extra burden. Patients may already experience high work demands from pursuing their own health goals or may experience challenges with both physical and cognitive functions that act as barriers to participation in human factors and ergonomics research or design processes (Holden et al., 2020). Mackrill et al. (2017) described how digital technologies may be used to encourage user participation in healthcare environment design. Martin et al. (2012) described a user-centered approach that can be used to elicit design requirements to support medical device development. Physical patient ergonomics is needed to develop more robust and comprehensive user descriptors of patient characteristics (anthropometric measurements, strength measurements, visual perception abilities, range of motion, and so on) that can address some of the potential challenges with engaging patients in design processes and support improved medical device design. Additional work is needed to continue to identify innovative mechanisms for supporting human factors and ergonomics work with patient and caregivers.

3.3 INTERSECTION OF PHYSICAL PATIENT ERGONOMICS WITH OTHER DOMAINS OF PATIENT ERGONOMICS

Work is rarely purely physical. The cognitive and organizational ergonomics aspects of patient work are often intertwined with physical ergonomics aspects (see Chapters 2 and 4 in this volume). For example, modern medication management involves cognition and there is a growing area of cognitive patient ergonomics research related to this topic (Chapter 2 in this volume). However, accessing pills as a part of medication administration is a physical task and methods and knowledge from physical ergonomics literature may be important for improving the design, performance, and safety related to this aspect of medication management. In contrast, safe patient ambulation may be considered a primarily physical task that can be supported by physical patient ergonomics, but ambulation can also be influenced by dual-task performance or distractions that must be processed, and which would benefit from cognitive patient ergonomics approaches.

Physical ergonomics offers one perspective, or slice, of understanding patient needs and capabilities to guide design to support patient work. There is a small core of primarily physical patient ergonomics work (e.g., patient ambulation, product package design for physical access, design of the built environment to accommodate assistive devices) that intersects with other domains of patient ergonomics, including cognitive and organizational ergonomics as well as other fields of usability and product design. The patient work system is often shared with or intersecting with the work system of healthcare professionals (Chapter 4 in this volume). Therefore, physical patient ergonomics is also one component of physical ergonomics applications in healthcare more generally (e.g., physical fatigue in healthcare professionals, design of tools, and the built environment to reduce postural demands and work-related musculoskeletal issues in surgeons).

3.4 CASE STUDY: HOME HEALTHCARE DEVICE DESIGN

Home-based rehabilitation systems have the potential to support self-management of physical improvement following impairment. Designing such systems with consideration for physical ergonomics can provide more effective designs and enhance the user experience. This has been evaluated in the case of a home-based stroke rehabilitation system.

Over 700,000 new cases of stroke occur each year (Rosamond et al., 2007). This number is projected to increase with the aging of the baby boomer generation. Despite evidence that improvements can still be made several years post stroke (Wolf et al., 2010), rehabilitation options in chronic stages of stroke are not well established and only a small fraction of individuals with stroke have continued interactions with a therapist 1 year after their stroke (Langhammer & Stanghelle, 2003). Although written home exercise programs are commonly prescribed at the end of formal therapies (Jeanne Langan et al., 2018), they do not support self-management for home rehabilitation. Continued practice is key to improvement of functional ability (Novak, 2011). Hence, developing approaches to better support patients with their home program is important. To address this need, the mRehab system was designed and developed (Bhattacharjya et al., 2019; Cavuoto et al., 2018; Lin et al., 2016; Nwogu et al., 2017).

mRehab is a home-rehabilitation portable measurement system that can provide critical data on upper extremity functional mobility across a range of capabilities (Figure 3.1, top). The system contains a smartphone embedded in 3D-printed objects and a mobile application (app) that provides the user with instructions, activity tracking, and performance feedback. The capabilities supported include general arm movement, forearm pronation and supination, scapula/arm stabilization, elbow extension, power grip, and fine motor control. These motions are important for performing essential ADLs and improving chronic post-stroke quality of life. The use of 3D printing provides a customizable option. A smartphone is embedded in a 3D-printed mug with a handle that can be adapted to an individual user's hand function, a 3D-printed bowl for bimanual rehabilitation, or a 3D-printed box-mounted holder for key and doorknob turning tasks. The 3D-printed items were designed to securely hold a smartphone or to hold the smartphone in place while a 3D-printed lever swiped across the smartphone screen recording touch data capturing rotational movements for the key or doorknob. The smartphone's built-in sensors capture activity data, including movement time, acceleration, orientation, and smoothness (jerk), during the prescribed home rehabilitation activities. The app allows patients to select the rehabilitation activity to be trained, set goals, track task completion (through

FIGURE 3.1 (Top) mRehab system: a smartphone with application, 3D-printed objects, and box. (Bottom) Design iterations for the mug design.

real-time auditory feedback), and monitor performance on movement time and smoothness compared to prior performance.

Physical ergonomics considerations were critical to the design of an effective system that meets the unique needs of the target population. The applications of physical ergonomics to this design process are highlighted below.

3.4.1 Physical Device Design

The use of 3D printing technology allows for customization of the objects used for training activities. In the current system, the mug, bowl, key, and doorknob were selected to support upper extremity functional activity rehabilitation. In the future, an expanded set of objects could be designed and printed to provide individualized rehabilitation based on the specific needs and goals of the patient. For the current objects, the size and shape of the mug handle, bowl, and key can be adjusted. Participants in both in-lab and in-home usability testing of the system had generally positive feedback on the 3D-printed objects (Bhattacharjya et al., 2019; Cavuoto et al., 2018).

A patient-centered design process was followed to arrive at the final object design. Feedback from individuals with stroke and older adults was solicited at each stage of the design process. An example of the design modifications for the handle of the 3D-printed mug is shown in Figure 3.1, bottom. For many individuals with stroke, the impact on their neurological system can result in a clenched hand on their affected side, and an inability to grasp in a traditional manner. As seen in the leftmost figure, the rounded handle can be challenging to use with this condition. Users preferred the "D" shape that is seen in the rightmost figure. This allowed enough space for varied hand sizes and could accommodate overlapping fingers, but also had the closure at the bottom of the handle to allow users to rest their hand and not fear the mug sliding out of their hand during use.

The patient-centered design approach affected other physical aspects of the design. For the mug, right- and left-handed versions were made to make the phone screen visible to patients depending on their affected side. For the bowl (shown in Figure 3.1, top), an edge was added along the top so that patients could hold onto the bowl without worrying about it accidentally slipping out of their hands. The bowl sides also have a shallower slope to encourage forearm supination while holding it.

3.4.2 Application of Ergonomics Methods for Performance Evaluation

Receiving feedback can assist rehabilitation progress by providing patients a better understanding of their abilities, building self-efficacy, and fostering the ability to set appropriate goals for themselves (Dobkin et al., 2010; Liu & Chan, 2014; Novak, 2011). However, most objective feedback is currently limited to task completion success/failure and completion time. Although in general, a faster completion time can indicate improved ability, completion time may plateau while performance gains may still be noticeable. In addition, task control is often equally important for successful performance of ADLs. For example, when drinking from a cup, if the cup is lifted quickly, but without control, liquid may spill over the cup edges. A trained therapist could assess quality by visual inspection, as an ergonomist might assess injury risk in the workplace. However, this becomes challenging as 1) individuals with chronic stroke are not regularly assessed by a therapist, 2) changes may become more subtle as a patient progresses, and 3) a clinic visit only provides a single time-point of observation. In a recent survey of practicing occupational and physical therapists, although only 11% of respondents reported using objective feedback more than 75% of the time, a majority prioritized the need for a means to provide objective feedback to therapists and patients on completion time, smoothness of movement, and symmetry of movement (Langan et al., 2018).

In developing valid metrics for performance quality for monitoring rehabilitation progress and providing feedback, designers can draw from kinematic analysis measured that has been used in physical ergonomics to evaluate ergonomic risk and characterize expert task performance. Kinematic measures such as smoothness of movement (jerk score) have been used in research to provide information on the efficiency of a movement (Langan et al., 2013). The utility of jerk as a metric of motor skill, physical exertion, and fatigue development has been documented in the ergonomics literature (Maman et al., 2017; Zhang et al., 2019). Whereas kinematic analysis had traditionally required a motion capture laboratory facility for accurate assessment, recent advances in sensor design and analysis approaches for inertial measurement units (IMUs) have allowed for accurate field-based measurement (Schall et al., 2016).

3.5 CASE STUDY: PATIENT PHYSICAL FUNCTION ACROSS SETTINGS

Up to 60% of adults aged 65 years and older will lose independence in one or more ADLs or the ability to independently walk across a small room during their hospital stay (Hastings et al., 2018). Loss of physical function in older adults during hospitalization has been documented in the literature for over 30 years and has been identified using multiple terms including *Functional Decline during Hospitalization, Hospital-Associated Disability,* and, more recently, *Trauma of Hospitalization.* Multiple comorbid conditions and the impact of an acute illness place older adults at high risk for a functional injury. However, imposed bed rest or limited ambulation and forced dependency in completing own ADLS are identified as the most predictable and preventable causes of loss of ADL ability in older adults during a hospital stay (C. J. Brown et al., 2004). Interventions to prevent loss of ADL independence in older patients have been targeted at an organizational level with little understanding of the patient perspective or barriers that prevent patient work to maintain their ability to remain functionally independent.

Only a few studies captured what it is like for older adults to be hospitalized and the impact of the environment on whether or not they get up to walk or are encouraged to be independent during their stay. Findings indicate that older adult patients view the ability to physically function as critical to their sense of wellbeing and expect they will be discharged with improved ADLs and able to go home (Boltz et al., 2010). Others identified that older patients desire to be involved in completing their everyday personal care and to independently walk during their stay, but are often denied the opportunity to work (perform own ADL care) and are restricted in how often or how much they are allowed to walk during hospitalization (Boltz et al., 2012; Penney & Wellard, 2007). Patient work, defined as the exertion of effort and investment of time on the part of patients to accomplish something (Strauss, 1993), has not been well studied in the context of hospital settings. Only one study (Bodden & King, 2018) captured hospital barriers from a patient perspective that prevent older adults from engaging in work to ambulate during a hospital stay. In hospitals, older adults often experience uncertainty about whether or not they can get up to walk. Uncertainty about the work of walking is related to poor communication or lack of approval from the nursing staff as to when, where, or how far patients can walk; lack of awareness that walking is beneficial to recovery; or who to ask for help when they want to get out of bed. The concept of poor communication as a barrier to patient work is consistent with other study findings (Penney & Wellard, 2007) (see also Chapter 6 in this volume). Older adults also receive a strong message from hospital clinicians (nurses and physicians) that they cannot walk without assistance or supervision because of concerns for falls. Older adults identified that because they are restricted in movement, walking is of low priority during the hospital stay or not important for maintaining their independence in ADLs after discharge (Bodden & King, 2018; C. J. Brown et al., 2007). Older adults also see hospitals as non-welcoming spaces for walking due to physical layouts that are confusing and lacking in signage for wayfinding, cluttered patient rooms and hallways, and seating (chairs, couches) that make it difficult to get in and out of (Bodden & King, 2018). The complexity of hospital environments adds to high task difficulty for older adult patients to initiate walking during hospitalization. If the work of walking (such as walking to and from the bathroom or out in the hallway, or completing a physical therapy session) becomes too challenging, patients' ability to work to complete the task of walking is significantly influenced, further jeopardizing maintenance of their functional independence at discharge.

Additionally, few studies have investigated work older adults engage in to return to their normal state of functional ability or what strategies they used to regain their functional independence after a hospital stay. Only one study identified that older adults post discharge engage in two types of work to resume their normal physical functional status, *Working to Regain* or *Working to Maintain* (Liebzeit, Bratzke, Boltz, et al., 2019). Which type of work older adults engaged in is dependent upon the person's physiologic and physical reserve and whether or not they have sufficient support from others (Liebzeit, Bratzke, Boltz, et al., 2019). Older adults who are *Working to Regain* often look for body cues, such as increased energy and strength, that they are ready to begin the work to get back to normal. Strategies older adults use to complete their work include pushing themselves to accomplish more in terms of how long they engage in exercise or walking and the intensity (how many minutes) they maintain of the physical activity. Older adults also monitor themselves for improvements (being able to climb a flight of stairs) as a sign that they are progressing and prevented setbacks by going to see their health provider sooner so their progress is not impacted. Those who are *Working to Maintain* describe their health as compromised, which impacts their ability to complete the physical work of getting stronger. Older adults who are *Working to Maintain* are forced to redefine what normal life now looks like for them and have to accept new restrictions such as what they can do for themselves, depending on others for assistance with ADLs, getting outside of their home to socialize or purchase groceries, and giving up hobbies or paid employment (Liebzeit, Bratzke, Boltz, et al., 2019).

The presence of reliable supporters (unpaid family members and friends or paid healthcare professionals [e.g., physicians, rehabilitation therapists, home health nursing assistance]) is critically important for older adults to engage in *Working to Regain* function. (Liebzeit, Bratzke, & King, 2019). Supporters provide an important source of encouragement for older adults when the physical work to get their function back becomes difficult. Supporters also fill in temporarily to help with household tasks (cooking, cleaning) so the older persons can focus their energy and efforts on getting better. Paid supporters (physical therapists) provide important information for older adults who are *Working to Regain* on how to perform exercises correctly and serve as another source for monitoring their health status and progression for getting back to a normal life (Liebzeit, Bratzke, & King, 2019). The concept of patient work after a hospital stay has also been described by others who identify that older adults transition through phases post discharge (Werner et al., 2019). These phases include changes in older adults' roles and responsibilities, establishing and integrating new self-care routines, and addressing new challenges and need for resources. The process of patient work after a hospital stay requires a high workload demand on part of the older adult and often occurs over months to years (Liebzeit, Bratzke, Boltz, et al., 2019; Werner et al., 2019).

3.6 DISCUSSION AND CONCLUSIONS

Physical patient ergonomics is a growing area of research and application that can help improve the design of tools and environments to support the work activities and health goals for patients, such as Mr. A mentioned at the start of the chapter.

Although examples of physical patient ergonomics research in the literature span different patient physical work activities and patient characteristics, as well as use a range of physical ergonomics approaches, this domain of patient-centered human factors and ergonomics is less developed than cognitive patient ergonomics. This may be partially attributed to the challenges to defining clear boundaries for physical patient ergonomics and the intersection of patient physical function and work goals with cognition and work organization.

The summary of current physical patient ergonomics literature and the two case studies illustrate the value of physical patient ergonomics for improving patient safety, performance of physical work activities, comfort, and satisfaction. Specifically, as highlighted with the first case study, in the design of health-related devices for home use, integration of user anthropometry and the variability that accompanies a specific patient population are necessary for designing devices to support patient work activities. As shown in the second case study, addressing ergonomic concerns for the design of the physical environment is critical for supporting mobility and resumption of independent mobility when transitioning from a hospital to other settings.

More research is needed to expand our understanding of the unique needs and capabilities of diverse patient populations as they complete multidimensional work tasks related to their health and well-being. Given current population aging trends, as well as increasing numbers of patients managing chronic and complex medical conditions in home and community environments (i.e., outside of formal healthcare settings), we need to recognize and support physical function, safety, and performance. Aging is not only associated with changes in cognitive function, the changes in sensory ability and physical function that occur as a natural byproduct of aging need to be considered in patient ergonomics research and practice. Other disciplines, such as rehabilitation sciences, nursing, and industrial design, are working to support the universal design and the functional performance of different patient populations. The discipline of human factors and ergonomics can offer a framework for integrating the physical function with the cognitive and sensory changes, whereas most of the other disciplines primarily focus on the physical alone. Physical ergonomics approaches that seek to understand and, when relevant, quantify patient work demands and patient capacity to safely and successfully complete physical work offer the value that can complement the perspectives of other fields aiming to support patient health and function. Furthermore, understanding how the physical impacts (either supporting or impairing) cognitive function and vice versa can allow for healthcare systems designs that support the patient as a whole. There are opportunities for human factors and ergonomics to explore whether our understanding of physical work and methods used in industrial ergonomics can translate to understanding and supporting patient work. Do we need different methods or tools, for example, to assess the postural demands of patient work as opposed to the existing tools that are used to assess occupational work? Future research should consider the need for and value of new technologies, new analysis methods, or new design approaches that support patient work. As new approaches and technologies are developed, these must also be validated to determine effectiveness for the challenges faced by patients.

Lastly, much of the research on physical patient ergonomics has focused on the experiences of patients, but non-professional caregivers are also facing challenges with physical tasks and work related to caring for their loved ones. Few studies have directly addressed the physical needs and capabilities of informal caregivers. Additional research is needed to understand the needs and design interventions to support physical work activities or interactions with physical tools and environments for both patients and caregivers.

REFERENCES

Bhattacharjya, S., Stafford, M. C., Cavuoto, L. A., Yang, Z., Song, C., Subryan, H., . . . Langan, J. (2019). Harnessing smartphone technology and three dimensional printing to create a mobile rehabilitation system, mRehab: Assessment of usability and consistency in measurement. *Journal of Neuroengineering and Rehabilitation, 16*(1), 127.

Bodden, J., & King, B. (2018). How task complexity and barriers influence older adult engagement in ambulation during hospitalization. *Innovation in Aging, 2*(Suppl 1), 949.

Boltz, M., Capezuti, E., Shabbat, N., & Hall, K. (2010). Going home better not worse: Older adults' views on physical function during hospitalization. *International Journal of Nursing Practice, 16*(4), 381–388.

Boltz, M., Resnick, B., Capezuti, E., Shuluk, J., & Secic, M. (2012). Functional decline in hospitalized older adults: Can nursing make a difference? *Geriatric Nursing, 33*(4), 272–279.

Bonenberg, A., Branowski, B., Kurczewski, P., Lewandowska, A., Sydor, M., Torzyński, D., & Zabłocki, M. (2019). Designing for human use: Examples of kitchen interiors for persons with disability and elderly people. *Human Factors and Ergonomics in Manufacturing & Service Industries, 29*(2), 177–186.

Boute, B., Veldeman, L., Speleers, B., Van Greveling, A., Van Hoof, T., Van de Velde, J., . . . Detand, J. (2018). The relation between patient discomfort and uncompensated forces of a patient support device for breast and regional lymph node radiotherapy. *Applied Ergonomics, 72*, 48–57.

Brown, C., Kitchen, K., & Nicoll, K. (2012). Barriers and facilitators related to participation in aquafitness programs for people with multiple sclerosis: A pilot study. *International Journal of MS Care, 14*(3), 132–141.

Brown, C. J., Friedkin, R. J., & Inouye, S. K. (2004). Prevalence and outcomes of low mobility in hospitalized older patients. *Journal of the American Geriatrics Society, 52*(8), 1263–1270.

Brown, C. J., Williams, B. R., Woodby, L. L., Davis, L. L., & Allman, R. M. (2007). Barriers to mobility during hospitalization from the perspectives of older patients and their nurses and physicians. *Journal of Hospital Medicine: An Official Publication of the Society of Hospital Medicine, 2*(5), 305–313.

Carlson, D., & Myklebust, J. (2002). Wheelchair use and social integration. *Topics in Spinal Cord Injury Rehabilitation, 7*(3), 28–46.

Cavuoto, L. A., Subryan, H., Stafford, M., Yang, Z., Bhattacharjya, S., Xu, W., & Langan, J. (2018). *Understanding user requirements for the design of a home-based stroke rehabilitation system.* Paper presented at the Proceedings of the Human Factors and Ergonomics Society Annual Meeting.

Definition and Domains of Ergonomics. (2019). Retrieved from https://www.iea.cc/whats/

Dobkin, B. H., Plummer-D'Amato, P., Elashoff, R., Lee, J., & Grp, S. (2010). International randomized clinical trial, stroke inpatient rehabilitation with reinforcement of walking speed (SIRROWS), improves outcomes. *Neurorehabilitation and Neural Repair, 24*(3), 235–242.

Dueñas, L., i Bernat, M. B., del Horno, S. M., Aguilar-Rodriguez, M., & Alcántara, E. (2016). Development of predictive models for the estimation of the probability of suffering fear of falling and other fall risk factors based on posturography parameters in community-dwelling older adults. *International Journal of Industrial Ergonomics, 54*, 131–138.

Ehmen, H., Haesner, M., Steinke, I., Dorn, M., Gövercin, M., & Steinhagen-Thiessen, E. (2012). Comparison of four different mobile devices for measuring heart rate and ECG with respect to aspects of usability and acceptance by older people. *Applied Ergonomics, 43*(3), 582–587.

Fausset, C. B., Mitzner, T. L., Price, C. E., Jones, B. D., Fain, B. W., & Rogers, W. A. (2013). *Older adults' use of and attitudes toward activity monitoring technologies.* Paper presented at the Proceedings of the Human Factors and Ergonomics Society Annual Meeting.

Gonzalez, E. T., Jones, A. M., Harley, L. R., Burnham, D., Choi, Y. M., Fain, W. B., & Ghovanloo, M. (2014). *Older adults' perceptions of a neckwear health technology.* Paper presented at the Proceedings of the Human Factors and Ergonomics Society Annual Meeting.

Gyi, D. E., Sims, R., Porter, J. M., Marshall, R., & Case, K. (2004). Representing older and disabled people in virtual user trials: Data collection methods. *Applied Ergonomics, 35*(5), 443–451.

Harper, C., Avera, A., Crosser, A., Jefferies, S., & Duke, T. (2017). *An exploration of interactive wayfinding displays in hospitals: Lessons learned for improving design.* Paper presented at the Proceedings of the Human Factors and Ergonomics Society Annual Meeting.

Hastings, S. N., Choate, A. L., Mahanna, E. P., Floegel, T. A., Allen, K. D., Van Houtven, C. H., & Wang, V. (2018). Early mobility in the hospital: Lessons learned from the STRIDE Program. *Geriatrics, 3*(4), 61.

Healthy People 2030 Framework. (2014). Retrieved from https://www.healthypeople.gov/2020/About-Healthy-People/Development-Healthy-People-2030/Framework

Hensler, S., Herren, D. B., & Marks, M. (2015). New technical design of food packaging makes the opening process easier for patients with hand disorders. *Applied Ergonomics, 50*, 1–7.

Holden, R. J., Schubert, C. C., & Mickelson, R. S. (2015). The patient work system: An analysis of self-care performance barriers among elderly heart failure patients and their informal caregivers. *Applied Ergonomics, 47*, 133–150.

Holden, R. J., Toscos, T., & Daley, C. N. (2020). Researcher reflections on human factors and health equity. In R. Roscoe, E. K. Chiou, & A. R. Wooldridge (Eds.), *Advancing Diversity, Inclusion, and Social Justice through Human Systems Engineering* (pp. 51–62). Boca Raton, FL: CRC Press.

Hossain, M. A. (2014). Perspectives of human factors in designing elderly monitoring system. *Computers in Human Behavior, 33*, 63–68.

Howard, S., Lang, A., Sharples, S., & Shaw, D. (2017). See I told you I was taking it!–Attitudes of adolescents with asthma towards a device monitoring their inhaler use: Implications for future design. *Applied Ergonomics, 58*, 224–237.

International Organization for Standardization (ISO). (2005). *7730: 2005. Ergonomics of the thermal environment-analytical determination and interpretation of thermal comfort using calculation of the PMV and PPD indices and local thermal comfort criteria.*

Kabel, A., Dimka, J., & McBee-Black, K. (2017). Clothing-related barriers experienced by people with mobility disabilities and impairments. *Applied Ergonomics, 59*, 165–169.

Kabra, C., Jaiswal, R., Arnold, G., Abboud, R., & Wang, W. (2015). Analysis of hand pressures related to wheelchair rim sizes and upper-limb movement. *International Journal of Industrial Ergonomics, 47*, 45–52.

Kayes, N. M., McPherson, K. M., Schluter, P., Taylor, D., Leete, M., & Kolt, G. S. (2011). Exploring the facilitators and barriers to engagement in physical activity for people with multiple sclerosis. *Disability and Rehabilitation, 33*(12), 1043–1053.

Kenward, M. G. (1971). An approach to the design of wheelchairs for young users. *Applied Ergonomics, 2*(4), 221–225.

Kloosterman, M. G., Buurke, J. H., Schaake, L., Van der Woude, L. H., & Rietman, J. S. (2016). Exploration of shoulder load during hand-rim wheelchair start-up with and without power-assisted propulsion in experienced wheelchair users. *Clinical Biomechanics, 34*, 1–6.

Krapp, K. (2007). Activities of daily living evaluation encyclopedia of nursing & allied health. *Enotes Nursing Encyclopedia*.

Lang, A. R., Martin, J. L., Sharples, S., & Crowe, J. A. (2013). The effect of design on the usability and real world effectiveness of medical devices: A case study with adolescent users. *Applied Ergonomics, 44*(5), 799–810.

Langan, J., DeLave, K., Phillips, L., Pangilingan, P., & Brown, S. H. (2013). Home-based telerehabilitation shows improved upper limb function in adults with chronic stroke: A pilot study. *Journal of Rehabilitation Medicine, 45*(2), 217–220.

Langan, J., Subryan, H., Nwogu, I., & Cavuoto, L. (2018). Reported use of technology in stroke rehabilitation by physical and occupational therapists. *Disability and Rehabilitation: Assistive Technology, 13*(7), 641–647.

Langhammer, B., & Stanghelle, J. K. (2003). Bobath or motor relearning programme? A follow-up one and four years post stroke. *Clinical Rehabilitation, 17*(7), 731–734.

Lee, C.-L. (2019). An evaluation of tactile symbols in public environment for the visually impaired. *Applied Ergonomics, 75*, 193–200.

Lee, C.-L., Chen, C.-Y., Sung, P.-C., & Lu, S.-Y. (2014). Assessment of a simple obstacle detection device for the visually impaired. *Applied Ergonomics, 45*(4), 817–824.

Lemlouma, T., & Chalouf, M. A. (2012). *Smart media services through tv sets for elderly and dependent persons*. Paper presented at the International Conference on Wireless Mobile Communication and Healthcare.

Lewis, L., Sharples, S., Chandler, E., & Worsfold, J. (2015). Hearing the way: Requirements and preferences for technology-supported navigation aids. *Applied Ergonomics, 48*, 56–69.

Li, J., Ma, Q., Chan, A. H., & Man, S. (2019). Health monitoring through wearable technologies for older adults: Smart wearables acceptance model. *Applied Ergonomics, 75*, 162–169.

Liebzeit, D., Bratzke, L., Boltz, M., Purvis, S., & King, B. (2020). Getting back to normal: A grounded theory study of function in post-hospitalized older adults. *The Gerontologist*, 60(4), 704–714.

Liebzeit, D., Bratzke, L., & King, B. (2020). Strategies older adults use in their work to get back to normal following hospitalization. *Geriatric Nursing*, 41(2), 132–138.

Lin, F., Ajay, J., Langan, J., Cavuoto, L., Nwogu, I., Subryan, H., & Xu, W. (2016). *A portable and cost-effective upper extremity rehabilitation system for individuals with upper limb motor deficits*. Paper presented at the 2016 IEEE Wireless Health (WH).

Liu, K. P. Y., & Chan, C. C. H. (2014). Pilot randomized controlled trial of self-regulation in promoting function in acute poststroke patients. *Archives of Physical Medicine and Rehabilitation, 95*(7), 1262–1267.

Mackrill, J., Marshall, P., Payne, S. R., Dimitrokali, E., & Cain, R. (2017). Using a bespoke situated digital kiosk to encourage user participation in healthcare environment design. *Applied Ergonomics, 59*, 342–356.

Maman, Z. S., Yazdi, M. A. A., Cavuoto, L. A., & Megahed, F. M. (2017). A data-driven approach to modeling physical fatigue in the workplace using wearable sensors. *Applied Ergonomics, 65*, 515–529.

Martin, J. L., Clark, D. J., Morgan, S. P., Crowe, J. A., & Murphy, E. (2012). A user-centred approach to requirements elicitation in medical device development: A case study from an industry perspective. *Applied Ergonomics, 43*(1), 184–190.

Mshali, H., Lemlouma, T., Moloney, M., & Magoni, D. (2018). A survey on health monitoring systems for health smart homes. *International Journal of Industrial Ergonomics, 66*, 26–56.

Novak, I. (2011). Effective home programme intervention for adults: A systematic review. *Clinical Rehabilitation, 25*(12), 1066–1085.

Nwogu, I., Jha, S., Cavuoto, L., Subryan, H., & Langan, J. (2017). *A portable upper extremity rehabilitation device.* Paper presented at the Workshops at the Thirty-First AAAI Conference on Artificial Intelligence.

Penney, W., & Wellard, S. J. (2007). Hearing what older consumers say about participation in their care. *International Journal of Nursing Practice, 13*(1), 61–68.

Preusse, K. C., Mitzner, T. L., Fausset, C. B., & Rogers, W. A. (2017). Older adults' acceptance of activity trackers. *Journal of Applied Gerontology, 36*(2), 127–155.

Rangel, M., & Mont'Alvão, C. (2011). *Color and wayfinding: A research in a hospital environment.* Paper presented at the Proceedings of the Human Factors and Ergonomics Society Annual Meeting.

Roda-Sales, A., Vergara, M., Sancho-Bru, J. L., Gracia-Ibáñez, V., & Jarque-Bou, N. J. (2019). Effect of assistive devices on hand and arm posture during activities of daily living. *Applied Ergonomics, 76*, 64–72.

Rosamond, W., Flegal, K., Friday, G., Furie, K., Go, A., Greenlund, K., . . . Amer Heart, A. (2007). Heart disease and stroke statistics -2007 update - A report from the American Heart Association Statistics Committee and Stroke Statistics Subcommittee. *Circulation, 115*(5), E69–E171.

Rousek, J., & Hallbeck, M. (2011). The use of simulated visual impairment to identify hospital design elements that contribute to wayfinding difficulties. *International Journal of Industrial Ergonomics, 41*(5), 447–458.

Schall, M. C., Jr., Fethke, N. B., Chen, H., Oyama, S., & Douphrate, D. I. (2016). Accuracy and repeatability of an inertial measurement unit system for field-based occupational studies. *Ergonomics, 59*(4), 591–602.

Sheikhzadeh, A., Yoon, J., Formosa, D., Domanska, B., Morgan, D., & Schiff, M. (2012). The effect of a new syringe design on the ability of rheumatoid arthritis patients to inject a biological medication. *Applied Ergonomics, 43*(2), 368–375.

Sherif, S., Hasan, S., Arnold, G., Abboud, R., & Wang, W. (2016). Analysis of hand pressure in different crutch lengths and upper-limb movements during crutched walking. *International Journal of Industrial Ergonomics, 53*, 59–66.

Steinfeld, E., & Mullick, A. (1990). Universal design: The case of the hand. *Innovation, Fall*, 27–31.

Story, M. F., Luce, A. C., & Rempel, D. M. (2009). *Accessibility and usability of diabetes management devices for users with vision disabilities.* Paper presented at the Proceedings of the Human Factors and Ergonomics Society Annual Meeting.

Story, M. F., Winters, J. M., Lemke, M. R., Barr, A., Omiatek, E., Janowitz, I., . . . Rempel, D. (2010). Development of a method for evaluating accessibility of medical equipment for patients with disabilities. *Applied Ergonomics, 42*(1), 178–183.

Strauss, A. (1993). *Continual Permutations of Action Aldyne de Gruyter.* New York.

Suryadevara, N. K., Mukhopadhyay, S. C., Wang, R., & Rayudu, R. (2013). Forecasting the behavior of an elderly using wireless sensors data in a smart home. *Engineering Applications of Artificial Intelligence, 26*(10), 2641–2652.

Tsai, K.-H., Yeh, C.-Y., & Lo, H.-C. (2008). A novel design and clinical evaluation of a wheelchair for stroke patients. *International Journal of Industrial Ergonomics, 38*(3–4), 264–271.

Vincent, C. J., & Blandford, A. (2014). The challenges of delivering validated personas for medical equipment design. *Applied Ergonomics, 45*(4), 1097–1105.

Wang, Y., Wu, D., Zhao, M., & Li, J. (2014). Evaluation on an ergonomic design of functional clothing for wheelchair users. *Applied Ergonomics, 45*(3), 550–555.

Ward, J., Buckle, P., & Clarkson, P. J. (2010). Designing packaging to support the safe use of medicines at home. *Applied Ergonomics, 41*(5), 682–694.

Werner, N. E., Tong, M., Borkenhagen, A., & Holden, R. J. (2019). Performance-shaping factors affecting older adults' hospital-to-home transition success: A systems approach. *The Gerontologist, 59*(2), 303–314.

White, M. M., Morejon, O. N., Liu, S., Lau, M. Y., Nam, C. S., & Kaber, D. B. (2017). Muscle loading in exoskeletal orthotic use in an activity of daily living. *Applied Ergonomics, 58*, 190–197.

Wilkinson, A., Kanik, M., O'Neill, J., Charoenkitkarn, V., & Chignell, M. (2017). *Ambient activity technologies for managing responsive behaviours in dementia.* Paper presented at the Proceedings of the International Symposium on Human Factors and Ergonomics in Health Care.

Wolf, S. L., Thompson, P. A., Winstein, C. J., Miller, J. P., Blanton, S. R., Nichols-Larsen, D. S., Morris, D. M., Uswatte, G., Taub, E., Light, K. E., & Sawaki, L. (2010). The EXCITE stroke trial: comparing early and delayed constraint-induced movement therapy. *Stroke, 41*(10), 2309-2315.

World Health Organization (WHO). (2014). *Constitution of the World Health Organization. 2006.* Available from: http://www. who. int/governan ce/eb/who_constitution_en. pdf.

Zhang, L., Diraneyya, M. M., Ryu, J., Haas, C. T., & Abdel-Rahman, E. M. (2019). Jerk as an indicator of physical exertion and fatigue. *Automation in Construction, 104*, 120–128.

Zulas, A. L., Crandall, A. S., & Schmitter-Edgecombe, M. (2014). *Caregiver needs from elder care assistive smart homes: Children of elder adults assessment.* Paper presented at the Proceedings of the Human Factors and Ergonomics Society Annual Meeting.

4 Macroergonomics of Patient Work

Engaging Patients in Improving Sociotechnical Context of Their Work

Pascale Carayon
University of Wisconsin-Madison

Armagan Albayrak and Richard Goossens
Delft University of Technology

Peter Hoonakker and Bat-Zion Hose
University of Wisconsin-Madison

Michelle M. Kelly
University of Wisconsin School of Medicine and Public Health

Marijke Melles
Delft University of Technology

Megan E. Salwei
Vanderbilt University Medical Center

CONTENTS

Researchers and practitioners in the human factors and ergonomics (HFE) discipline aim to improve "work." We are often reminded of the Greek origin of the term "ergonomics": in Greek "ergo" means work and "nomos" means law. Work is performed by a range of people who are paid or unpaid individuals. Patients (i.e., unpaid) do work (Strauss, 1993; Strauss et al., 1982). HFE researchers and practitioners are increasingly paying attention to work performed by patients and their informal caregivers. An informal caregiver, often a family member, provides care, typically unpaid, to someone with whom they have a personal relationship (National Research Council, 2010). In this chapter, we focus on macroergonomics of patient work and describe conceptual and methodological approaches to analyze and improve patient work systems. We will refer to informal caregivers (family members, friends, and neighbors) as caregivers in the rest of the chapter.

4.1 PATIENT WORK: IMPORTANCE OF THE CONTEXT/SYSTEM

Some patient work is recognized, but other work may not be visible or recognized by clinicians as well as patients themselves. Visible work of patients is, for example, taking the appropriate medications on a set time, exercising, wound dressing at home, following a diet, scheduling appointments with their clinicians, and so on. Strauss and colleagues (1982) described "invisible work" done by patients; for instance, hospitalized patients manage their diabetes (secondary condition) whereas physicians and nurses treat their main illness (primary condition). Gorman et al. (2018) identified several instances of work done by breast cancer survivors that were not visible to their physicians; for instance, a patient went online to identify side effects of a medication and communicated this to her doctor who, after becoming aware of this invisible work, decided to change the prescription. Visible or invisible work done by patients occurs in various contexts and settings such as home, hospital, emergency room, or primary care clinic. In addition, the recent movement toward patient-centered care has produced various organizational approaches and practices, such as shared decision-making (Hargraves et al., 2016; Stiggelbout et al., 2012) and patient-facing technologies (Ahern et al., 2011), which require active participation of patients and their caregivers. These strategies and practices involve *work* that patients need to perform, including personal health management work needed to deal with their health or chronic condition(s), such as taking medications and managing personal health information. From a macroergonomics perspective, we need to understand the context or system in which various forms of patient work occur and how the broader "work system" can be designed to support the work of patients and improve patient outcomes, including outcomes important to patients themselves.

Patients do work on their own or with their caregivers. They also do some work in collaboration with healthcare staff. For instance, patients may do the "mirror image" work of clinicians' work: a patient keeps a diary about their daily activities related to a medical condition (e.g., migraine) and the clinician looks for patterns in these data (Strauss et al., 1982). Collaborative work among patients, caregivers, and clinicians involves physical activities (e.g., physical exam and daily living activities), cognitive activities (e.g., sharing information about a medication regimen), and psychosocial activities (e.g., conversation about disposition during an encounter in the emergency department) (see Chapters 2 and 3 in this volume). This collaborative work engages patients (and caregivers and families) at various system levels: (1) direct care; (2) organizational design and governance; and (3) policy making (Carman et al., 2013). At the direct care level, patients may go online to look for information about treatment options after receiving a new diagnosis (see Chapter 7 in this volume). This work may be in preparation for a follow-up visit and discussion with a specialist about the care plan. At the organizational level, patients may do some work in engaging with healthcare organizations, such as patient/family advisory groups or quality improvement teams. At the policy level, patients and their communities (e.g., geographic/living community or patient advocacy community) may participate in developing and implementing programs. There is also an increasing movement toward the active participation of patients in research (Greenhalgh et al., 2019), which involves work such as making decisions about research questions and methods (e.g., measurement of patient-centered outcomes) and analyzing research data (see Chapter 13 in this volume).

In summary, patient work is done individually or collaboratively, includes various physical, cognitive, and psychosocial activities, and involves engagement at multiple system levels. It is important to recognize that patient work takes place in a context or system, which influences that work, facilitates it, or hinders it. Understanding the system in which patient work occurs is the purpose of macroergonomics applied to patient work: the macroergonomics perspective allows one to describe, assess, and improve the "work system" in which patients do their various types of work. This has the potential to lead to improvements in outcomes, patient experience, and patient-centered health policies. Patient work is often done in concert with caregivers (e.g., family members, friends, neighbors, and community members). Work done by caregivers is another form of "unpaid work" that occurs in an organizational or sociotechnical work system. In the rest of the chapter, we will use the term of "patient work" as an umbrella term to describe the individual work of patients as well as the collaborative work of patients and their caregivers.

4.2 WHAT IS MACROERGONOMICS?

Macroergonomics is a branch of HFE that espouses a systems approach to understanding, assessing, and improving interactions of people with tasks, technologies, the physical environment, and the psychosocial and organizational work setting (Hendrick & Kleiner, 2001). A critical aspect of macroergonomics is the emphasis on the broader organizational and sociotechnical context in which "work" is performed. According to macroergonomics experts (Hendrick, 1991; Hendrick & Kleiner, 2001; Zink, 2000), it is not sufficient for HFE to consider the micro-level interactions, such as physical or cognitive activities performed by workers with various tools (i.e., the so-called micro-ergonomic triad of "person-task-tool"). Macroergonomics specifies that higher-level organizational factors need to be considered, in particular, when redesigning work and improving outcomes for both workers and their organizations (Hendrick, 2007).

Macroergonomics researchers have proposed systems models of the organizational context of work. For instance, Kleiner (2006, 2008) defines the work system as a combination of three sub-systems influenced by the *external environment*: *personnel sub-system*, *technological sub-system*, and *organizational sub-system*. Smith and Carayon's model of the work system is another macroergonomics model that points to the broad context in which work is performed (Carayon, 2009; Smith & Carayon-Sainfort, 1989; Smith & Carayon, 2001). The work system model is centered around the "worker" who performs tasks and activities using tools and technologies; this work occurs in a physical environment and in a specific organizational context. This work system model has five elements: (1) the *person* ("worker" who could be the patient), (2) *tasks*, (3) *tools and technologies*, (4) *physical environment*, and (5) *organization*. In a macroergonomics approach, work system elements interact with each other (Wilson, 2014); therefore, changing one element will have an impact on other elements of the work system. This has implications for design and redesign, including the need for a systems-based participatory approach to improvement.

Similar to other branches of HFE, macroergonomics has traditionally focused on paid work; therefore, the organizational context is often conceptualized as the company that hires workers and provides the physical, informational, and organizational resources (and constraints) necessary to do the work. According to Strauss (1993), patient work involves "exertion of effort and investment of time on the part of patients or family members to produce or accomplish something" (pp. 64–65). In this chapter, we extend the concept of macroergonomics to the unpaid work of patients and their informal caregivers.

Carayon and colleagues (2013) defined five key elements of macroergonomics, which distinguish macroergonomics from other HFE approaches to improving health care and patient safety: (1) *systems approach*, (2) *joint optimization of performance and well-being*, (3) *consideration of organizational and sociotechnical context*,

TABLE 4.1
Macroergonomic Elements of Patient Work

Macroergonomic Elements	Application to Patient Work
1. Systems approach	Patient work occurs in a system. There are various models of the patient work system.
2. Joint optimization of performance and well-being	Various patient outcomes need to be considered (e.g., care effectiveness, care efficiency, and patient safety), as well as patient experience and other outcomes that matter to patients.
3. Consideration of organizational and sociotechnical context	Patient work occurs in varied contexts: e.g., home, work, school, clinical setting, and patient/clinician encounter.
4. System interactions and levels	Patient work occurs at multiple system levels and involves interactions between multiple system elements.
5. Implementation process	Patient participation is a strategy to foster patient-centered care and promote patient empowerment.

(4) *system interactions and levels*, and (5) *implementation process*. Table 4.1 lists the five macroergonomic elements and their implications for patient work.

First, because macroergonomics is anchored in the Sociotechnical Systems Theory (Pasmore, 1988; Trist, 1981; Trist & Bamforth, 1951), analysis and improvement of work are based on a systems approach (Wilson, 2014); this is the essence of the work system models of Kleiner (2006, 2008) and Smith and Carayon (Carayon, 2009; Smith & Carayon-Sainfort, 1989, 2001). Therefore, patient work occurs in a "work system" with elements that interact with each other. In the following section, we describe various models of the patient work system.

Second, in line with the dual objective of HFE (Dul et al., 2012), macroergonomics aims to improve both performance and well-being. In the context of patient work, this translates to the need to consider multiple patient-related outcomes, including healthcare quality and patient safety as well as patient experience and other outcomes that matter to patients. See below the discussion of the patient journey mapping approach that integrates consideration of patient experience.

Third, patient work occurs in a broader organizational or sociotechnical context; understanding the system (or context) is critical for redesign or improvement. For instance, approaches to "patient compliance" (e.g., compliance with medication regimen) tend to focus on changing the behavior or attitude of the patient, but miss opportunities to address the context or environment that can either hinder or facilitate the patient's behavior, as well as the emotional and subjective aspect of patient work (e.g., taking medications may signal that something is 'wrong' with the person) (Valdez et al., 2015).

Fourth, because patient work occurs in a system, we need to understand the various elements of the patient work system and the system levels at which patients work or engage in their health care. As noted above, the conceptual framework on patient engagement by Carman and colleagues (2013) identifies three system levels: direct care, organizational, and policy level.

Finally, macroergonomics fully embraces the philosophy and methods of participation (Noro & Imada, 1991). In the context of patient-centered care, designing and redesigning the work system of patients should involve patients in a meaningful and impactful manner. This is in line with other human-centered design approaches, such as experience-based design (Bate & Robert, 2006), which integrates the actual experience of patients in co-designing health services. The proliferation of terms connected to patient-centered care shows the increasing importance of the concept; however, it can add to confusion. A concept analysis based on an extensive literature review by Castro et al. (2016) helps to clarify the connected concepts of *patient participation*, *patient-centered care*, and *patient empowerment*. Patient participation can be considered a key strategy to foster patient-centered care and, therefore, promotes patient empowerment. Patient participation occurs either at an individual level or collective level. At the individual level, patient participation "revolves around a patient's rights and opportunities to influence and engage in the decision-making about his care through a dialogue attuned to his preferences, potential, and a combination of his experiential and the professional's expert knowledge" (Castro et al., 2016, p. 1929). At the collective level, patient participation is the "contribution of patients or their representing organizations in shaping health and social care services by means of active involvement in a range of activities at the individual, organizational, and policy level that combine experiential and professional knowledge" (Castro et al., 2016, p. 1929). In subsequent sections, we describe various examples of collective patient participation in the design and implementation of sociotechnical system improvements for healthcare quality and patient safety.

4.3 MACROERGONOMIC MODELS OF PATIENT WORK

Macroergonomic models of the work system have been applied and adapted to patient work. In this section, we review the original SEIPS (Systems Engineering Initiative for Patient Safety) model (Carayon et al., 2006) and its adaptation, SEIPS 2.0 (Holden et al., 2013), which clearly describe the individual work of clinicians, the individual work of patients, and the collaboration between clinicians and patients. Since a lot of patient work occurs in the home environment, we also describe HFE models of home health care that adopt a macroergonomics perspective.

4.3.1 SEIPS Models

The original SEIPS model of work system and patient safety (Carayon et al., 2006) integrates the macroergonomic work system model of Smith and Carayon (Carayon, 2009; Smith & Carayon-Sainfort, 1989) with the Structure-Process-Outcome model of Donabedian (1978, 1988). According to the SEIPS model (Carayon et al., 2006, 2014), the work system can be the system of clinicians or the system of patients. When analyzing the work system of clinicians, we describe tasks performed by clinicians, some of them involving patients; for instance, a nurse administering medications to patients. In the clinicians' work system, patients may influence system elements, for instance as a source of distraction or interruption in the physical environment. The original SEIPS model also describes the *"person"* in the work system as a patient

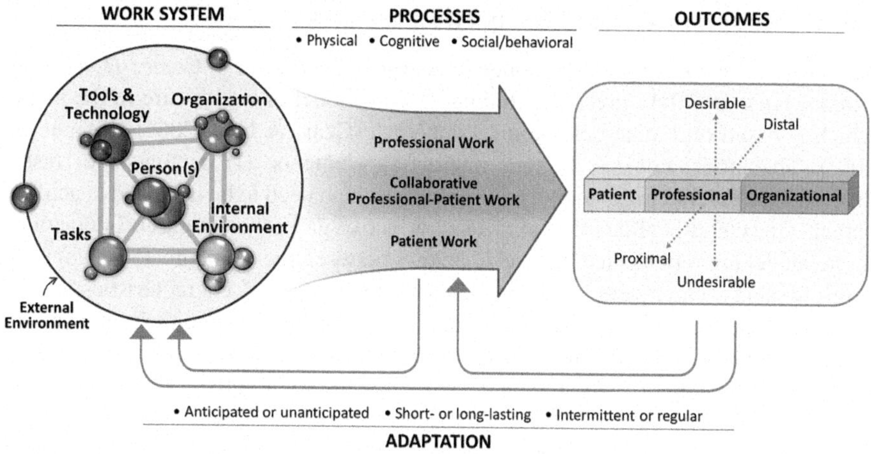

FIGURE 4.1 SEIPS 2.0 Model (Holden et al., 2013).

who, for instance, takes medications (*task*) using a medication dispenser box (*tool*) at home (*physical environment*) on a specific schedule (*organization*). SEIPS 2.0 went one step further and clearly delineated the work done by clinicians, the work done by patients, and the work done in collaboration between clinicians and patients (Holden et al., 2013) (see Figure 4.1).

4.3.2 Model of the Patient Work System

A patient's management of their health is a complex process embedded in a specific context. The patient work framework (Holden, Schubert, et al., 2015; Valdez et al., 2015) is related to two system models: SEIPS 2.0 (Holden et al., 2013) and the Human Factors Model of Health Care in the Home (see below) (Henriksen et al., 2009), and takes into account the larger process and context (a key element of macroergonomics; see Table 4.1) as well as patient perspectives. The patient work framework describes elements of a patient's work system: person, tasks, tools and technologies, physical environment, and social-organizational environment; these elements interact with each other and influence patient outcomes (e.g., patient health). It can be used in combination with user-centered design approaches to design health information technologies (see Chapters 5 and 10 in this volume) and improve integration of technologies in patients' lives, consequently enhancing patient health management (Valdez et al., 2015).

Holden et al. (2015) applied the Patient Work System model to analyze barriers to self-care work experienced by patients with heart failure (Holden, Schubert, et al., 2015). Using interviews, observation, and survey data collected from 30 patients with heart failure, researchers identified several barriers in each of the work system elements. For example, a *person* barrier was a patient's knowledge about heart failure self-care, and a *tools and technology* barrier was the unavailability of computers. Researchers emphasized that the context and work system interactions need to be addressed in the design of interventions to support patient work.

4.3.3 Models of Health Care in Home Settings

The landmark report by the National Academies, *Health Care Comes Home: The Human Factors* (2011), presents a human factors model of health care in the home, which is based on a macroergonomics approach (Czaja & Nair, 2006; Czaja et al., 2001). The model consists of four interacting elements: (1) person(s), (2) tasks, (3) equipment/technology, and (4) environment, which need to be designed to achieve patient safety and healthcare quality (e.g., effective and efficient care). Interactions in the model are represented by double arrows between the elements. Person(s) (e.g., care recipients, caregivers, and clinicians) bring a range of characteristics, skills/abilities, education, health conditions, preferences, and attitudes to the home healthcare experience. Person(s) interactions with tasks and equipment/technology vary based on their cognitive, perceptual, and physical capabilities. The interactions between tasks and equipment/technology vary based on cognitive, sensory, and physical demands placed on people related to their own capabilities. Person(s), tasks, and equipment/technology interact with each other in multiple environments (i.e., health policy environment, community environment, social environment, and physical environment), which are represented as concentric circles in the model. The interactions and "fit" among the system elements influence the efficiency, effectiveness, and safety of home health care.

Henriksen et al. (2009) developed a macroergonomic model of home health care that highlights factors specific to home health care that contribute to patient safety and quality. The model distinguishes between active errors, likely made by caregivers responding to urgent patient needs, and latent conditions (i.e., potential contributing factors that can go unnoticed in the home care delivery system). The seven contributing factors are (1) patient characteristics; (2) provider characteristics; (3) nature of home healthcare tasks; (4) design features of the physical environment; (5) the medical devices and technologies used; (6) the social and community environments; and (7) the external environment. Each contributing factor includes five to eight sub-factors (e.g., a patient's health literacy and knowledge have implications for self-care management). Home healthcare stakeholders should consider the design of these components and their interactions with one another to avoid threats to safety and quality.

Using this human factors model of home health care, McGraw (2019) identified factors contributing to the development of community-acquired pressure ulcers, which were identified in interviews of nurses working in home healthcare settings. For example, patients with dementia *(patient characteristic)* were often unable to participate in pressure area care. Home healthcare providers' deficits in health knowledge *(clinician characteristic)* were attributed to their limited education and training. McGraw defined four categories of interventions based on the identified risks and contributing factors for pressure ulcer: (1) behavioral interventions (e.g., patient education); (2) technical interventions (e.g., digital screens providing regular personalized prompts for pressure ulcer prevention); (3) safeguarding interventions (e.g., when caregivers failed to escalate skin changes); and (4) initiatives to promote better integration between health, local authorities, and families (e.g., one-to-one teaching at the bedside).

4.3.4 Summary and Synthesis of Macroergonomic Models of Patient Work

Patient work is a combination of *tasks* performed by a *person* using various *tools*, which occurs in a broader *organizational and sociotechnical context*. The macroergonomic models of patient work reviewed in this section propose *different system elements*: person; tasks; tools; physical environment; organization; social environment; community; and external environment. The models combine *multiple system levels*, such as the home environment, community, and external environment. In an effort to propose a generic macroergonomic model of the patient work system, Holden and colleagues (2017) conducted a secondary data analysis of three separate studies of work performed by chronically ill patients and their informal caregivers. The researchers analyzed data from the following three studies: The Asthma and Technology Study of individuals using a new asthma management technology; the Keystone Beacon Project that developed a community-based care management program for older patients with heart failure and chronic obstructive pulmonary disease (Carayon et al., 2012; Carayon, Hundt, et al., 2015); and the Caring Hearts Study of older adults with heart failure and their informal caregivers (Srinivas et al., 2017). The consolidated model includes the micro-ergonomic triad of person-tasks-tools at its core; the triad is embedded in the system levels of home setting and community (see Figure 4.2) and is influenced by factors in three domains: physical context, social context, and organizational context. The researchers identified 17 physical, social, and organizational context (i.e., macroergonomic) factors with positive, negative, or mixed impact on patient health and health behavior. For example, patients living

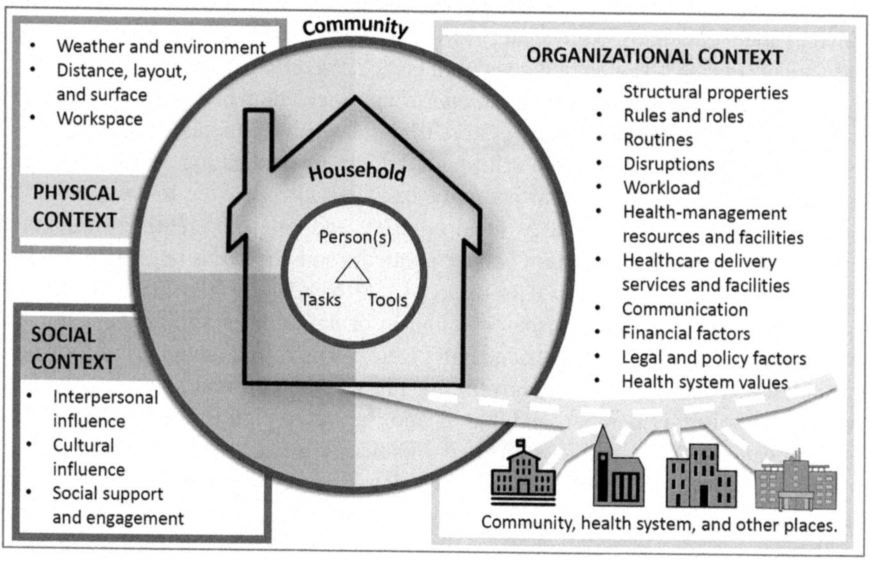

FIGURE 4.2 Consolidated model of macroergonomic patient work system (Holden et al., 2017).

in rural areas, large distances from their primary care clinic or grocery store (i.e., physical context), may be less likely to take their medications on time or visit their primary care physician if a problem occurs due to the physical barrier of distance.

4.4 MACROERGONOMICS OF PATIENT-CENTERED CARE

Macroergonomic (sociotechnical) approaches to work system design can be applied to improve patient work and enhance patient-centered care. In this section, we describe two examples of the design and implementation of health information technologies in sociotechnical systems that can foster patient-centered care (i.e., care that is respectful of and responsive to individual patient preferences, needs, and values and ensuring that patient values guide all clinical decisions) (Institute of Medicine [IOM], 2001).

4.4.1 Supporting Patient Participation and Empowerment through Health Information Technologies

Various health information technologies have been developed to enhance the work of patients and their interactions with clinicians (see Chapter 5 in this volume). One example is patient portals. Linked to an electronic health record system, patient portals allow patients to view online information about their care, including information about their medical conditions, medications, test results, and the clinicians who care for them. Patient portals can also be used to schedule appointments, pay medical bills, and send secure messages to clinicians. Patient portals have evolved to support the work of patients and clinicians in different contexts. Patient portals were initially designed to support the care of patients in the ambulatory setting and to facilitate their interactions with outpatient clinicians (Kelly et al., 2018). Newer inpatient or acute care portals support patients during hospitalization and will be discussed in this section.

Inpatient portals are online applications, often provided on bedside tablet computers to hospitalized patients (Kelly et al., 2018). They give patients or their caregivers real time, secure access to their hospital diagnoses, medication information, test results, and clinician names and photographs. They also provide tools to enhance patients' hospital experience (e.g., meal ordering and entertainment) and facilitate communication and collaboration between patients and clinicians (e.g., goal for discharge and secure messaging) (Collins et al., 2016; Kelly et al., 2018). By fostering increased information transparency and information sharing, inpatient portals are intended to empower patients and caregivers during vulnerable periods of acute hospitalization. Patients and caregivers use the information contained in inpatient portals to monitor, remember, and understand their care plan (patient work) and to communicate and collaborate with their inpatient clinician (patient and clinician collaborative work) (Kelly et al., 2019; Woollen et al., 2016). This technology also allows patients to play a more active role in improving the safety of care (Prey et al., 2018), such as facilitating patient reconciliation of medications during transitions to inpatient care or identification and reporting of errors.

Despite these potential benefits, effectively integrating this technology into the work system of patients and their inpatient healthcare teams can be challenging.

Multiple work system barriers may hinder their implementation. For instance, some clinicians worry that releasing test results in real time may confuse hospitalized patients and increase their anxiety (Caligtan et al., 2012). These concerns are heightened in the context of offering inpatient portals to patients with lower health or computer literacy or among clinicians who practice a more paternalistic style of medicine, both being *person* barriers. Increased access to information may lead to patients asking a greater number of questions (either in person or through the secure-messaging functionality of the inpatient portal). Although this information may enhance patient participation, it could also increase clinician workload (Caligtan et al., 2012; Kelly et al., 2017; O'Leary et al., 2016). Other important but competing clinical care activities act as *task* barriers to effectively using inpatient portals to support patient engagement.

A macroergonomics approach is necessary to ensure that the needs of both patients and clinicians are considered in the (re)design, implementation, and evaluation of technologies such as inpatient portals; this will help to support care processes and improve outcomes for both patients and clinicians (Carayon, 2012; Collins et al., 2018; Holden et al., 2013). Design efforts should employ user-centered design approaches (see Chapter 10 in this volume) to determine the optimal content and design requirements for patient portals and their integration across care settings (Collins et al., 2018).

4.4.2 Design of Patient-centered Decision Support

Although most clinical decision support (CDS) technologies are designed to be clinician facing, patient decision aids can support collaborative work between patients and clinicians (see Chapter 6 in this volume). This includes support to patient decision-making by assessing patient values, goals, and preferences in relation to evidence-based medicine and potential risks, benefits, and outcomes of care. Melnick and colleagues (2015) applied a macroergonomic approach to design a patient-centered CDS for patients with minor head injury in the emergency department (ED), called 'Concussion or Brain Bleed.' The CDS utilized the Canadian Computed Tomography (CT) Head Rule, a clinical prediction rule designed to reduce unnecessary imaging for minor head injuries. Patients, ED clinicians, experts in clinical informatics, and health services researchers developed the CDS utilizing user-centered design methods (e.g., critical decision method) and HF principles (e.g., user involvement) through four iterative steps: (1) initial prototype development, (2) usability assessment, (3) field testing, and (4) beta testing (Melnick et al., 2015, 2017).

The technology was designed not only to support clinician decision-making for CT scans but also to support communication between patients and clinicians. Concussion or Brain Bleed starts with a patient-focused CDS including a welcome screen, injury evaluator, and risk visualization. Next, the CDS facilitates risk discussions between patients and clinicians, including potential patient-specific considerations (e.g., radiation and claustrophobia). Finally, a paper handout is given to patients summarizing the discussion and care decision.

Researchers implemented the CDS in an ED and evaluated its impact (Singh et al., 2017). Forty-one patients and 29 clinicians in the ED participated in the study. Research assistants trained patients and clinicians on how to use Concussion or Brain Bleed; the patient and clinician then used the CDS in the ED and completed surveys to evaluate the patient experience (patient knowledge on injury and risks, satisfaction, decisional conflict, and trust in physician), clinician experience (acceptability and usability), and healthcare utilization (CT obtained in the ED and admits to hospital). The researchers conducted follow-up calls with patients seven days later to determine patient outcomes, such as return ED or primary care visit, use of neuroimaging, and diagnosed brain injury. They compared the number of patient knowledge questions (e.g., available diagnostic options and patient individual risk) answered correctly between pre- and post-visit surveys, finding that the CDS increased patient knowledge by 42% (as measured by nine questions on, e.g., concussion and treatment options); 85%–88% of patients were satisfied with the clarity, amount, and helpfulness of information provided by the technology. Clinicians also reported positive results from the CDS, with 85% of clinicians rating the information as somewhat or extremely helpful to patients with high usability ratings based on the System Usability Scale (SUS) (Brooke, 1996) (score of 85 out of 100). They found no missed brain injuries at the seven-day follow-up. These results demonstrate a key element of macroergonomics, joint optimization (Hendrick, 1991), in which work system elements (e.g., CDS technology) positively impact clinician well-being as well as patient outcomes (i.e., knowledge) and safety (i.e., appropriate CT use).

4.5 PATIENT JOURNEY

Patient work happens over time and space; this is the essence of the patient journey (Carayon & Wood, 2009; Carayon & Wooldridge, 2019; Carayon et al., 2020). In a macroergonomics approach to patient work, we need to pay attention to the whole patient journey, not just specific encounters between patients and healthcare services. The patient journey approach emphasizes the transitions that occur between steps or phases of the journey and that either build or hinder patient safety (Carayon et al., 2020; Vincent & Amalberti, 2016; Vincent et al., 2017). The human-centered design method of patient journey mapping can be considered a macroergonomics approach to analyze and improve the various systems in which patients work (McCarthy et al., 2016; Simonse et al., 2019). It is a method from the human-centered design sciences "to see and understand the patient's experience by separating the management of a specific condition or treatment into a series of consecutive events or steps" (Trebble et al., 2010). The patient journey mapping approach is a well-established method to comprehensibly represent a distributed healthcare system, including all actors and processes that take place in that system, as well as experiences from a patient's perspective (Simonse et al., 2019). The ultimate aim of the patient journey mapping method is to identify design opportunities for new product and service systems aimed at improving entire sociotechnical care systems. In addition to identifying design opportunities, a patient journey clearly highlights the system boundaries within which the new service or product–service combination needs to function.

4.5.1 Visualizing the Patient Journey

A patient journey is a visual representation of the distributed healthcare system that a patient is part of. It includes the different stages of a care pathway and the different factors that affect the patient experience, directly or indirectly, during their journey. Activities in time or interactions of the actors in the system are referred to as "touchpoints." The emotions of patients along their journey are also mapped to gain insight into how patients experience their care throughout the journey. Reviewing emotions and touchpoint opportunities can provide useful insights for improving the distributed healthcare system.

The patient journey visualizes health care as "work-as-done" from the perspective of patients, meaning that it is based on actual processes and interactions. Data collection methods include participatory ergonomics methods such as observations, interviews, focus group sessions, generative design techniques, and co-design methods (see Chapters 9 and 10 in this volume). Ideally, all of the different actors involved in the patient journey, in particular patients, are involved in the design or redesign of the journey (see also Chapter 13 in this volume on patient participant involvement).

4.5.2 Mapping the Patient Journey

One way to understand patient work holistically in all of its complexity is by researching and constructing the patient journey. Constructing a patient journey consists of five steps with the active participation of patients.

4.5.2.1 Step 1: Define Design Goal and Research Aim

The first essential step is to define the purpose of the journey to be analyzed. First, a (preliminary) design goal needs to be defined. Following the definition of the design goal, the research aim of the specific patient journey needs to be defined. For example, in a study by de Ridder et al. (2018), the design goal was defined as improving the perioperative patient experience in an outpatient setting. The research aim of the patient journey mapping process was to map the patient experience before, during, and after surgery. Note that a journey often takes a broader perspective than the design goal (i.e., the perioperative phase in the study of de Ridder et al. (2018) vs. the pre-, peri-, and postoperative phases mapped in the journey). This is because the problem, or touchpoint, that needs improvement might be affected by events or opportunities that happen in time or place other than the touchpoint that is the focus of the design goal. The research aim determines what actors should be included in the journey and which phases should be covered by the journey. Note that the design goal might be redefined based on insights gathered throughout the patient journey mapping process.

4.5.2.2 Step 2: Identify Actors and Their Relations

The next step is to define the main actor and identify all actors that directly or indirectly affect the work or experience of the main actor; the relationships between all of the actors, in particular with the main actor, also need to be defined. When the design goal is to improve patient work or patient experience, the main actor is the patient. Often a mind map is constructed to represent the actors and their interdependencies.

To construct the mind map and define the actors and interdependencies, different methods are used, such as observational research, interviews, shadowing, literature review, and analysis of clinical protocols, and current workflows (see Chapter 9 and 12 in this volume). Since the different actors are all part of the patient journey, their interactions take place in multiple physical contexts and over time. Actors involve multiple people (e.g., patient, caregivers, physician, and nurse) using various tools and technologies (e.g., electronic health record, measuring devices, rehabilitation aids, and wayfinding systems in hospitals). For example, in a study to enhance parental involvement in pediatric oncology care (Kleinsmann et al., 2018), 25 actors were identified ranging from oncologist and nurse to physiotherapist and school staff. This high number of actors involved illustrates the immense complexity of pediatric oncology. Through iterative steps, the design goal was refined as parental involvement in medication adherence, and the number of most relevant actors within the design space reduced to eight. Note that, in this project, the tools and technologies were not initially considered. Later in the project, tools and technologies were added to the patient journey map and included technologies such as electronic health records, written reports, medication packaging, and food (used by parents to persuade their child to take their medications).

4.5.2.3 Step 3: Map the Journey: Rank Actors, Define Phases, and Identify Touchpoints

When all relevant actors and technologies and their connections are defined, the journey can be constructed. Actors are listed in the left column, at the vertical axis, starting with the main actor (i.e., the patient). The other actors are then listed in order of relevance or appearance in the clinical pathway. Next, the different phases of the patient journey are listed on the horizontal axis from left to right. Simonse et al. (2019) used the patient journey mapping approach to improve satisfaction of patients undergoing gastrointestinal diagnosis by means of video capsule endoscopy (VCE) technology. The patient journey included multiple phases: consultation explaining the procedure; the intake procedure at hospital; connecting a recorder to the patient to be able to see the images broadcasted by the VCE technology; swallowing the video capsule; waiting and walking around the hospital; discharge to home; and recovery at home. The duration of each phase was recorded, which could range from weeks, to days, to minutes. Once the axes are set, the touchpoints are defined. A touchpoint describes the activity of a specific actor in a specific phase. Activities can be direct interactions with the patient (e.g., a consultation between a patient and a clinician to discuss treatment options) or activities that do not involve direct interaction with the patient but still affect the patient experience (e.g., a surgeon preparing the video endoscopy service). Often quotes are used to describe touchpoints.

4.5.2.4 Step 4: Map Emotions

In this step, the patient experience is summarized by mapping their emotions throughout the journey. Patient experience is defined as "the sum of all interactions, shaped by an organization's culture, that influence patient perceptions across the continuum of care" (The Beryl Institute, 2019). Mapping patient emotions over time helps the designer to identify possible problems and friction points in the care process.

A patient-reported experience questionnaire could be used to assess the patient experience with respect to several factors, such as communication, involvement, and pain control (Black et al., 2014). Alternative data collection methods include observations or interviews using tools to measure emotions, such as the toolkit of Desmet (2002) that describes 24 positive and 24 negative emotions regarding product use. The use of quotes can help to illustrate and better understand the emotions that are mapped in the journey and, therefore, help to identify potential problems. In the study of de Ridder et al. (2018) on the perioperative experience of hand and wrist surgical patients, strong negative patient emotions were found during receiving anesthesia and recovery in the recovery room. The emotions were based on data collected via observations and interviews with patients and staff.

4.5.2.5 Step 5: Identify Problem Areas and Design Opportunities

Negative emotions can be associated with dissatisfaction with a current situation. Positive emotions with procedures that go well and affect the health system in a positive way should be preserved or even extended to other procedures. The journey map can help to identify the so-called friction points (e.g., patients who do not "adhere" to their rehabilitation plan) as well as possible causes of these problems (e.g., one-time instruction about their rehabilitation during a consultation) and possible solutions (e.g., information provision in various formats and at various points in time). It is important to consider that problems are relative, depend on perspective, and are time-related. Once a negative emotion is identified, that particular touchpoint can be analyzed in more depth. Redesigning these touchpoints by taking into consideration all processes and actor activities that affect this touchpoint plays a key role in improving healthcare processes. The patient journey map developed by de Ridder and colleagues (2018) describes the perioperative experience of hand and wrist surgical patients and was used to identify eight factors that affect the patient experience in negative and positive ways. Four factors were time-independent and present over the whole patient journey: insecurity; loneliness; lack of information; and lack of reassurance by the staff. In addition to these emotions experienced throughout the journey, four time-dependent factors were identified: lack of control; acceptance; curiosity; and relief. Given the identified factors, two paths for redesign were defined. The first design path involved interventions aimed at improving the overall patient experience such as a digital platform to educate and prepare patients for their surgery. The second design path consisted of interventions aimed at a specific moment or experience, such as redesigning the surgical cloth and patient-specific lighting during surgery.

The concept of patient journey and the associated methodology of patient journey mapping create a holistic overview of the healthcare system(s) involved in a certain clinical procedure or protocol or a specific group of patients, while providing the opportunity to identify and zoom in on specific friction points. Patient journey mapping is an iterative human-centered design process where the visualizations can inspire and challenge patients and healthcare staff to rethink how the current health system is functioning, therefore leading to changes in communication, interactions, and touchpoints with the patient.

4.6 RECOMMENDATIONS AND IMPLICATIONS FOR RESEARCH AND PRACTICE OF MACROERGONOMICS OF PATIENT WORK

Analyzing and improving patient work systems pose multiple challenges to macroergonomics researchers and practitioners. Collecting data from patients, in particular in the home environment or community settings, requires attention to the following issues (Holden, McDougald Scott, et al., 2015):

- Development of partnership with patients. This involves consideration for issues such as different priorities between researchers or practitioners and patients, and limited trust or mistrust.
- Addressing patient characteristics. Patients may have cognitive or perceptual limitations that reduce their ability to communicate and engage in a participatory redesign process.
- Logistics and procedures. For instance, access to patients can be challenging because of physical distance, scheduling of home visit, and burden associated with data collection.
- Data quality, interpretation, and integration. In macroergonomic projects on analysis and improvement of patient work, multiple and mixed data are often collected, which raises issues of potential conflict between data sources (e.g., patient experience versus clinician's perspective versus data recorded in medical records).
- Patient sampling. Another major challenge for both research and practice in macroergonomics of patient work is related to the participation of patients in the redesign and improvement of patient work systems. For instance, it is not clear how many patients and what type of patients should be involved. To ensure that the patient journey map is of high-quality, it is important to involve as many patients as possible to understand their experiences; this will help to answer multiple questions: What do patients experience that is common across patients? What are the unique experiences? What are the common and unique steps of the patient journey? Saturation is probably a good criterion to use in this process (Sandelowski, 1995); however, researchers and practitioners still need to identify the range of patients targeted for sampling. Patients and other actors (e.g., healthcare staff) should also be involved in the validation and interpretation of results to create a shared view on the redesign. Having shared insights yields strong and effective design directions for the improvement of patient-centered service delivery.
- Use of mixed methods to collect relevant data. Research on macroergonomics of healthcare quality and patient safety is likely to rely on mixed methods to collect and analyze data (Carayon, Kianfar et al., 2015); this is also true for research on macroergonomics of patient work. For instance, the patient journey mapping methodology described above involves multiple data collection methods, including qualitative research methods with often limited sample sizes. The reflective and iterative design process, the participatory

approach with multiple actors and perspectives, and insights from different sources make macroergonomic approaches, such as the patient journey mapping, valuable to analyze, understand, and redesign health systems.

Improving patient work is likely to be a multi-factorial problem: meeting needs of multiple people (i.e., those of patients, but also needs of informal caregivers, and healthcare staff) and addressing multiple outcomes of patients and other "workers" (e.g., caregivers and healthcare staff). This requires consideration of multiple perspectives in the human-centered participatory design process of improving patient work (Détienne, 2006; Xie et al., 2015). Macroergonomic approaches and methods need to be further developed to effectively consider these multiple perspectives in improving patient work; this includes developing methods to identify and manage trade-offs in the design process.

4.7 CONCLUSION

Patients perform work to manage or maintain their health, which involves multiple tasks performed in physical, social, and organizational environments, using tools and technologies. The tasks are organized in a certain manner; for example, in the case of a patient with diabetes, she or he monitors glucose levels and then administers insulin. Macroergonomic models have many elements in common as described above. The variations across the models include differences such as the distinction between an internal and external environment. Models may also differ in how they relate system elements to other factors or system levels. For example, the SEIPS models link the work system to process and outcomes. Czaja and colleagues (Czaja & Nair, 2006; Czaja et al., 2001) relate the core micro-ergonomic triad to multiple system levels.

The macroergonomic models of patient work provide deep and broad opportunities to understand and, therefore, improve the system(s) in which patient work occurs. For instance, there may be a misfit between different system elements (e.g., a person does not have the adequate technologies to manage their health information and communicate with their clinician), which may lead to a negative impact on care processes or patient outcomes. If this misfit or barrier is addressed, outcomes may improve. Such improvement needs to consider how the patient's work system interacts with the work systems of their care team members, such as informal caregivers and clinicians. This requires an appreciation for the various work systems involved in patient care over time and space, which is the core of the patient journey concept.

ACKNOWLEDGMENTS

This work was supported by the Clinical and Translational Science Award (CTSA) program, through the NIH National Center for Advancing Translational Sciences (NCATS), Grant Number: 1UL1TR002373. The content is solely the responsibility of the authors and does not necessarily represent the official views of the NIH.

REFERENCES

Ahern, D. K., Woods, S. S., Lightowler, M. C., Finley, S. W., & Houston, T. K. (2011). Promise of and potential for patient-facing technologies to enable meaningful use. *American Journal of Preventive Medicine, 40*(5, Supplement 2), S162–S172.

Bate, P., & Robert, G. (2006). Experience-based design: From redesigning the system around the patient to co-designing services with the patient. *Quality & Safety in Health Care, 15*(5), 307–310.

Black, N., Varaganum, M., & Hutchings, A. (2014). Relationship between patient reported experience (PREMs) and patient reported outcomes (PROMs) in elective surgery. *The BMJ Quality & Safety, 23*(7), 534–542.

Brooke, J. (1996). SUS: A "quick and dirty" usability scale. In P. W. Jordan, B. A. Werdmeester, & A. L. McClelland (Eds.), *Usability evaluation in industry*. London: Taylor & Francis, 189–194.

Caligtan, C. A., Carroll, D. L., Hurley, A. C., Gersh-Zaremski, R., & Dykes, P. C. (2012). Bedside information technology to support patient-centered care. *International Journal of Medical Informatics, 81*(7), 442–451.

Carayon, P. (2009). The balance theory and the work system model... Twenty years later. *International Journal of Human-Computer Interaction, 25*(5), 313–327.

Carayon, P. (2012). Sociotechnical systems approach to healthcare quality and patient safety. *Work: A Journal of Prevention, Assessment and Rehabilitation, 41*(0), 3850–3854.

Carayon, P., Alyousef, B., Hoonakker, P., Hundt, A. S., Cartmill, R., Tomcavage, J., . . . Walker, J. (2012). Challenges to care coordination posed by the use of multiple health IT applications. *Work: A Journal of Prevention, Assessment and Rehabilitation, 41*(0), 4468–4473.

Carayon, P., Hundt, A. S., Hoonakker, P., Kianfar, S., Alyousef, B., Salek, D., . . . Tomcavage, J. (2015). Perceived impact of care managers' work on patient and clinician outcomes. *European Journal of Patient-Centered Healthcare, 3*(2), 158–167.

Carayon, P., Hundt, A. S., Karsh, B.-T., Gurses, A. P., Alvarado, C. J., Smith, M., & Brennan, P. F. (2006). Work system design for patient safety: The SEIPS model. *Quality & Safety in Health Care, 15*(Supplement I), i50–i58.

Carayon, P., Karsh, B.-T., Gurses, A. P., Holden, R. J., Hoonakker, P., Hundt, A. S., . . . Wetterneck, T. B. (2013). Macroergonomics in health care quality and patient safety. *Review of Human Factors and Ergonomics, 8*, 4–54.

Carayon, P., Kianfar, S., Li, Y., Xie, A., Alyousef, A., & Wooldridge, A. (2015). A systematic review of mixed methods research on human factors and ergonomics in health care. *Applied Ergonomics, 51*, 291–321.

Carayon, P., Wetterneck, T. B., Rivera-Rodriguez, A. J., Hundt, A. S., Hoonakker, P., Holden, R., & Gurses, A. P. (2014). Human factors systems approach to healthcare quality and patient safety. *Applied Ergonomics, 45*(1), 14–25.

Carayon, P., & Wood, K. E. (2009). Patient safety: The role of human factors and systems engineering. In W. B. Rouse & D. A. Cortese (Eds.), *Engineering the system of healthcare delivery* (pp. 23–46). IOS Press, Amsterdam, The Netherlands.

Carayon, P., & Wooldridge, A. (2019). Improving patient safety in the patient journey: Contributions from human factors engineering. In A. E. Smith (Ed.), *Women in industrial and systems engineering: Key advances and perspectives on emerging topics*. Springer Nature, Cham, Switzerland.

Carayon, P., Wooldridge, A., Hoonakker, P., Hundt, A. S., & Kelly, M. M. (2020). SEIPS 3.0: Human-centered design of the patient journey for patient safety. *Applied Ergonomics, 84*.

Carman, K. L., Dardess, P., Maurer, M., Sofaer, S., Adams, K., Bechtel, C., & Sweeney, J. (2013). Patient and family engagement: A framework for understanding the elements and developing interventions and policies. *Health Affairs, 32*(2), 223–231.

Castro, E. M., Van Regenmortel, T., Vanhaecht, K., Sermeus, W., & Van Hecke, A. (2016). Patient empowerment, patient participation, and patient-centeredness in hospital care: A concept analysis based on a literature review. *Patient Education and Counseling, 99*(12), 1923–1939.

Collins, S. A., Dykes, P., Bates, D. W., Couture, B., Rozenblum, R., Prey, J., . . . Dalal, A. K. (2018). An informatics research agenda to support patient and family empowerment and engagement in care and recovery during and after hospitalization. *Journal of the American Medical Informatics Association, 25*(2), 206–209.

Collins, S. A., Rozenblum, R., Leung, W. Y., Morrison, C. R., Stade, D. L., McNally, K., . . . Dalal, A. K. (2016). Acute care patient portals: A qualitative study of stakeholder perspectives on current practices. *Journal of the American Medical Informatics Association, 24*(e1), e9–e17.

Czaja, S. J., & Nair, S. N. (2006). Human factors engineering and systems design. In G. Salvendy (Ed.), *Handbook of human factors and ergonomics* (3rd ed.). New York: John Wiley & Sons, 32–49.

Czaja, S. J., Sharit, J., Charness, N., Fisk, A. D., & Rogers, W. (2001). The center for research and education on aging and technology enhancement (CREATE): A program to enhance technology for older adults. *Gerontechnology, 1*(1), 50–59.

Desmet, P. M. A. (2002). *Designing emotions*. Delft, The Netherlands: Delft University of Technology.

Détienne, F. (2006). Collaborative design: Managing task interdependencies and multiple perspectives. *Interacting with Computers, 18*(1), 1–20.

Donabedian, A. (1978). The quality of medical care. *Science, 200*, 856–864.

Donabedian, A. (1988). The quality of care. How can it be assessed? *Journal of the American Medical Association, 260*(12), 1743–1748.

Dul, J., Bruder, R., Buckle, P., Carayon, P., Falzon, P., Marras, W. S., . . . van der Doelen, B. (2012). A strategy for human factors/ergonomics: Developing the discipline and profession. *Ergonomics, 55*(4), 377–395.

Gorman, R. K., Wellbeloved-Stone, C. A., & Valdez, R. S. (2018). Uncovering the invisible patient work system through a case study of breast cancer self-management. *Ergonomics, 61*(12), 1575–1590.

Greenhalgh, T., Hinton, L., Finlay, T., Macfarlane, A., Fahy, N., Clyde, B., & Chant, A. (2019). Frameworks for supporting patient and public involvement in research: Systematic review and co-design pilot. *Health Expectations, 0*(0), 1–17.

Hargraves, I., LeBlanc, A., Shah, N. D., & Montori, V. M. (2016). Shared decision making: The need for patient-clinician conversation, not just information. *Health Affairs, 35*(4), 627–629.

Hendrick, H. W. (1991). Human factors in organizational design and management. *Ergonomics, 34*, 743–756.

Hendrick, H. W. (2007). Macroergonomics: The analysis and design of work systems. *Reviews of Human Factors and Ergonomics, 3*(1), 44–78.

Hendrick, H. W., & Kleiner, B. M. (2001). *Macroergonomics - An introduction to work system design*. Santa Monica, CA: The Human Factors and Ergonomics Society.

Henriksen, K., Joseph, A., & Zayas-Caban, T. (2009). The human factors of home health care: A conceptual model for examining safety and quality concerns. *Journal of Patient Safety, 5*(4), 229–236.

Holden, R. J., Carayon, P., Gurses, A. P., Hoonakker, P., Hundt, A. S., Ozok, A. A., & Rivera-Rodriguez, A. J. (2013). SEIPS 2.0: A human factors framework for studying and improving the work of healthcare professionals and patients. *Ergonomics, 56*(11), 1669–1686.

Holden, R. J., McDougald Scott, A., Hoonakker, P. L. T., Hundt, A. S., & Carayon, P. (2015). Data collection challenges in community settings: Insights from two studies of patients with chronic disease. *Quality of Life Research, 24*, 1043–1055.

Holden, R. J., Schubert, C. C., & Mickelson, R. S. (2015). The patient work system: An analysis of self-care performance barriers among elderly heart failure patients and their informal caregivers. *Applied Ergonomics, 47*(0), 133–150.

Holden, R. J., Valdez, R. S., Schubert, C. C., Thompson, M. J., & Hundt, A. S. (2017). Macroergonomic factors in the patient work system: Examining the context of patients with chronic illness. *Ergonomics, 60*(1), 26–43.

Institute of Medicine (IOM). (2001). *Crossing the quality chasm: A new health system for the 21st century*. Washington, DC: National Academies Press.

Kelly, M. M., Coller, R. J., & Hoonakker, P. L. (2018). Inpatient portals for hospitalized patients and caregivers: A systematic review. *Journal of Hospital Medicine, 13*(6), 405–412.

Kelly, M. M., Dean, S. M., Carayon, P., Wetterneck, T. B., & Hoonakker, P. L. T. (2017). Healthcare team perceptions of a portal for parents of hospitalized children before and after implementation. *Applied Clinical Informatics, 26*(01), 265–278.

Kelly, M. M., Thurber, A. S., Coller, R. J., Khan, A., Dean, S. M., Smith, W., & Hoonakker, P. L. T. (2019). Parent perceptions of real-time access to their hospitalized child's medical records using an inpatient portal: A qualitative study. *Hospital Pediatrics, 9*(4), 273–280.

Kleiner, B. M. (2006). Macroergonomics: Analysis and design of work systems. *Applied Ergonomics, 37*(1), 81–89.

Kleiner, B. M. (2008). Macroegonomics: Work system analysis and design. *Human Factors, 50*(3), 461–467.

Kleinsmann, M., Sarri, T., & Melles, M. (2018). Learning histories as an ethnographic method for designing teamwork in healthcare. *CoDesign, 16(2)*, 1–19.

McCarthy, S., O'Raghallaigh, P., Woodworth, S., Lim, Y. L., Kenny, L. C., & Adam, F. (2016). An integrated patient journey mapping tool for embedding quality in healthcare service reform. *Journal of Decision Systems, 25*(suppl1), 354–368.

McGraw, C. A. (2019). Nurses' perceptions of the root causes of community-acquired pressure ulcers: Application of the model for examining safety and quality concerns in home healthcare. *Journal of Clinical Nursing, 28*(3–4), 575–588.

Melnick, E. R., Hess, E. P., Guo, G., Breslin, M., Lopez, K., Pavlo, A. J., . . . Post, L. A. (2017). Patient-centered decision support: Formative usability evaluation of integrated clinical decision support with a patient decision aid for minor head injury in the emergency department. *Journal of Medical Internet Research, 19*(5), e174.

Melnick, E. R., Lopez, K., Hess, E. P., Abujarad, F., Brandt, C. A., Shiffman, R. N., & Post, L. A. (2015). Back to the bedside: Developing a bedside aid for concussion and brain injury decisions in the emergency department. *EGEMS (Wash DC), 3*(2), 1136.

National Research Council. (2010). *The role of human factors in home health care: Workshop summary*. Washington, DC: The National Academies Press.

National Research Council. (2011). *Health care comes home: The human factors*. Washington, DC: National Academies Press.

Noro, K., & Imada, A. (1991). *Participatory ergonomics*. London: Taylor & Francis.

O'Leary, K. J., Sharma, R. K., Killarney, A., O'Hara, L. S., Lohman, M. E., Culver, E., . . . Cameron, K. A. (2016). Patients' and healthcare providers' perceptions of a mobile portal application for hospitalized patients. *BioMed Central Medical Informatics and Decision Making, 16*(1), 123.

Pasmore, W. A. (1988). *Designing effective organizations: The sociotechnical systems perspective*. New York: John Wiley & Sons.

Prey, J. E., Polubriaginof, F., Grossman, L. V., Masterson Creber, R., Tsapepas, D., Perotte, R., Vawdrey, D. K. (2018). Engaging hospital patients in the medication reconciliation process using tablet computers. *Journal of the American Medical Informatics Association, 25*(11), 1460–1469.

Ridder, E. d., Dekkers, T., Porsius, J. T., Kraan, G., & Melles, M. (2018). The perioperative patient experience of hand and wrist surgical patients: An exploratory study using patient journey mapping. *Patient Experience Journal, 5*, 97–107.
Sandelowski, M. (1995). Sample size in qualitative research. *Research in Nursing & Health, 18*(2), 179–182.
Simonse, L., Albayrak, A., & Starre, S. (2019). Patient journey method for integrated service design. *Design for Health, 3*(1): 82–97.
Singh, N., Hess, E., Guo, G., Sharp, A., Huang, B., Breslin, M., & Melnick, E. (2017). Tablet-based patient-centered decision support for minor head injury in the emergency department: Pilot study. *Journal of Medical Internet Research mHealth and uHealth, 5*(9), e144.
Smith, M. J., & Carayon-Sainfort, P. (1989). A balance theory of job design for stress reduction. *International Journal of Industrial Ergonomics, 4*(1), 67–79.
Smith, M. J., & Carayon-Sainfort, P. (2001). Balance theory of job design. In W. Karwowski (Ed.), *International Encyclopedia of ergonomics and human factors* (pp. 1181–1184). London: Taylor & Francis.
Srinivas, P., Cornet, V., & Holden, R. (2017). Human factors analysis, design, and evaluation of Engage, a consumer health IT application for geriatric heart failure self-care. *International Journal of Human-Computer Interaction, 33*(4), 298–312.
Stiggelbout, A. M., Weijden, T. V., Wit, M. P., Frosch, D., Legare, F., Montori, V. M., . . . Elwyn, G. (2012). Shared decision making: Really putting patients at the centre of healthcare. *The BMJ, 344*, e256.
Strauss, A. L. (1993). *Continual permutations of action*. New York, NY: Aldine Transaction.
Strauss, A. L., Fagerhaugh, S., Suczek, B., & Wiener, C. (1982). The work of hospitalized patients. *Social Science & Medicine, 16*(9), 977–986.
The Beryl Institute. (2019). Defining patient experience. Retrieved from https://www.theberylinstitute.org/page/DefiningPatientExp
Trebble, T. M., Hansi, N., Hydes, T., Smith, M. A., & Baker, M. (2010). Process mapping the patient journey: An introduction. *The BMJ, 341*, 394–401.
Trist, E. L. (1981). *The evolution of socio-technical systems*. Toronto, ON: Quality of Working Life Center.
Trist, E. L., & Bamforth, K. (1951). Some social and psychological consequences of the long-wall method of coal getting. *Human Relations, 4*, 3–39.
Valdez, R. S., Holden, R. J., Novak, L. L., & Veinot, T. C. (2015). Transforming consumer health informatics through a patient work framework: Connecting patients to context. *Journal of the American Medical Informatics Association, 22*(1), 2–10.
Vincent, C., & Amalberti, R. (2016). *Safer healthcare - Strategies for the real world*. Springer Open, Cham, Switzerland.
Vincent, C., Carthey, J., Macrae, C., & Amalberti, R. (2017). Safety analysis over time: Seven major changes to adverse event investigation. *Implementation Science, 12*(1), 151.
Wilson, J. R. (2014). Fundamentals of systems ergonomics/human factors. *Applied Ergonomics, 45*(1), 5–13.
Woollen, J., Prey, J., Wilcox, L., Sackeim, A., Restaino, S., Raza, S. T., . . . Vawdrey, D. (2016). Patient experiences using an inpatient personal health record. *Applied Clinical Informatics, 7*(2), 446–460.
Xie, A., Carayon, P., Cartmill, R., Li, Y., Cox, E. D., Plotkin, J. A., & Kelly, M. M. (2015). Multi-stakeholder collaboration in the redesign of family-centered rounds process. *Applied Ergonomics, 46*(Part A), 115–123.
Zink, K. (2000). Ergonomics in the past and the future: From a German perspective to an international one. *Ergonomics, 43*(7), 920–930.

Section III

Patient Ergonomics Domains

5 Consumer Health Information Technology
Integrating Ergonomics into Design, Implementation, and Use

Teresa Zayas-Cabán
Office of the National Coordinator for Health Information Technology

P. Jon White
Veterans Affairs Salt Lake City Health Care System

CONTENTS

Managing one's health, which includes both general wellness and health care, encompasses a complex set of activities. Successfully doing so requires, among many other things, that individuals or patients and their informal caregivers, who manage or support health care on someone's behalf, manage their health information. Whether it is information that they know, such as symptoms, copies of records they keep and take with them, or a bag of prescription bottles taken to a doctor's appointment, patients or their caregivers are generally the primary source of information about their health and the care that has been provided to them across space and time.

Effectively managing health information, however, is increasingly complex. Individuals no longer receive care and medicine at only the local general practitioner's office and the neighborhood drug store. Healthcare delivery in the U.S. is a $3 trillion industry (Centers for Medicare & Medicaid Services, 2018) and includes several different types of organizations. Interaction with all these organizations and services results in the generation of health information that is critical for effective health management by patients and caregivers, and healthcare delivery by clinicians. Unfortunately, detailed notes about appointments or test results created by physicians and other care professionals can be difficult for patients or caregivers to obtain, despite the clearly established legal right to one's own health information (U.S. Department of Health & Human Services, 2016b). Beyond clinical care, health insurance requirements bewilder doctors and patients alike.

The reality is that most health management activities take place in homes and communities outside of the formal healthcare delivery settings (i.e., hospitals, clinics, laboratories, and other facilities), where patients and caregivers shoulder the responsibility of managing medication regimens and care plans, nutrition, wellness activities such as exercise or meditation, and coordinating care. For those with chronic or acute health issues, managing care usually involves coordination across providers and settings (see Chapter 7 in this volume). Since the healthcare system often does not appropriately share health information needed for effective diagnosis and treatment among different clinicians, there is increased responsibility and pressure for information management on patients or caregivers. Clearly, health management comprises a challenging set of tasks for patients and caregivers.

5.1 A DIGITAL APPROACH TO MANAGING HEALTH INFORMATION

By its nature, health management is information intensive. In recent decades, health care has followed the broader societal trend toward the use of digital information systems. Health information technology (IT), or "the use of information and communication technology in health care to support the delivery of patient or population care or to support patient self-management" (Agency for Healthcare Research and Quality, 2013), is seen as a way to support clinicians, patients, and caregivers alike in management of healthcare activities.

Implementation and use of health IT in healthcare delivery increased dramatically between 2009 and 2015 across the U.S., and a clear majority of U.S. doctors and hospitals now use electronic health record (EHR) systems (Henry et al., 2016; Office of the National Coordinator for Health Information Technology, 2016), a type of clinical health IT system. Many of those applications provide patients and caregivers direct access to some of the health information in their doctors' or hospitals' EHR systems. However, clinical health IT systems used by doctors and hospitals do not comprehensively cover all the information important for patients' improvement and maintenance of their health and well-being. Doctors and other clinicians have also encountered significant issues with the usefulness of their health IT systems and its impact on their productivity (Friedberg et al., 2013; Khairat et al., 2018; Ratwani et al., 2019; Sinsky et al., 2016). Moreover, due to technological and policy barriers, patients and caregivers still struggle to access electronic health data stored in their care providers' health IT systems (Office of the National Coordinator for Health Information Technology, 2017a).

To manage their health information, people have, for many years, used different approaches, such as having notebooks of records that they compile and bring to all their appointments (Moen & Brennan, 2005). Some patients or caregivers have found that the simplest way to summarize their current condition is to bring all the present prescriptions to their appointments. Others try compiling copies of medical records from across different clinicians, laboratories, and other healthcare services. In recent decades, however, there has been growth in health IT intended for use by consumers of health care (Demiris, 2016; Eysenbach, 2000). Collectively referred to as consumer health IT, patient or caregiver use of information and communication technologies is seen as a way "to improve their medical outcomes and/or participate in [the] healthcare decision-making process" (Jimison et al., 2008, p. 9). Policymakers and advocates see consumer health IT as a potential way to support patients and caregivers with health information and care management. Federal policy initiatives have incented their adoption and use (Centers for Medicare & Medicaid Services, 2019; Office of the National Coordinator for Health Information Technology, 2019). Funding for research, provided by the Agency for Healthcare Research and Quality (2019), the National Institutes of Health (2018), the National Science Foundation (2019), and the Robert Wood Johnson Foundation (2015), has further fueled interest in consumer health IT applications. These trends have resulted in increased availability of consumer health IT applications, which, if well designed, implemented, and used, should effectively support patient and caregiver health information and health management work and result in improved health outcomes.

5.2 COMMON CONSUMER HEALTH IT APPLICATIONS AND THEIR USES

The broad term "consumer health IT" encompasses many different types of applications and functionalities, which are described in more detail below.

5.2.1 Types of Consumer Health IT Applications

Table 5.1 summarizes the types of tools currently available in the market or provided to patients or caregivers by their doctors or hospitals.

Perhaps the most prevalent types of consumer health IT applications are *personal health records* (PHRs) and *patient portals*. PHRs were originally envisioned as standalone applications that would store all of an individual's health records from multiple sources (Tang et al., 2006). Several commercial systems were launched, and some eventually shut down due to barriers to easily extracting and integrating health data from multiple information systems and sources (*The end of Google Health*,

TABLE 5.1
Consumer Health IT Applications and Their Functionalities

Type	Typical Functionalities
Personal health records	Integrate clinical notes, laboratory results, prescription records, claim records, and benefit information
Patient portals	*Clinician portal:* Shares clinician-generated health information and allows appointment scheduling, prescription refill or renewal requests, secure messaging with clinicians, and billing and payment
	Health insurance portal: Shares insurer-generated information, such as claims records and benefits information
Diagnostic tools	Provide diagnostic information based on consumer input
Care management tools	Record treatment plan, track symptoms, provide decision support, and aid medication management
Telehealth	Delivers remote care delivery
Shared decision-making tools	Provide risk appraisal and decision support and document patient values, wishes, and decisions
Communication tools	Provide communication with or from clinicians, other patients, or caregivers
Health and wellness tools	Provide informational or educational resources as well as fitness or dietary support

2011; Truong, 2019). Patient portals, sometimes referred to as tethered PHRs, on the other hand, are typically offered by direct care providers (i.e., doctors, hospitals, and other care professionals) and are a feature of their EHR systems (Office of the National Coordinator for Health Information Technology, 2017b). These usually give patients or caregivers access to information such as current medications, laboratory test results, after-care visit (outpatient) or discharge (inpatient) summaries, and immunization records. They also include other functionalities, such as secure messaging with clinicians, appointment scheduling, prescription renewal or refill requests, or billing and payment. In addition to portals offered by care providers, health insurers often provide portals with individually relevant information, such as claims and benefits information (J. M. Grossman et al., 2009). Insurer portals can also include health assessments and other tools. Ancillary service providers, such as pharmacies (Ching & Kapoor, 2017) and laboratories (Hale, 2018; Rappleye, 2019), also have patient portal offerings that may, for example, provide prescription records and allow for refill or renewal requests or share laboratory results.

Many patients and caregivers desire faster and easier diagnosis of their symptoms than can be obtained through traditional means. *Diagnostic tools* utilize not only clinical data such as laboratory values and vital signs, but also broadly available consumer technology data, such as smartphone pictures, audio input from microphones, and accelerometer data, in increasingly sophisticated analytic capabilities (Hernández-Neuta et al., 2019).

Patients and caregivers also frequently use tools to support care management activities by monitoring their symptoms, recording their own information, or taking action on health-related tasks through *care management tools* (Finkelstein et al., 2012) (Chapter 7 in this volume). Functionalities of these tools include tracking of symptoms and relevant health measures, decision support on seeking or performing care, medication management, and managing claims and other health-related financial information. To the extent that diagnostic and care management tools are taking the place of a human care professional, regulation of such tools or some of their functions by the Food and Drug Administration (FDA) has been a policy topic of discussion.

Increasingly, patients and caregivers are seeking care from clinicians remotely through *telehealth* applications. These tools include video teleconferencing with care professionals (American Telemedicine Association, 2019). Telehealth can also be offered remotely at dedicated locations (e.g., at schools), at specific facilities (e.g., nursing home receiving specialty care), or in patients' homes. The use of telehealth has been increasing, partially driven by improved health insurance coverage of those services and also driven by the COVID-19 pandemic (Barnett et al., 2018).

Clinical decision-making is one important aspect of care delivery because it involves the individual receiving care or their caregivers. *Shared decision-making tools* encourage and facilitate communication between clinician and patient and can document how the decision was reached (Finkelstein et al., 2012). These tools can increase consumer understanding of a condition or treatment and improve adherence to recommended management.

Communication is an essential component of health management (Chapter 6 in this volume). Communication occurs from clinician to patient, but can also entail patient-to-patient, caregiver-to-patient, caregiver-to-caregiver, and caregiver-to-clinician. *Communication tools* such as secure messaging are providing new channels for patients and caregivers to be in more frequent touch with their clinical team (Finkelstein et al., 2012; Goldzweig et al., 2012). However, tools are increasingly available to connect patients to "peers" who may share a diagnosis or treatment, as well as to connect caregivers to each other (Falisi et al., 2017). Patients and caregivers are employing social networking and social media to stay informed about new developments in their area of interest or connect to their peer groups (Smailhodzic et al., 2016).

Finally, *health and wellness tools* that are not directly associated with medical care and conditions comprise a significant component of available consumer health IT. Whether used to monitor and improve diet, track steps and exercise, or guide meditation, these tools are widely used by consumers and can have a significant impact on overall health (Muoio, 2019).

5.2.2 Mobile Applications and Consumer Health IT

Broader trends in consumer IT use have also recently begun to affect the development and use of different kinds of consumer health IT applications. There has been an evolution in the use of mobile applications (apps) instead of a personal computer or internet browser technology to manage a host of information, including health-related information. Increasingly, consumers are also gathering health information about themselves through sensors, such as those available via smartphones, and wearable digital devices, such as activity trackers (see Chapters 7, 9, and 12 in this volume).

Application programming interfaces, or APIs, are particular sets of rules and specifications that software programs can follow to communicate with each other directly (Stack Overflow, 2011). The use of APIs has made it easier to exchange many different types of data across devices and platforms (e.g., from a wearable device to a smartphone). This has created opportunities for the development of consumer health IT apps, also referred to as mobile health apps or mHealth apps. There are currently more than 300,000 consumer health IT apps available to patients or caregivers (Research2Guidance, 2017). In the pursuit of increased interoperability, defined as the ability of a system to exchange electronic health information with and use electronic health information from other systems without special effort on the part of the user (adapted from the Institute for Electrical and Electronics Engineering (1990)), between health IT systems, the healthcare sector has been developing and implementing standardized APIs. The rate of implementation is expected to increase due to the passage of the 21st Century Cures Act of 2016 (Cures Act), which requires that patients be given access to their digital health information *without special effort* through an API (21st Century Cures Act, 2016). To implement relevant provisions of the Cures Act, the Office of the National Coordinator for Health Information Technology and the Centers for Medicare & Medicaid Services (CMS) released regulations that require API development and use (21st Century Cures Certification, 2020; CMS Interoperability and Patient Access, 2020).

These policy and technological advances continue to take place within the broader policy and regulatory environment. Health care is a heavily regulated economic sector at both the federal and state levels, and those regulations have a significant impact on the kinds of technology and capabilities available to consumers and how health information is shared. In particular, privacy and security must be considered when developing, implementing, and using consumer health IT. The Health Insurance Portability and Accountability Act (HIPAA) Privacy Rule guarantees individuals the right of access to their health data, but also governs the privacy of those data at the federal level (U.S. Department of Health & Human Services, 2015). Issues with security, which is regulated at the federal level by the HIPAA Security Rule (U.S. Department of Health & Human Services, 2017), have also become increasingly important in light of massive breaches of health information, which have affected tens of millions of patients (*10 largest data breaches of 2019,* 2019). Furthermore, consumer health IT applications may also be regulated by other federal agencies, such as the Federal Trade Commission or Federal Communications Commission (U.S. Department of Health & Human Services, 2016a), depending on their functionality. In addition, applications that fall under the FDA's definition of a medical device may require FDA approval before they can be offered to patients or caregivers (Larson, 2018).

5.2.3 Effectiveness and Prevalence of Consumer Health IT

While the reasons why people use consumer health IT vary, at least some use can be attributed to the hope or belief that using these tools will improve health and well-being. The impact of consumer health IT has been a subject of study by the health services research community. As shown in Table 5.2, published research

TABLE 5.2
Types of Health Outcomes Typically Studied When Evaluating Consumer Health IT

Types of Health Outcomes	Sample Outcomes
Clinical	• Quality of life[a, b, c] • Disease status[a, b, c] • Functional status[a, b]
Process	• Receiving appropriate treatment[a, b, c] • Healthcare utilization[a, b, c] • Therapeutic adherence[a, b, c]
Intermediate	• Satisfaction[c] • Self-management[a, b] • Safety[c] • Health knowledge[a, b, c] • Health behaviors[a, b, c]
Economic	• Healthcare costs[a, b] • Access to care[a, c]

Notes:
[a] Jimison et al. (2008).
[b] Gibbons et al. (2009).
[c] Finkelstein et al. (2012).

has tended to focus on health outcomes, such as clinical, process, intermediate, and economic (Finkelstein et al., 2012; Gibbons et al., 2009; Jimison et al., 2008; Zayas-Cabán & Marquard, 2012).

Systematic reviews have shown that health outcomes across many conditions, such as heart disease, cancer, and diabetes mellitus, generally improve when consumer health IT is employed (Finkelstein et al., 2012; Gibbons et al., 2009; Jimison et al., 2008). Several reviews have also found that the most effective consumer health IT solutions were those that had a "closed feedback loop" that not only monitored or transmitted and interpreted patient data, but also provided the patient or caregiver with a response (Gibbons et al., 2009; Jimison et al., 2008; Or & Karsh, 2009). Reviews have also noted that the usefulness and usability of consumer health IT could be significantly improved and that study of usefulness or usability-related outcomes was minimal. More recently published scientific literature reflects increased recognition of patient ergonomics issues and their relevance to consumer health IT (Marquard & Zayas-Cabán, 2012; Pavlas et al., 2018; Tarver & Haggstrom, 2019; Valdez et al., 2015). The focus on ergonomics in the literature, however, is primarily on aspects of usability, with less attention to the usefulness of the applications. Although usability is critical to ensure that end users can easily and efficiently interact with a given consumer health IT application, applications must also be useful to effectively support patient and caregiver work.

While consumer health IT has much promise, the overall adoption and use remain low, which could be considered a proxy measure of the usefulness and usability of existing applications. Patient portals are widely offered by doctors and hospitals and are required under federal incentive programs meant to spur adoption and use of health IT and programs designed to regulate health IT systems (Centers for Medicare & Medicaid Services, 2019; Office of the National Coordinator for Health Information Technology, 2019; Turner et al., 2019). However, extensive availability does not necessarily result in extensive use, and surveys have shown low use levels across the entire patient population (Anthony et al., 2018; Patel & Johnson, 2019). Consumer health IT or mobile health app offerings have increased significantly as well, but their overall use remains low compared to the use of other mobile apps, which are used to manage personal information, such as financial or social, though it is expected to increase (Clement, 2020; HealthWorksCollective, 2018; Muoio, 2019). The causes of low use are many, but a significant barrier is poorly addressed ergonomic issues, which are particularly important considerations for underserved, elderly, and chronically ill individuals (Jimison et al., 2008). Notably, use appears to differ across sociodemographic characteristics. Studies show that, generally, high consumer health IT use is skewed to individuals who are younger, healthier, and have higher education, health literacy, and income (Anthony et al., 2018; Bol et al., 2018; Carroll et al., 2017; Robbins et al., 2017; Strekalova, 2019). Barriers to use include limited IT knowledge or computer literacy, lack of perceived benefits, and concerns about privacy and security (Anthony et al., 2018; De La Cruz Monroy & Mosahebi, 2019; Gibbons et al., 2009; Jimison et al., 2008; Robbins et al., 2017; Strekalova, 2019). In addition, studies have shown that consumer health IT applications that do not fit into patient, caregiver, or provider workflow are not readily adopted or used (De La Cruz Monroy & Mosahebi, 2019; Gibbons et al., 2009; Jimison et al., 2008).

5.2.4 Patient Ergonomics and Consumer Health IT

Ergonomics has been broadly shown to be critical to other technology design, information systems used in work office settings, and other consumer products (National Research Council, 2011b). Increased prioritization of patient and caregiver access to health information, increased opportunities for health apps, and current low usage levels create an excellent opportunity to incorporate ergonomics in future applications and avoid repeating mistakes made with clinical health IT systems. Indeed, based on outcomes in other similar domains, the application of ergonomics to the understanding of patients' and caregivers' health information management work is essential for effective design, implementation, and use of consumer health IT. The stakes are higher—consumer health IT is meant to support laypeople with a critical need for access to information to take care of themselves and others in managing their health and health information. These individuals will have varying cognitive and physical characteristics, will perform this *work* within different organizational contexts, and will have little-to-no IT support and varying IT resources (see Chapters 2, 3, and 4 in this volume).

5.3 PATIENT ERGONOMICS ISSUES AND CONSIDERATIONS IN DESIGN, IMPLEMENTATION, AND USE OF CONSUMER HEALTH IT

To be effective, consumer health IT applications must function within a patient's broader "work system" and support information exchange and integration. Choices made regarding interfaces with other information systems and the ease with which patients or caregivers can obtain needed health information will enhance or limit a given solution's utility. Moreover, since applications found to be most effective are the ones that share information and feedback between patients or caregivers and healthcare professionals, these applications may need to account for the needs of patients, caregivers, and/or clinicians. This may create challenges in design, implementation, and use—including carefully balancing the needs, preferences, and constraints of these three user groups.

It is useful to consider the whole design lifecycle (Austin & Boxerman, 1997) and, to the extent possible, use iterative and ongoing processes to inform design and implementation with lessons learned from preceding and following phases (see Chapter 10 in this volume). It is also important to pay attention to all three domains of ergonomics during the design lifecycle: (1) cognitive, (2) physical, and (3) organizational (International Ergonomics Association, 2019). This chapter will not review the three ergonomics domains in depth—those are discussed in Chapters 2, 3, and 4 in this volume—but these serve as a good guiding frame to address the complex issues that arise in designing, implementing, and using high-quality consumer health IT. To illustrate the range of ergonomics issues that must be considered throughout the consumer health IT design lifecycle, we employ two vignettes of a patient and a caregiver and how they interact with patient portals offered by healthcare professionals (see Boxes 5.1 and 5.2).

5.3.1 Ergonomics Considerations for Consumer Health IT Design

Human-centered design "aims to make interactive systems more usable by focusing on the use of the system and applying human factors/ergonomics and usability knowledge and techniques" (International Organization for Standardization, 2019, p. 2) (see Chapter 10 in this volume). Such design has been found to be effective at improving use and desired outcomes of consumer health IT (Karsh et al., 2011). Good system design begins with careful consideration of the existing need or problem that health IT will address. The need or problem can include supporting a specific task or process (e.g., managing a complex medication regimen) or focusing on a patient population with high need (e.g., oncology patients). Consumer health IT applications should ideally extend (i.e., provide new needed functionality), optimize (i.e., conduct a task efficiently or effectively), or enhance (i.e., essentially handle a task "for" the user) a user's performance of health information management or health management tasks. Designers must then identify the user base for the intended health IT and consider the attributes of those users. This includes understanding current information needs and related processes; user needs, goals, and motivations; and the environments and context in which the health IT application will be used (National Research Council, 2011a, 2011b; National Library of Medicine, 2018). For example, if the user base is a

BOX 5.1 USING A PATIENT PORTAL FOR A PHYSICIAN OFFICE VISIT

Susie has been dealing with asthma ever since she was young. She has been doing well for the last couple of years, but recently a bout with the flu and the cold weather have made it difficult to breathe, especially when she climbs the stairs at work, and will wake her up during the night. Her primary care doctor has recently given her a course of prednisone, which made her feel better, but now that the medicines are done she is back to feeling bad.

She calls her primary care doctor, and they agree it is time to see the pulmonologist. When she calls the pulmonologist, they ask her to come in and bring records of her recent doctor visits, the most recent X-ray report, pulmonary function tests, and current medications. Susie remembers that a couple of months ago, her primary doctor's office signed her up to use its patient portal so she can access her records and make appointments online. She decides that would be the easiest way to get her records.

When she goes to sign in, though, Susie does not remember her username or password. To get those, she has to call the number on the portal website to have her password reset. Susie leaves a voicemail and waits for several hours. Her call is returned in the afternoon, and she is finally able to log in.

Within the portal, Susie sees where she can pay her bill, but cannot find where her records are located. She calls back the help number on the website and waits again. After holding for 20 minutes, the support person walks her through the menu of options and Susie eventually finds the doctor's note from her last visit, which has her breathing tests and medications.

Unfortunately, her most recent X-ray tests were done at urgent care, and whereas her primary care doctor has a copy of the radiology report in the EHR, the report is not available to Susie via the patient portal. Frustrated, Susie prints out what she has and hurries over to the pulmonologist. They squeeze her in, but because it is near the end of the day, Susie has to wait for an hour and a half. The pulmonologist's office sends Susie to radiology to get new X-rays because neither Susie nor the office had access to her previous X-rays. After another 90 minutes at radiology, Susie heads back to the pulmonologist's office, which is now closed. Susie heads home, feeling tired, defeated, and worse than when she started the day.

pediatric population, it will have a substantially different set of cognitive and physical attributes than, say, an older adult population (Safran, 2019). Some of these attributes can be garnered from the existing literature, but the reality is that ergonomics practitioners will have to reach out to the target population to assess its needs and capabilities. Practically speaking, this means the practitioner will need to determine what the target population needs to accomplish and how it will go about accomplishing it.

BOX 5.2 USING A PATIENT PORTAL FOR AN EMERGENCY ROOM VISIT

Luis takes care of his sister Cristina, who has a genetic condition that causes heart problems. All their lives, Luis has kept meticulous track of Cristina's medications in a physical binder. While this required a lot of effort, Luis knew that all the information was correct because he always double-checked the information. Recently, Cristina's cardiologist switched to using an EHR system with a patient portal. Luis likes using technology and has also been using the patient portal to keep track of Cristina's heart medications.

A few months ago, something changed in Cristina's heart rhythm. She was getting tired more easily. After seeing the cardiologist, one medication was stopped, and she was prescribed a new medication. Cristina started the medication using samples from the doctor's office, but when Luis tried to fill a prescription at the pharmacy for Cristina to continue taking the medication regularly, he was told that the copay was several hundred dollars. Unsure what to do, he sent a message about the expensive copay to the cardiologist using the patient portal app on his phone. At the cardiologist's practice, a nurse saw Luis' message and spoke to the cardiologist, who sent in a new prescription for a generic medication. However, the pharmacy did not have the medication readily available and had to order it.

Cristina ran out of the medicine Saturday morning. By Sunday morning, she was feeling weak, and Luis took her to the emergency room. During the intake process, Luis told the triage staff the name of Cristina's cardiologist and they obtained her medication list and medical history from the local health information exchange (HIE). When the HIE was implemented, the system was set to update information once per week on Mondays, rather than every time new information was entered. Consequently, the medication list included a drug Cristina was no longer taking, but no one asked Luis if the list was current as she had been started on the new medicine just that week.

The emergency room physician took one look at Cristina's electrocardiogram (EKG) and decided to admit her for an abnormal heart rhythm. He ordered several medications and admitted her to the cardiology service. One of the medicines he ordered would have made Cristina's heart problem worse unless she was also taking the medication she had stopped taking. Fortunately, the nurse reviewed the emergency room physician's medication orders with Luis and Cristina. Luis noticed her new medication was not on the list and showed the nurse the current medication list available through the portal app. The nurse found the physician and discussed the order for the problematic medication. The physician then ordered Cristina's new medication, avoiding any potential adverse impact on her health. Understanding these issues can help establish functional requirements and usage scenarios, as well as develop and test system specifications and designs.

There are many ergonomics methods that can be employed to understand these design considerations (see, e.g., Chapters 9–12 in this volume). As a result of the complexities of health information and health management, a combination of iterative design approaches should be used to learn about patients and caregivers and their needs (International Organization for Standardization, 2019; Montague, 2012). As noted earlier, an initial literature review of both the ergonomics literature and the specific content domain may provide insights into the relevant user population, problem of interest, and relevant ergonomics considerations (Holden et al., 2020). Interviews, surveys, and focus groups are often used to understand user characteristics, their abilities, and environments (Agency for Healthcare Research and Quality, n.d.; Holden et al., 2020; International Organization for Standardization, 2019; Montague, 2012; Zayas-Cabán & Valdez, 2012). Observations in the setting in which activities of interest are carried out can be extremely helpful in understanding which tasks are undertaken, technologies used, environmental constraints, and organizational issues (Montague, 2012). Observations can range from more ethnographic activities to use of, or adaptation of, specific techniques such as checklists, work sampling, time and motion studies, and task analysis (Agency for Healthcare Research and Quality, n.d.; Holden et al., 2020; National Research Council, 2011b; Zayas-Cabán & Valdez, 2012). Findings from these earlier activities can be used to develop personas or flowcharts to aid in design, as well as result in requirements and prototypes (Agency for Healthcare Research and Quality, n.d.; Montague, 2012). These results need to be validated using some of the previously mentioned techniques. Prototypes can also be validated using heuristic reviews by experts and usability testing (Agency for Healthcare Research and Quality, n.d.; International Organization for Standardization, 2019; Montague, 2012; National Research Council, 2011b; Zayas-Cabán & Valdez, 2012).

Patient portals were designed to be used by essentially all patients (or proxies in the case of pediatric populations and other patients with unique circumstances) with a specific set of features, which could include (1) access to some health record information, such as immunization records, laboratory and other diagnostic test results, after-care visit or discharge summaries, and medications; (2) appointment scheduling; (3) secure messaging with clinicians; and (4) other basic information, such as clinic contact information and patient contact information. It is unclear how these functions were selected when these systems were designed and whether these functions would support or meet critical health information needs or health management tasks. A recent systematic review (Dendere et al., 2019) provides examples of distinct approaches used when making design choices for patient portals. Some healthcare organizations employed an iterative design approach, where patients could inform some of the functionality in resulting portal designs through pilot-testing, observations, and interviews. For example, patient-requested features included use of games, access to all health information, individualized content, and ability to communicate with clinicians, which were used to inform the portal's design. At other institutions, however, it was healthcare organization leaders who collaborated with developers to design the systems. By not engaging end users in the design process, particularly patients and caregivers, it is unlikely all user needs were reflected in the resultant design (Eyasu et al., 2019; Portz et al., 2019).

Designing for an entire patient population also means accounting for a wide range of cognitive and physical abilities. Effective systems would need to reach individuals with varied literacy, health literacy, numeracy, and computer literacy (Chapter 2 in this volume). They would also need to account for a range of differences in mobility, vision, and other physical abilities (Chapter 3 in this volume). Portals would need to work across a range of possible evolving technologies and platforms that patients would use. For example, the first patient portals were available before the availability of smartphones and mobile apps in the consumer product market (Wang et al., 2004; Weingart et al., 2006) and might have required redesign for use on a patient's or caregiver's smartphone via an app. They would also need to accommodate patient and caregiver access needs and preferences within the broader legal and regulatory landscape as well as within the provider organization's policies (Baldwin et al., 2017). One large academic medical center describes that the policies they used to register patients and provide access to information allow "proxy" (e.g., caregivers, guardians, and others) access to the portal on behalf of a patient, implement messaging between users and care professional staff, determine the availability of laboratory test results through the portal, and implement appointment and bill management functionalities (Osborn et al., 2011). These policies may have impacted patient or caregiver ability or willingness to sign up for and use a portal account. Across institutions, clinicians have expressed concerns about the impact patient or caregiver use of the patient portal may have on their workload (Dendere et al., 2019). Clinicians' concerns may impact both their willingness to promote portal use to patients or caregivers and using the technology themselves, which may impact patients' or caregivers' use of the technology.

In Susie's case (see Box 5.1), the portal was designed so that she would have access to some, but not all, of her health information via her primary care doctor's EHR system. In this case, not only would having access to her X-ray reports have enabled her to be seen by her pulmonologist quickly, but would have also likely reduced the need for duplicate testing. Had the developer of Susie's doctor's EHR system conducted human-centered design to understand what information patients such as Susie might need available through the patient portal, they might have uncovered the fact that having access to X-ray reports is extremely valuable. In Luis' and Cristina's case (see Box 5.2), the vendor that developed Cristina's cardiologist's EHR system had conducted focus groups with several patients and caregivers to understand what information they would need access to via the patient portal. The resulting portal included needed information, making it easier for Luis to monitor Cristina's medications and communicate with her clinicians.

5.3.2 Ergonomics Considerations for Consumer Health IT Implementation

After design and development, it is time to deploy the health IT. Implementation includes the steps needed to make the consumer health IT application operational, including conducting training and/or testing before launch, deployment, or making the application available in the market. This is a critical phase of the lifecycle for several reasons. Choices will be made during implementation about what functions are available, how the users will be supported, and what resources will be available for the operational system, all of which will affect the user experience and utility of the consumer IT application. If these choices are not made with the end user in mind, suboptimal outcomes are likely to result. Since it is the first time users will encounter the technology, their initial experience will influence their attitudes and future use. These implementation choices may also have long-term impacts that may be difficult to reverse and may impact future decisions. As with the design phase, ergonomic domains are useful constructs to illustrate a range of considerations when undertaking the implementation of consumer health IT.

In the cognitive ergonomics domain, implementers should consider outreach to potential users before implementation, engagement of users, and training before launch. A recent systematic review found that several different training approaches have been employed with varying levels of success, which may impact patients' or caregivers' abilities to effectively use a given portal (Dendere et al., 2019). The physical ergonomics aspects of implementation include understanding the infrastructure base of consumer technology and testing and monitoring successful install and account signups. Organizational ergonomics issues to consider are interactions with the broader work system, including patients' caregivers and healthcare professionals, adequate implementation resources, and policies developed and put in place to oversee the technology's operation. Dendere et al. (2019) found that healthcare provider organization leadership was critical to portal implementation across a range of institutions, including the development of training and decisions regarding how to integrate the portal into clinical care. Implementation approaches and choices can impact patients' and caregivers' abilities to interact with the technology (Baldwin et al., 2017). Studies have found that the interpretation of relevant legislation and regulation has resulted in the varying implementation of policies governing sharing of health information. Some under-resourced provider organizations have had difficulties with implementation (Dendere et al., 2019).

Several ergonomics methods can be used to ensure that relevant issues are accounted for in the implementation of consumer health IT. Heuristic reviews and usability testing are useful in evaluating a developed consumer health IT application before launching it (Agency for Healthcare Research and Quality, n.d.; Montague, 2012; National Research Council, 2011b; Zayas-Cabán & Valdez, 2012). Surveys and interviews can be used to elicit feedback during the implementation process to ensure that the final consumer health IT application adequately addresses user needs and to identify any potential problems early on (Agency for Healthcare Research and Quality, n.d.; Holden et al., 2020). These approaches can be effectively integrated into the implementation phase to ensure that the desired outcomes are achieved for both developers and users.

In Susie's case, as shown in Box 5.1, rather than employing an automated user name and password retrieval system, her primary care doctor's office decided to employ a phone number. The number of staff available to answer the phone may not have been sufficient to provide a timely response to patients. Patient portals and other consumer health IT must be implemented so that minimal training and IT support are required. Individuals may not have the time to contact a support person or wait for a response. In Luis' and Cristina's case, as described in Box 5.2, Cristina's cardiologist office had a process in place to monitor incoming messages through the portal. Medication-related messages were routed to a nurse for quick review and possible action. The cardiologist was able to respond quickly and send a new prescription to the pharmacy.

5.3.3 Ergonomics Considerations for Consumer Health IT Use

Once implementation has occurred, evaluators and system administrators can determine whether the intended purpose of the health IT is being fulfilled. It is vital to systematically assess performance, use, and utility after implementation and during the use of consumer health IT using ergonomics approaches, paying particular attention to unintended outcomes (Baldwin et al., 2017; International Organization for Standardization, 2019). For example, some studies have found differential use of portal features (e.g., low use of the messaging feature), whereas others have found that users want additional functionality (e.g., to be able to send a message to a given staff member, be alerted that a message had been read, and know when staff would respond), and some have found functionality inadequate (e.g., could only access selected health information instead of all their health information) (Dendere et al., 2019).

Particular cognitive ergonomics issues that need to be addressed after the application is implemented include whether the consumer health IT solution is being used as intended, what kinds of difficulties users are encountering, in what unintended ways it is being utilized, whether characteristics of users and non-users differ, and whether actual use suggests future development needs. For example, several studies have found differences between portal users and non-users. Patient portal users are typically white, younger, female, and have higher income, education, and health literacy (Anthony et al., 2018; L. V. Grossman et al., 2019). Concerns regarding privacy and security can also impact portal use, as well as perceived portal benefits (Dendere et al., 2019). Studies also continue to find that some interfaces are not well-designed, inhibiting patients' or caregivers' abilities to effectively use them (Dendere et al., 2019). Regarding physical ergonomics, it is also important to determine if there are barriers to use for different populations, if any accessibility issues are not being met, whether the consumer health IT application can be effectively used across patients' differing physical infrastructures, whether the infrastructure has unintended effects on use and outcome, and how the solution will adapt to changing infrastructure requirements. Organizational ergonomics aspects include paying attention to how the application's use aligns with structures, policies, and processes; whether access rights need to be reassessed; and whether they fit with the patient's or caregiver's broader work system. Since these portals are offered by care provider organizations, the relationship with clinicians has been shown to influence use. If clinicians do not encourage patients to use the portal or they do not want to use the portal themselves, this will impact patient and caregiver use (Dendere et al., 2019). In particular, research has shown that clinicians may not offer all their patients access to their organization's patient portal (Anthony et al., 2018). There may also be disparities in different provider organization's abilities to engage patients or caregivers through their portals (Dendere et al., 2019).

Human-centered design also requires evaluating the use and system performance (International Organization for Standardization, 2019). It is important to have processes in place to elicit user feedback after implementation and establish clear performance metrics (Holden et al., 2020; Montague, 2012). Interviews and surveys can be used to continually gather feedback about the system once it is in use (Holden et al., 2020). Summative usability evaluations can be conducted to ensure that the system is easy to use on an ongoing basis (Agency for Healthcare Research and Quality, n.d.; Holden et al., 2020; International Organization for Standardization, 2019; Montague, 2012; Zayas-Cabán & Valdez, 2012). Log data can be used to understand feature use and inform the need for reviewing performance and functionality (Holden et al., 2020). If appropriate preparations have been made, data such as login frequency, time spent using an application, functionality used, and instances of malicious attacks, such as unauthorized access or availability attacks, can be derived from the information system. Data collected using these methods can be used to improve existing services, but importantly can also be used in future design and development efforts to deliver better products and outcomes.

In the vignettes, Susie needed to be able to quickly use the system to obtain the information she needed for her specialist appointment. Unfortunately, she had difficulty finding the information she needed because some of it came from another provider and was not shared via her portal. This caused her to have to go to the radiologist and, in the end, she was unable to get the care she so desperately needed. If the primary care doctor's portal developer has a process in place to receive user feedback on an ongoing basis, they may be able to ensure that Susie and other patients have all the information they need through the portal. Luis, on the other hand, was able to use the information on the portal app to show the nurse administering Cristina's medications the current medication list. Having the application available on his phone with current and relevant information made it easier to communicate with hospital staff and care for Cristina.

5.4 CONCLUSION AND RECOMMENDATIONS FOR ERGONOMICS PRACTITIONERS

Patients and caregivers manage a multitude of complex and information-intensive health activities. There is increased development of consumer health IT to support their efforts. This increased development has, unfortunately, been characterized by inconsistent attention to ergonomics issues, with resultant suboptimal outcomes. As the demand for consumer health IT is expected to grow, the field has a unique opportunity to embed ergonomics in design, implementation, and use.

This chapter lays out considerations throughout the design lifecycle and across ergonomics domains. Based on these considerations, we offer four main recommendations for incorporating ergonomics into the design, implementation, and use of consumer health IT:

1. *Approach design, implementation, and use of consumer health IT to improve the work of patients and caregivers.* It is critical that consumer health IT applications are designed and implemented to effectively support

health information and health management activities. Ergonomics principles and methods described in this chapter should be employed during design to establish how the proposed application will support patient or caregiver work. That is the principal goal against which success or failure must be measured. Otherwise, the resultant consumer health IT application may inadvertently create more work for patients and caregivers, which may result in little to no use and limited impact on outcomes.

2. *Engage users throughout the design lifecycle of consumer health IT.* Users should be integral parts of the teams that design and implement consumer health IT, and actual use should be carefully evaluated (see also Chapter 13 in this volume). Data about user health information management tasks should be gathered using ergonomics methods, and those data should be used to inform design. Once designed, the health IT should be tested by potential users in their actual health information management settings. Users are the principal target of implementation, which should be executed with them primarily in mind. Finally, actual use must be evaluated since broad deployment often reveals uses and outcomes not previously identified in the design lifecycle.
3. *Use existing ergonomics and human factors knowledge and resources, such as models and techniques to improve the design, implementation, and use of consumer health IT.* There is an extant and extensive knowledge base and set of ergonomics principles, theories, and methods that can be leveraged for effective design and implementation of consumer health IT. Relevant methods include literature reviews, observation, interviews, surveys, task analysis, heuristic reviews, usability testing, log data analysis, and summative usability evaluations. A combination of methods should be used to learn about users and their needs. While the extant knowledge base should be used in combination with field-based methods to understand user needs and evaluate the technology, they provide a strong foundation from which to work.
4. *Iterate cyclically on design, implementation, and use to achieve better usefulness and usability of consumer health IT.* Use ergonomics methods and approaches to ensure consumer health IT continually meets patients' and caregivers' needs. Employing an iterative design approach with pilot testing of early prototypes using ergonomic techniques will help ensure that the consumer health IT application more accurately reflects user needs. Even after accounting for ergonomics considerations, assessing user needs, and engaging users throughout the design lifecycle, unanticipated issues are likely to emerge during implementation and use. In addition, user needs may change, the tasks may change, the technology or supporting infrastructure may change, and new organizational considerations, such as changing policies or regulations, may need to be considered. Ergonomics techniques should be used to continually assess the consumer health IT application during use to inform redesign and implementation. Mechanisms must be in place to obtain feedback, monitor and observe use, and then iterate on the application's design or implementation. These mechanisms should include

techniques that continually identify ergonomics issues through both system performance and use monitoring and soliciting user feedback through interviews and surveys.

Attention to these issues will create more effective, efficient, and useful tools, with the goal of better supporting patients and caregivers in their health information management and health management work using health IT, leading to improved outcomes.

ACKNOWLEDGMENTS

The authors would like to thank Palladian Partners and Jesse Zarley for copy editing support and reference formatting assistance. The findings and conclusions in this chapter are those of the authors and do not necessarily represent the views of the Office of the National Coordinator for Health Information Technology, the U.S. Department of Health and Human Services, or the Veterans Affairs Salt Lake City Health Care System.

REFERENCES

21st Century Cures Act, H.R. 34, 114th Congress, Pub. L. No. 114–255 § 1001–18001, 1033 Stat. 130 (2016).

21st Century Cures Act: Interoperability, information blocking, and the ONC Health IT Certification Program, § 170 and 171 (2020).

Agency for Healthcare Research and Quality. (2013). *Exploratory and developmental grant to improve health care quality through health information technology (IT) (R21)*. Retrieved from https://grants.nih.gov/grants/guide/pa-files/PA-14-001.html

Agency for Healthcare Research and Quality. (2019). *AHRQ digital healthcare research funding opportunities*. Retrieved from https://digital.ahrq.gov/ahrq-digital-healthcare-research-funding-opportunities

Agency for Healthcare Research and Quality. (n.d.). *Workflow assessment for health IT toolkit*. Retrieved from https://digital.ahrq.gov/workflow

American Telemedicine Association. (2019). *Telehealth basics*. Retrieved from https://www.americantelemed.org/resource/why-telemedicine/

Anthony, D. L., Campos-Castillo, C., & Lim, P. S. (2018). Who isn't using patient portals and why? Evidence and implications from a national sample of US adults. *Health Affairs, 37*, 1948–1954.

Austin, C. J., & Boxerman, S. B. (1997). *Information systems for health services administration* (5 ed.). Ann Arbor, MI: Health Administration Press.

Baldwin, J. L., Singh, H., Sittig, D. F., & Giardina, T. D. (2017). Patient portals and health apps: Pitfalls, promises, and what one might learn from the other. *Healthcare (Amsterdam, Netherlands), 5*(3), 81–85.

Barnett, M. L., Ray, K. N., Souza, J., & Mehrotra, A. (2018). Trends in telemedicine use in a large commercially insured population, 2005–2017. *Journal of the American Medical Association, 320*(20), 2147–2149.

Bol, N., Helberger, N., & Weert, J. C. M. (2018). Differences in mobile health app use: A source of new digital inequalities? *The Information Society, 34*(3), 183–193.

Carroll, J. K., Moorhead, A., Bond, R., LeBlanc, W. G., Petrella, R. J., & Fiscella, K. (2017). Who uses mobile phone health apps and does use matter? A secondary data analytics approach. *Journal of Medical Internet Research, 19*(4), e125.

Centers for Medicare & Medicaid Services. (2018). *NHE fact sheet.* Retrieved from https://www.cms.gov/research-statistics-data-and-systems/statistics-trends-and-reports/nationalhealthexpenddata/nhe-fact-sheet.html

Centers for Medicare & Medicaid Services. (2019). *Promoting interoperability (PI).* Retrieved from https://www.cms.gov/Regulations-and-Guidance/Legislation/EHRIncentivePrograms/index.html

Ching, J., & Kapoor, K. (2017). Benefits, barriers, and impact of patient portals and mHealth applications. *PharmacyToday, 23*(10), 42.

Clement, J. (2020). *Number of mobile app downloads worldwide from 2016-2019 (in billions).* Medicare and Medicaid programs; Patient Protection and Affordable Care Act; Interoperability and patient access for Medicare advantage organization and Medicaid managed care plans, state Medicaid agencies, CHIP agencies, and CHIP managed care entities, issuers of qualified health plans in the federally-facilitated exchanges and health care providers, § 406, 407, 422, 423, 431, 438, 457, 482, 485, 156 (2020). Retrieved from https://www.statista.com/statistics/271644/worldwide-free-and-paid-mobile-app-store-downloads/

De La Cruz Monroy, M. F. I., & Mosahebi, A. (2019). The use of smartphone applications (apps) for enhancing communication with surgical patients: A systematic review of the literature. *Surgical Innovation, 26*(2), 244–259.

Demiris, G. (2016). Consumer health informatics: Past, present, and future of a rapidly evolving domain. *Yearbook of Medical Informatics,* (Suppl 1), S42–47.

Dendere, R., Slade, C., Burton-Jones, A., Sullivan, C., Staib, A., & Janda, M. (2019). Patient portals facilitating engagement with inpatient electronic medical records: A systematic review. *Journal of Medical Internet Research, 21*(4), e12779.

The end of Google Health. (2011). *HITECH answers.* Retrieved from https://www.hitechanswers.net/the-end-of-google-health/

Eyasu, T., Leung, K., & Strudwick, G. (2019). Guiding improvements in user experience: Results of a mental health patient portal user interface assessment. *Studies in Health Technology and Informatics, 257,* 110–114.

Eysenbach, G. (2000). Consumer health informatics. *The BMJ, 320*(7251), 1713–1716.

Falisi, A. L., Wiseman, K. P., Gaysynsky, A., Scheideler, J. K., Ramin, D. A., & Chou, W. S. (2017). Social media for breast cancer survivors: A literature review. *Journal of Cancer Survivorship: Research and Practice, 11*(6), 808–821.

Finkelstein, J., Knight, A., Marinopoulos, S., Gibbons, M. C., Berger, Z., Aboumatar, H., . . . Bass, E. B. (2012). Enabling patient-centered care through health information technology. *Evidence Report/Technology Assessment,* (206), 1–1531.

Friedberg, M. W., Chen, P. G., Van Busum, K. R., Aunon, F., Pham, C., Caloyeras, J. P., . . . Tutty, M. (2013). *Factors affecting physician professional satisfaction and their implications for patient care, health systems, and health policy.* Santa Monica, CA: RAND Corporation.

Gibbons, M. C., Wilson, R. F., Samal, L., Lehman, C. U., Dickersin, K., Lehmann, H. P., . . . Bass, E. B. (2009). Impact of consumer health informatics applications. *Evidence Report/Technology Assessment,* (188), 1–546.

Goldzweig, C. L., Towfigh, A. A., Paige, N. M., Orshansky, G., Haggstrom, D. A., Beroes, B. S., . . . Shekelle, P. G. (2012). *Systematic review: Secure messaging between providers and patients, and patients' access to their own medical record: Evidence on health outcomes, satisfaction, efficiency and attitudes.* Washington, DC: Department of Veterans Affairs.

Grossman, J. M., Zayas-Cabán, T., & Kemper, N. (2009). Information gap: Can health insurer personal health records meet patients' and physicians' needs? *Health Affairs, 28*(2), 377–389.

Grossman, L. V., Masterson Creber, R. M., Brenda, N. C., Wright, D., Vawdrey, D. K., & Ancker, J. S. (2019). Interventions to increase patient portal use in vulnerable populations: A systematic review. *Journal of the American Medical Informatics Association, 26*(8–9), 855–870.

Hale, C. (2018). Quest to allow patients to order lab testing from home. *FierceBiotech*. Retrieved from https://www.fiercebiotech.com/medtech/quest-to-allow-patients-to-order-lab-testing-direct-from-home

HealthWorksCollective. (2018). *Mobile medical apps: A game changing healthcare innovation*. Retrieved from https://www.healthworkscollective.com/mobile-medical-apps-a-game-changing-healthcare-innovation/

Henry, J., Pylypchuk, Y., Searcy, T., & Patel, V. (2016). *Adoption of electronic health record systems among U.S. non-federal acute care hospitals: 2008–2015*. (ONC Data Brief, no.35). Washington, DC: Office of the National Coordinator for Health Information Technology.

Hernández-Neuta, I., Neuman, F., Brightmeyer, J., Ba Tis, T., Madaboosi, N., Wei, Q., . . . Nilsson, M. (2019). Smartphone-based clinical diagnostics: Towards democratization of evidence-based health care. *Journal of Internal Medicine, 285*(1), 19–39.

Holden, R. J., Cornet, V. P., & Valdez, R. S. (2020). Patient ergonomics: 10-year mapping review of patient-centered human factors. *Applied Ergonomics, 82*(2020), doi:10.1016/j.apergo.2019.102972

IEEE. (1990). *IEEE standard computer dictionary: A compilation of IEEE standard computer glossaries*. New York: IEEE.

International Ergonomics Association. (2019). *Definitions and domains of ergonomics*. Retrieved from https://www.iea.cc/whats/index.html

International Organization for Standardization. (2019). Ergonomics of human-system interaction — Part 210: Human-centred design for interactive systems.

Jimison, H., Gorman, P., Woods, S., Nygren, P., Walker, M., Norris, S., & Hersh, W. (2008). Barriers and drivers of health information technology use for the elderly, chronically ill, and underserved. *Evidence Report/Technology Assessment*, (175), 1–1422.

Karsh, B.-T., Holden, R. J., & Or, C. K. (2011). Human factors and ergonomics of health information technology implementation. In P. Carayon (Ed.), *Handbook of human factors and ergonomics in health care and patient safety* (2nd ed., pp. 249–264). Boca Raton, FL: CRC Press.

Khairat, S., Burke, G., Archambault, H., Schwartz, T., Larson, J., & Ratwani, R. M. (2018). Perceived burden of EHRs on physicians at different stages of their career. *Applied Clinical Informatics, 9*(2), 336–347.

10 largest data breaches of 2019. (2019). Becker's Healthcare. Retrieved from https://www.beckershospitalreview.com/cybersecurity/10-largest-data-breaches-of-2019.html.

Larson, R. S. (2018). A path to better-quality mhealth apps. *JMIR mHealth and uHealth, 6*(7), e10414.

Marquard, J. L., & Zayas-Cabán, T. (2012). Commercial off-the-shelf consumer health informatics interventions: Recommendations for their design, evaluation and redesign. *Journal of the American Medical Informatics Association, 19*(1), 137–142.

Moen, A., & Brennan, P. F. (2005). Health@Home: The work of health information management in the household (HIMH): Implications for consumer health informatics (CHI) innovations. *Journal of the American Medical Informatics Association, 12*(6), 648–656.

Montague, E. (2012). *Designing consumer health IT: A guide for developers and systems designers* (AHRQ Publication No. 12–0066-EF). Rockville, MD: Agency for Healthcare Research and Quality.

Muoio, D. (2019). *Global medical app downloads exceeded 400M in 2018. MobiHealth News*. Retrieved from https://www.mobihealthnews.com/content/global-medical-app-downloads-exceeded-400m-2018

National Library of Medicine. (2018). *NLM research grants in biomedical informatics and data science (R01 clinical trial optional)*. Retrieved from https://grants.nih.gov/grants/guide/pa-files/par-18-896.html.

National Research Council. (2011a). *Consumer health information technology in the home: A guide for human factors design considerations.* Washington, DC: The National Academies Press.

National Research Council. (2011b). *Health care comes home: The human factors.* Washington, DC: The National Academies Press.

National Science Foundation. (2019). *Smart and connected health (SCH): Connecting data, people and systems.* Retrieved from https://www.nsf.gov/funding/pgm_summ.jsp?pims_id=504739

Office of the National Coordinator for Health Information Technology. (2016). *Office-based physician electronic health record adoption: 2004–2015* (Health IT Quick-Stat 50). Washington, DC. Retrieved from https://dashboard.healthit.gov/quickstats/pages/-physician-ehr-adoption-trends.php

Office of the National Coordinator for Health Information Technology. (2017a). *Improving the health record request process for patients: Insights from user experience research.* Washington, DC. Retrieved from https://www.healthit.gov/sites/default/files/onc_records-request-research-report_2017-06-01.pdf

Office of the National Coordinator for Health Information Technology. (2017b). *What is a patient portal?* Retrieved from https://www.healthit.gov/topic/certification-ehrs/about-onc-health-it-certification-program

Office of the National Coordinator for Health Information Technology. (2019). *About the Health IT Certification Program.* Retrieved from https://www.healthit.gov/topic/certification-ehrs/about-onc-health-it-certification-program

Or, C. K., & Karsh, B. T. (2009). A systematic review of patient acceptance of consumer health information technology. *Journal of the American Medical Informatics Association, 16*(4), 550–560.

Osborn, C. Y., Rosenbloom, S. T., Stenner, S. P., Anders, S., Muse, S., Johnson, K. B., . . . Jackson, G. P. (2011). MyHealthAtVanderbilt: Policies and procedures governing patient portal functionality. *Journal of the American Medical Informatics Association, 18*(Suppl 1), i18–23.

Patel, V., & Johnson, C. (2019). *Trends in individuals' access, viewing and use of online medical records and other technology for health needs: 2017–2018* (ONC Data Brief, no.47). Washington, DC: Office of the National Coordinator for Health Information Technology.

Pavlas, J., Krejcar, O., Maresova, P., & Selamat, A. (2018). Prototypes of user interfaces for mobile applications for patients with diabetes. *Computers, 8*(1), 1.

Portz, J. D., Bayliss, E. A., Bull, S., Boxer, R. S., Bekelman, D. B., Gleason, K., & Czaja, S. (2019). Using the Technology Acceptance Model to explore user experience, intent to use, and use behavior of a patient portal among older adults with multiple chronic conditions: Descriptive qualitative study. *Journal of Medical Internet Research, 21*(4), e11604.

Rappleye, E. (2019). *LabCorp rolls out direct-to-consumer blood tests. Becker's Health IT & CIO Report.* Retrieved from https://www.beckershospitalreview.com/consumerism/-labcorp-rolls-out-direct-to-consumer-blood-tests.html

Ratwani, R. M., Reider, J., & Singh, H. (2019). A decade of health information technology usability challenges and the path forward. *Journal of the American Medical Association, 321*(8), 743–744.

Research2Guidance. (2017). *mHealth App Economics 2017/2018: Current status and future trends in mobile health.* Retrieved from Berlin, Germany. https://research2guidance.com/325000-mobile-health-apps-available-in–2017

Robbins, R., Krebs, P., Jagannathan, R., Jean-Louis, G., & Duncan, D. T. (2017). Health app use among US mobile phone users: Analysis of trends by chronic disease status. *JMIR mHealth and uHealth, 5*(12), e197.

Robert Wood Johnson Foundation. (2015). *Project HealthDesign: Rethinking the power and potential of personal health records.* Retrieved from https://www.rwjf.org/en/library/research/2010/10/project-healthdesign--rethinking-the-power-and-potential-of-pers.html

Safran, C. (2019). *InfoSAGE information sharing across generations and environments - Final report.* Rockville, MD: Agency for Healthcare Research and Quality. Retrieved from https://healthit.ahrq.gov/ahrq-funded-projects/infosage-information-sharing-across-generation-and-environments/final-report

Sinsky, C., Colligan, L., Li, L., Prgomet, M., Reynolds, S., Goeders, L., . . . Blike, G. (2016). Allocation of physician time in ambulatory practice: A time and motion study in 4 specialties. *Annals of Internal Medicine, 165*(11), 753–760.

Smailhodzic, E., Hooijsma, W., Boonstra, A., & Langley, D. J. (2016). Social media use in healthcare: A systematic review of effects on patients and on their relationship with healthcare professionals. *BioMed Central Health Services Research, 16*(1), 442.

Stack Overflow. (2011). *What exactly is the meaning of an API?* Retrieved from https://stackoverflow.com/questions/7440379/what-exactly-is-the-meaning-of-an-api

Strekalova, Y. A. (2019). Electronic health record use among cancer patients: Insights from the Health Information National Trends Survey. *Health Informatics Journal, 25*(1), 83–90.

Tang, P. C., Ash, J. S., Bates, D. W., Overhage, J. M., & Sands, D. Z. (2006). Personal health records: Definitions, benefits, and strategies for overcoming barriers to adoption. *Journal of the American Medical Informatics Association, 13*(2), 121–126.

Tarver, W. L., & Haggstrom, D. A. (2019). The use of cancer-specific patient-centered technologies among underserved populations in the United States: Systematic review. *Journal of Medical Internet Research, 21*(4), e10256.

Truong, K. (2019). Microsoft HealthVault is officially shutting down in November. *MedCity News.* Retrieved from https://medcitynews.com/2019/04/microsoft-healthvault-is-officially-shutting-down-in-november/

Turner, K., Hong, Y. R., Yadav, S., Huo, J., & Mainous, A. G. (2019). Patient portal utilization: Before and after stage 2 electronic health record meaningful use. *Journal of the American Medical Informatics Association, 26*(10), 960–967.

U.S. Department of Health & Human Services. (2015). *The HIPAA privacy rule.* Retrieved from https://www.hhs.gov/hipaa/for-professionals/privacy/index.html

U.S. Department of Health & Human Services. (2016a). *Examining oversight of the privacy & security of health data collected by entities not regulated by HIPAA.* Washington, DC: Author. Retrieved from https://www.healthit.gov/sites/default/files/non-covered_entities_report_june_17_2016.pdf

U.S. Department of Health & Human Services. (2016b). *Individuals' right under HIPAA to access their health information 45 CFR §164.524.* Retrieved from https://www.hhs.gov/hipaa/for-professionals/privacy/guidance/access/index.html

U.S. Department of Health & Human Services. (2017). *The security rule.* Retrieved from https://www.hhs.gov/hipaa/for-professionals/security/index.html

Valdez, R. S., Holden, R. J., Novak, L. L., & Veinot, T. C. (2015). Transforming consumer health informatics through a patient work framework: Connecting patients to context. *Journal of the American Medical Informatics Association, 22*(1), 2–10.

Wang, T., Pizziferri, L., Volk, L. A., Mikels, D. A., Grant, K. G., Wald, J. S., & Bates, D. W. (2004). Implementing patient access to electronic health records under HIPAA: Lessons learned. *Perspectives in Health Information Management, 1,* 11.

Weingart, S. N., Rind, D., Tofias, Z., & Sands, D. Z. (2006). Who uses the patient internet portal? The PatientSite experience. *Journal of the American Medical Informatics Association, 13*(1), 91–95.

Zayas-Cabán, T., & Marquard, J. L. (2012). Using human factors to guide the design and implementation of consumer health informatics interventions. In J. C. Augusto, M. Huch, A. Kameas, J. Maitland, P. McCullagh, J. Roberts, A. Sixsmith, & R. Wichert (Eds.), *Handbook of ambient assisted living* (Vol. 11, pp. 22–36). Amsterdam: IOS Press.

Zayas-Cabán, T., & Valdez, R. S. (2012). Human factors and ergonomics in home care. In P. Carayon (Ed.), *Handbook of human factors and ergonomics in health care and patient safety* (2nd ed., pp. 743–762). Boca Raton, FL: CRC Press.

6 Patient–Professional Communication

Onur Asan
Stevens Institute of Technology

Bradley H. Crotty
Medical College of Wisconsin

Avishek Choudhury
Stevens Institute of Technology

CONTENTS

Communication is defined in social science as "the production and exchange of information and meaning by use of signs and symbols" (Gerbnere, 1985). Patients communicate to doctors information about their symptoms and context; doctors communicate to patients educational points, diagnosis information, and instructions; professional teams communicate information to one another to carry out the care plans for patients; and patients communicate with one another as peer supports, such as within online or in-person support groups. Effective communication can make the difference between making and following through with correct diagnoses and treatment plans, or not, as seen in the case in Box 6.1.

BOX 6.1 A CASE STUDY ABOUT LEE, PART ONE

Lee is a 37-year-old woman who is looking to establish a primary care relationship. She recently moved into the area after taking a new job position. She began her search for a new doctor online, reviewing online videos and patient experience comments. Having Crohn's disease, a chronic illness affecting her digestive system, she has had her share of good and bad experiences receiving care.

When she initially developed symptoms, she did not think much of the lower abdominal discomfort but wanted to be checked out because she thought that something might not be right. After a normal pelvic examination, attention turned to her weight. Lee had been losing weight. The doctor asked about it, considering that she, like many young women, may have an eating disorder. Lee said she just was not hungry. She also had a tough time describing her symptoms, which seemed difficult to put into words. It did not help that her doctors and nurses seemed rushed with a backdrop of long lines of patients, a demanding computer that the doctors complained about, and frequent interruptions by the clinic staff.

After a few follow-up visits, each with different doctors who remained vigilant about the possibility of an eating disorder, she developed severe pain requiring an emergency room visit after her intestines became blocked. Further diagnostic inquiry and testing led to her diagnosis. Lee felt that the delay in her diagnosis was related to not being listened to, and she lost about 10% of her body weight during this time. It is important to her that she can find a doctor whom she can trust and communicate with well following that experience.

Patient–professional communication is considered the backbone of the healthcare visit and a crucial component of patient-centered care. Patient-centered care is defined as providing care that is respectful of, and responsive to, individual patient preferences, needs, and values and ensuring that patient values guide all clinical decisions, by the US Institute of Medicine (2001). Effective communication is required for attaining desirable healthcare outcomes (Maguire & Pitceathly, 2002); it affects patient satisfaction (Asan et al., 2014), adherence to treatment (Sawyer & Aroni, 2003), clinical outcomes (Arora, 2003), and patient trust (Asan, Tyszka, et al., 2018).

Patient–professional communication has two primary purposes: information sharing and establishing relationships between patients and care providers (Bylund et al., 2012), which are distinct but complementary and synergistic. Building the relationship leads to increased trust, which can facilitate more open and honest information sharing. Conversely, poor communication has been linked to medical errors, wrong or delayed diagnosis as shown in the case study in Box 6.1, and is associated with increased malpractice claims (Bari et al., 2016; Naughton, 2018).

6.1 COMMUNICATION: BACKGROUND AND RELEVANT THEORIES

Communication between people helps construct a shared understanding, which is critical to the healing professions (Figure 6.1). The word "communication" itself is

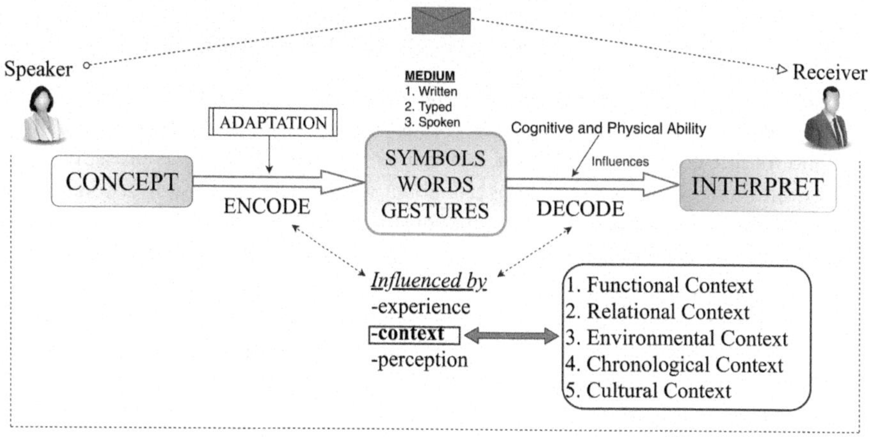

FIGURE 6.1 House of human communication.

derived from the Latin word "communis," meaning "to share." While we recognize that communication includes verbal messages, communication also encompasses nonverbal cues and signals, text, graphics, and electronic means of human interaction. Human communication is profoundly a collaborative meaning-making progression. We make meaning through a communal conception and elucidation of symbols, comprising words, gestures, images, sounds, and artifacts.

In 1948, Claude Shannon laid the groundwork for modeling communication in seminal articles in the *Bell Systems Technical Journal* (Shannon, 1948). Shannon, who had joined Bell Labs and had been working during World War II on cryptography, began more generically modeling communication to enhance the signal and reduce the noise, along channels. Shannon outlined the generic communication system as one where a reference message is encoded, transmitted, and then decoded (Slack, 1997) (Figure 6.1). Communication of a message, however, initiates with a thought (i.e., "reference" or "context"). In other words, every expression a speaker encodes is only an approximation of the thought the speaker has in mind (Shatz, 2006). If a person wants to convey that thought to others, he or she encodes the thought into symbols and behaviors. Using this concept of communication theory is not only of historical interest, but also of importance to understanding how communication can break down between patients and professionals.

As our case continues (Box 6.2), we see that Lee had a difficult time describing her symptoms, and the clinicians had a difficult time interpreting them. The symptoms were vague, and Lee used the term 'gnawing' rather than pain. In this example, 'gnawing' is a symbol or term to represent what Lee was experiencing, and the clinician must decode that through her past experiences and consider how that relates to a disease process. Even in face-to-face visits among patients and clinicians, the intended message may be received and interpreted differently depending on the context and experiences of the recipient (see Chapter 7 in this volume).

This challenge is further compounded when the patient and professional use different languages or have different cultural backgrounds. As a product of people's

BOX 6.2 A CASE STUDY ABOUT LEE, PART TWO

When Lee first developed symptoms, she felt a "gnawing" in her stomach. She had a hard time characterizing it, but "gnawing" was about the best word she could find. It was not quite a pain, nor would she characterize it as pressure—two common words that clinicians ask about when eliciting discomfort. When she was in the waiting room, other patients appeared to be more in pain than Lee, who appeared calm, without the usual grimace or tears that can be seen on people with abdominal pain. She was not hungry. She was tender when her abdomen was examined, which the emergency doctor identified only by looking closely at her face as she pushed with two hands in her left lower portion of Lee's abdomen. Lee had been under a lot of stress at that time, studying for her college examinations.

divergent personal and cultural experiences, the realization of an optimal shared understanding is an unlikely, fragile, and effortful inter-subjective task.

Beyond having a shared understanding, the communication process is further impacted by the varying physical and cognitive abilities of patients (Hannawa et al., 2017). As an example, relevant in health care, individuals with advanced age may become frail, during which people lose their physical (eyesight and hearing ability) and psychological (cerebral) reserves (Xue, 2011), which might hinder their ability to conceptualize, rationalize, and interpret information. This may require adaptation on the part of the doctor to ensure that concepts are adequately conveyed.

Communication must be tailored according to the patient's needs and ability to communicate. For instance, verbal communication accompanied by nonverbal physiognomies such as intensity, voice inflection, or facial expressions and gesticulations significantly alters the intentional meaning of the message. There are diverse communicative means to communicate with a patient to encourage follow-through with a therapeutic regimen or facilitate primary diagnosis. Reflecting on our case, the fact that Lee did not present with a tearful or pained facial expression may have led the doctors to consider alternative diagnoses such as an eating disorder for her weight loss rather than a digestive problem. At the same time, Lee's nonverbal facial expression during the examination was a significant indication of an abdominal problem.

As a result of important aspects of verbal and nonverbal elements of communication, and the importance of context (described further below), the medium or channel for patient–professional communication must be suited appropriately to the message. For example, when a complex message is being delivered that requires checking for understanding, a richer channel (such as in-person or through video) might be preferred over asynchronous messaging, which is not able to transmit nonverbal cues. In modern times, doctors and patients may use multiple channels for different elements of a transaction, moving from asynchronous message-based communication to voice to in-person as complexity rises. It perhaps would have been even more difficult to establish Lee's diagnosis without the richness of the examination and nonverbal communication.

BOX 6.3 A CASE STUDY ABOUT LEE, PART THREE

After Lee was treated for her intestinal blockage, she followed up with a gastroenterologist. She was particularly fond of him because he seemed very relatable to her, which increased her trust that he had her interests in mind. While he was about a generation older than her, he made strides to relate, or at least acknowledge, how a young adult would go through a new diagnosis of a chronic illness, one requiring injectable medicines on a regular basis and follow-up testing.

The communication accommodation theory (CAT) is a structure for understanding the interpersonal and intergroup dynamics of communicators (patient–professional) regulating their verbal and nonverbal configurations to each other (Asan, Kim, et al., 2018; Farzadnia & Giles, 2015). The CAT posits that the individual's perception of their communication partners has an impact on how they decide to behave in the interaction. CAT features individuals' beliefs and intentions underlying their communicative behavior in an immediate situation, oriented either convergently toward or divergently away from others present. CAT emphasizes how, when, and why people adapt their messages to match that of their interlocutors (accommodation) or not (non-accommodation) and how disagreements can be managed (Gasiorek & Giles, 2012). The theory argues that people accommodate those they admire (respect, trust) to attenuate social and communicative differences. As shown in the case study above, the communication style of the gastroenterologist increased Lee's trust in him (Box 6.3).

6.1.1 Context

The context in which the communication occurs has important implications for how the sender crafts the message, how the message is—or should be—conveyed, and how the recipient interprets the message. Depending on the situational and sociocultural complex, some particular way of communicating a message may be more effective. At the same time, a given approach that often works effectively may fail with a specific patient. Clinical communications are intensely context-dependent. Care professionals' failure to acknowledge the constraining and expediting effects of these contextual dimensions on establishing a shared understanding can compromise patient safety. The implication of a conversed message essentially depends on the perspective in which it is encoded and established. Such context contains multiple layers (Carayon, 2012) such as:

- Functional context: the goals people pursue in their interaction.
- Relational context: the professional–patient relationship, for example, how long they have known each other.
- Environmental context: factors such as comfort, accessibility, and physical safety.

- Chronological context: The sequence in which information is transmitted from the sender to the receiver.
- Cultural context: Ability to perceive and interpret information. The association of meaning to certain words or visuals depends on people's cultural background.

Throughout our case so far, the context has been incredibly important in Lee's story. Initially, she felt that the visits were rushed, in that she was trying to articulate what her symptoms were at the same time that many of her clinicians were being interrupted. In the context of young women, her symptoms were attributed to pelvic or gynecologic etiology. Furthermore, the physician most likely exhibited anchoring and availability bias given the surrounding context that eating disorders are prevalent among young women. With her pre-diagnosis follow-up visits, all with different doctors, those serving her each began building a relationship from scratch and lacked full chronological understanding of how her symptoms were evolving. Lee's initial story was interpreted through these lenses, which highlight environmental, relational, and chronological contexts.

6.1.2 The Patient's Context and Communication

The patient's context (Box 6.4) is an additional, yet, critical factor that can profoundly shape communication as well as health outcomes as patients work with their clinical professionals to meet their health objectives (see Chapter 4 in this volume). Patient context includes the health and social needs, personal values, and preferences of patients. The experiences of the patient, their literacy level, educational attainment, and health backgrounds also shape the ability to provide and use the information to and from doctors.

The issue of understanding the patient context and then being able to tailor messages appropriately (as related to CAT, described previously) is surfaced in the constraints imposed on most clinical encounters, namely limited time to gather information. Presently, the majority of ambulatory visits are less than 20 minutes (Lau et al., 2016). Under these constrained conditions, as well as other external

BOX 6.4 A CASE STUDY ABOUT LEE, PART FOUR

While Lee was beginning to gain some foothold over her Crohn's disease, she had to put her school on hold. She had been planning on graduating and then getting a job, not paying off emergency room bills. Her student insurance covered most of her expensive injectable treatments, but the out-of-pocket costs still added up. While she felt she worked well with her new gastroenterologist, she remained reserved around doctors whom she did not know well. She became more informed about Crohn's disease and entered into her follow-up appointments with more awareness and more of her perspective about her treatment.

pressures such as documentation requirements and quality of care measures that must be addressed, clinicians may not be able to adequately capture how the patient will be able to integrate the care plan into their life, or how different factors in their life might relate to their condition. Typically, patients would like to discuss more concerns than initially planned ones (Barry et al., 2000). However, only one in three visits employs agenda setting and prioritization of patient concerns, and one in ten visits covers the complete agenda (Barry et al., 2000; Robinson et al., 2016). These studies show missed opportunities to better understand the patient's context, including key details related to the patient's experience of their illness or treatment. These missed opportunities may undermine trust, connectedness, and information sharing. In turn, this can impact the clinician's ability to obtain an accurate diagnosis, as well as the patient's ability to contribute to, and follow through with, the treatment plan.

When clinicians have information about patient context (including needs, values, and preferences relevant to planning care), they can co-create care plans to be more reflective of patients' circumstances and increase the likelihood of the care plans being effective (Arora, 2003; Farzadnia & Giles, 2015; Roter, 1982; Sawyer & Aroni, 2003). In one research study where visits were audio-recorded, physicians being able to elicit contextual concerns and address them in care plans led to an overall 3.7 higher odds of a desired outcome (such as improved diabetes or hypertension control, adherence, follow through with treatment plan, keeping scheduled appointments, tests, or screenings) (Weiner et al., 2013).

Well-designed conversation aids and consumer and clinical informatics tools can increase the chance and the impact of effective context-based communication in everyday clinical encounters. Conversation aids prompt clinicians and patients to consider points that may not be adequately captured in more ad-hoc conversations (Montori et al., 2007; Zeballos-Palacios et al., 2019). These types of conversation aids, if used efficiently, might lead to improvements in quality, health, satisfaction, and reductions in both cost and healthcare disparities (Mayberry et al., 2006). Furthermore, using the electronic health records system (EHR) directly with patients in a collaborative fashion can also enhance the ability for patients and clinicians to ensure that they are checking for understanding and that the appropriate context is adequately conveyed. Finally, the use of EHRs also has a mixed impact on "patient-centered communication" including understanding patients' concerns, ideas, expectations, needs, problems, values, and feelings (Naughton, 2018).

6.2 MEASURING COMMUNICATION

Measurement approaches are required for either studying or improving communication in various healthcare settings. With the use of accepted measures, we know that patient–professional communication has been linked to patient satisfaction, trust, treatment adherence, and treatment outcomes (Eveleigh et al., 2012). We know that better communication between patients and professionals not only improves patient engagement (Asan, Tyszka, et al., 2018), but also helps acknowledge health issues and treatment selection (Travaline et al., 2005).

To measure communication adequately, we commonly need one or more methods sensitive to the aspects of communications described earlier. For example, when studying face-to-face communication, both verbal and nonverbal aspects can give different information of value to researchers measuring communication. Direct observation, video records, and transcripts, as well as interviews or surveys of participants, can be used to assess communication effectiveness. While transcripts of words are essential, any true in-depth understanding of this communication relies on a focus on nonverbal communication (D'Agostino & Bylund, 2014). Although perhaps initially counterintuitive, it is widely reported that face-to-face communication consists of three elements accounting for the overall message: words (the literal meaning) account for 7%; tone of voice accounts for 38%; and body language accounts for 55% of the message (Mehrabian, 1971).

6.2.1 Nonverbal Communication

Many studies report the importance of nonverbal communication between patients and clinicians and its link to several outcomes. Nonverbal communication consists of eye gaze, facial expression, gesturing, body postures, and positioning. Nonverbal communication helps to communicate care, concern, fear, respect, happiness, sadness, anger, surprise, fear, and disgust, which directly contribute to forming trust or mistrust. These messages do not stop when there is no verbal communication: even when people are silent, they are still communicating nonverbally.

Eye gaze has been reported as the most powerful component of nonverbal communication (Henry et al., 2012) and an important aspect of patient-centered communication (Gorawara-Bhat & Cook, 2011). Eye gaze is also used to understand the extent to which patients feel cared for by clinicians (Rose et al., 2014). Gaze provides an objective and measurable indication of attention and communication and can be an attribute that informs design guidelines (Asan, Tyszka, et al., 2018; Asan et al., 2015). Many studies have used eye gaze as a factor to quantitatively measure clinician–patient communication (Asan, Tyszka, et al., 2018; Asan et al., 2015; Gorawara-Bhat & Cook, 2011; Gorawara-Bhat et al., 2013; Montague & Asan, 2014; Ruusuvuori, 2001). The measurement of eye gaze, through video analysis, has been used and validated by previous studies (Asan et al., 2014; Asan & Montague, 2012; Montague & Asan, 2014) as a method to measure nonverbal communication between patient and physicians.

There is a strong relationship between nonverbal aspects of patient–professional communication and various health outcomes, systematically reviewed elsewhere (Henry et al., 2012). For instance, eye contact is found to influence patient-centeredness, rapport, physician awareness, and patients' physical and cognitive functioning (D'Agostino & Bylund, 2014). Body language such as head nodding, facial expression, body orientation, and postures are linked with patient satisfaction in many studies (Henry et al., 2012). Most studies of this kind collected video and subsequently analyzed recordings. Human factors researchers have also analyzed patient–professional nonverbal communication to understand needs for a better system and technology design to improve and facilitate nonverbal communication (Asan, 2017; Asan et al., 2014; Asan, Kim, et al., 2018; Asan & Montague, 2012; Asan, Tyszka, et al., 2018; Asan et al., 2015; Frankel et al., 2005).

6.2.2 Verbal Communication

Studies measure patient–professional verbal communication mainly using two approaches: surveys and coding of transcribed text. There are many psychometric survey instruments to measure patient–professional communication, with reported validity, consistency, reproducibility, responsiveness, and interpretability (Zill et al., 2014). Some of the widely used communication scales are Communication Assessment Tool (Makoul et al., 2007), Questionnaire on Quality of Physician–Patient Interaction (Bieber et al., 2010), Global Consultation Rating Scale (Burt et al., 2014), and the Measure of Patient-Centered Communication (Clayton et al., 2011). These surveys generally measure patients' perceptions of clinicians' communication skills, such as listening, involving patients in decisions, understanding, empathy, and respect. One study reviews 19 instruments assessing the patient–clinician relationship (Eveleigh et al., 2012). Beyond surveys, tools for making sense of transcriptions help assess communication in a more objective way by analyzing verbal communication patterns, used vocabularies, frequencies of words reflecting different emotions, and so on. Some of these approaches are the Four Habits Coding Scheme (Jensen et al., 2010; Krupat et al., 2006), Patient-Centered Behavior Coding Instrument (Zandbelt et al., 2005), and Roter Interaction Analysis System (Roter & Larson, 2002).

6.3 THE APPLICATION OF HUMAN FACTORS TO COMMUNICATION

Human factors and ergonomics (HFE), a science at the intersection of psychology and engineering, is devoted to designing all phases of a work system to improve human performance and safety. In the field of health care, the goals of HFE are to support the cognitive, physical, and social-behavioral work of patients and health professionals (Karsh et al., 2006), reduce risk and medical errors by managing hazards (Karsh et al., 2006), and foster high-quality experience for patients and health professionals (Saleem et al., 2009). Analyzing patient–professional communication using an HFE approach will yield a complete picture to see all factors from the individual, system, and organizational levels influencing this relationship and will better map out necessary interventions to improve patient–professional communication.

Using an HFE approach for analyzing healthcare interactions, we must consider all people and system components, including patients, clinicians, local environments and organization, tasks, tools, and technologies, which facilitate patient and professional work (Holden et al., 2013) (Figure 6.2). Patient–professional communication is influenced by each system component including organization (e.g., existing training program to improve professionals' communication), tools (e.g., portals to facilitate e-communication between patients and professionals), technologies (user-centered collaborative IT facilitating patient–professional communication during the visit), and local environment (ergonomically designed exam room layouts, which can improve face-to-face communication and minimize distraction during nonverbal communication) (see Chapters 2, 3, 4, 5, and 10 in this volume). Therefore, understanding the dynamics of patient–professional communication and the interaction with technologies and the environment is necessary to redesigning patient–professional work

BOX 6.5 A CASE STUDY ABOUT LEE, PART FIVE

After doing her online research, Lee identified a practice that she felt close to her. Looking on their website, she noticed a prominent "login" section for patients to access their health information and electronically communicate with the practice, curated digital tools for education and health management, and images of consultation rooms that looked more like small conference spaces rather than examination rooms.

Just before her first appointment, she received a text reminder on her phone, encouraging her to check-in online. There, she reviewed her medications (she had noticed that several were already imported from her last doctor), identified her top priorities for the visit, and completed a questionnaire about her perceived health status (a patient-reported outcome measure or PROM). She appreciated being able to complete this information without feeling pressured for time or other distractions of the usual waiting room.

When she arrived for her appointment, she was greeted and brought back to a consultation room. In the room, there was a semi-circular table with a large computer monitor at one end and a voice assistant next to it. Dr. Green followed her into the room and they both sat down at the table. The doctor had reviewed Lee's top priorities, which she had brought up on a tablet. That served as a starting point for their conversation. Tapping her badge to activate the computer, Dr. Green called up a graph of Lee's responses to the PROM questionnaire, and they both reflected on how she was doing with her Crohn's treatment. Wrapping up the visit, Dr. Green renewed Lee's medication, which Lee verified on the screen, ordered some additional lab tests, and "prescribed" a series of small check-ins for Lee to do electronically once every two weeks. They scheduled a virtual check-in for 3 months. Lee left feeling that she was heard, that she could trust Dr. Green, and that she had a very clear understanding of her next steps.

systems (Montague et al., 2011) because any change in the work system can cause changes in patient–professional communication patterns.

The Patient Work System, which is an application of human factors theories (Chapters 2 and 4 in this volume), has been getting attention from HFE researchers seeking to understand and improve the work done by patients (Holden et al., 2015). We believe that Patient Work System models (and patient ergonomics in general) guide researchers and practitioners to better understand patient–professional–technology interactions and their impact on outcomes in health care. The recognition that communication is associated with clinical outcomes has also elevated patient–professional communication as an important component of the care process. HFE approaches have been extensively used to design or evaluate collaborative communication technologies in health care. For instance, several studies systematically analyzed how to design and use telehealth to improve communication between patients and professionals as well as among professionals (McGuire et al., 2010) (see also Chapters 5 and 7 in this volume).

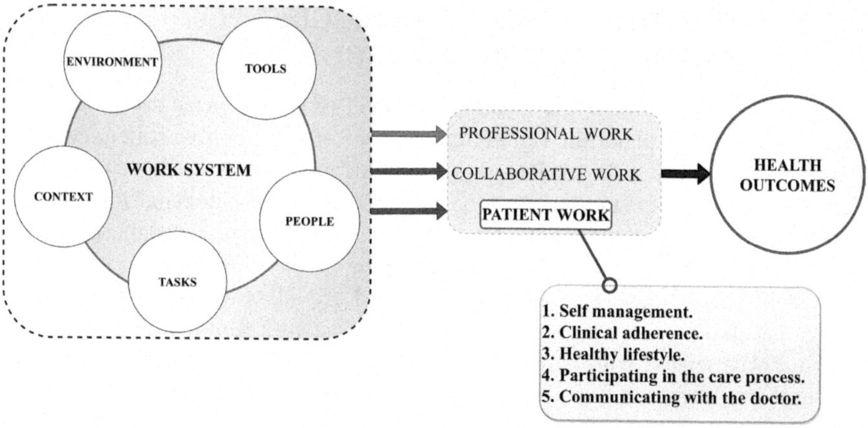

FIGURE 6.2 Patient–professional work model. (Adapted from Holden et al., 2013.)

Another HFE study explored how patient-centered communications occur through the hidden network of relationships linking patients and professionals in medicine (Zachary et al., 2012). Other HFE researchers explored and developed guidelines to improve patient–professional communications especially in geriatrics care (Hickman et al., 2006; Hickman et al., 2004). Furthermore, the design of new technologies and their impact on patient–professional communication, including technology-mediated communication, are also studied extensively by HFE researchers (Benda et al., 2017; Burnett et al., 2011; Wooldridge et al., 2016). Finally, as a part of the Patient Work System model, studies explored characteristics of patients, family members, their task, and environmental factors that contributed to older patients' self-care performance barriers, including ones related to communication (Holden et al., 2015, 2017). We can see from these studies that HFE methods are effective in evaluating, designing, and redesigning systems and technologies to improve patient–professional communication (for review of other patient ergonomics studies about communication, see Holden et al., 2020).

We can see how attention to HFE issues has enhanced communication and thus the visit experience of Lee (Box 6.5). The clinic Lee found was designed to address each of the Patient Work System components, optimizing the whole work system and its outcomes as a result: Dr. Green, Lee, the tasks to be done, the environment of care, the tools used, and the context of the visit were designed to improve Lee's performance, safety, and experience. Lee was able to prepare for her visit ahead of time and communicate her context and agenda at the outset of the visit. Lee was able to sit down and share her story in a more participatory and balanced setting than sitting on an examination table. Dr. Green was able to use a voice assistant to pull up functions of the record, and also use the technology collaboratively with Lee to review her self-reported health, order and verify medications, and set up the care plan. The work done by each of the actors, inclusive of professional work, patient work, and patient–professional collaborative work, was also demonstrated across the phases of care, including pre-appointment, appointment, and post-appointment with the check-in system. These designs led to increased trust and clarity of the care plan.

6.4 HEALTH INFORMATION TECHNOLOGIES AND PATIENT–DOCTOR INTERACTION

Emerging evidence shows mixed impacts of health IT (HIT) use on patient–professional communication, but under the right circumstances, it can improve patient-centered communication. The role of HIT in communication needs to be studied and improved, using HFE approaches. Moreover, the increasing demand for effective patient–professional communication has motivated the design of collaborative HIT and e-health tools. Software platforms driven by artificial intelligence (AI), telehealth, and secure messaging allow patients and providers to virtually communicate with each other, creating a broader healthcare network. Unfortunately, there are difficulties associated with patient-facing HIT, ranging from lack of awareness to structural challenges, to technology implementation (reviewed in Chapter 5 in this volume). For example, if HIT is not highly usable and resilient, it can lead to errors or distractions. Issues of data density, design, clarity regarding system status, and integration into workflow are all example factors affecting HIT usability.

In the literature, there is near-agreement that the EHR is not just a data repository, but a system in the clinical encounter that influences the dynamics of patient–professional interaction (Strowd, 2014; Ventres & Frankel, 2010). Despite potential benefits, the use of EHRs in primary care has also been accompanied by negative consequences, such as changes in the patient–professional dynamic, adverse impacts on patient–professional communication and patient-centered care, reduced physician attention to patient needs, and disengagement on the part of patients (Irani et al., 2009; White & Danis, 2013). Sample studies exploring the impact of EHR use on patient–professional communication in outpatient visits are summarized in Table 6.1.

It is generally recognized that EHRs were largely created to support billing and episodic transactions rather than longitudinal communication or clinical care. This leads to artifacts within EHRs, like concepts of bucketing communications and transactions into encounters rather than a more seamless longitudinal record. Recent studies focus on several ways to improve EHR-based patient-centered communication including (a) redesigning the interface and usability of EHRs and (b) adding functions to easily record patient contextualized data and make them available, such that the record could contain information about the patient as an individual, rather than a summary of their clinical data alone. Furthermore, because patients are the most underutilized member of the care team (deBronkart & Sands, 2018), well-designed informatics tools may overcome some of the barriers within current EHRs by directly inviting patients to share patient-centered data with clinician care team members.

The use of technologies to help continue the longitudinal relationship of clinical care continues to grow. Patient portals, which provide patients with a view of their EHR data and often enable a secure electronic messaging exchange between patients and the clinical office, can bring communication outside of the office. The use of patient portals and secure messaging could also serve to improve relationships. Related studies found that using such applications aids in increasing patient engagement and follow-through (Delbanco et al., 2012; Zhou et al., 2010).

As care activities move from in-person to virtual, richness is lost, which must be acknowledged and accounted for. As Lee interacts with Dr. Green over electronic

TABLE 6.1
Sample Studies Analyzing the Effect of Exam Room EHR on Patient–Doctor Communication

Studies	Outcome Measures	Data Collection Method
Hsu et al. (2005)	Physician's use of computer	Interviews or questionnaires
Lelievre and Schultz (2010); Rouf et al. (2007)	Patient preference for EHR and the effect of computer use on various aspects of patient–physician interaction	
Stewart et al. (2010)	Patient satisfaction on the quality of patient–clinician relationship	
Asan and Montague (2012); Margalit et al. (2006); Warshawsky et al. (1994)	Total encounter time, time of different phases of encounter, record use, and non-interaction time (NIT) between physician and patient	Quantitative behavioral analysis of video/audio recordings
Makoul et al. (2001)	Analysis of whether physicians accomplished communication tasks during encounters	
Als (1997); Frankel et al. (2005)	Impact of exam-room computers on communication between clinicians and patients	Qualitative behavioral analysis of video/audio recordings
Arar et al. (2005)	Process of care, themes discussed, names utilized for medications	

communication, Dr. Green may not be any longer able to perceive nonverbal communication that might be observed in-person. Checking for understanding is often lost in asynchronous transactions. As messages become even shorter (e.g., with text messages), the context important for deciphering meaning or understanding a broader picture may be lost. The level of concern or seriousness may also not be adequately conveyed. Lastly, in asynchronous messaging, checking for message receipt and understanding are warranted, given the finding that as many as half of the messages sent proactively by clinicians were unread (Crotty et al., 2015).

6.5 CASE STUDIES

We describe two interventional studies aiming to improve patient–professional communication in outpatient settings: a) using "mirrored" patient displays of EHRs and b) collecting and using HER-integrated contextualized data from patients ahead of outpatient visits.

6.5.1 The Impact of Double Screen EHR on Communication in Primary Care Settings

Emerging evidence shows that EHR use in the exam room can be complicated, but under the right circumstances, it can improve patient-centered communication

(Patel et al., 2017). We share an interventional study assessing patients' and physicians' perception of a second EHR screen-mirroring the EHR for patients, through which patients can see everything their physician sees and does on the main EHR screen, including the clinical documentation, review of data, and ordering of medications and tests (Asan, Tyszka, et al., 2018). This study explored the impact of the second screen on communication, patient education, patient engagement, and mutual trust.

The study was conducted in a general internal medicine clinic within an urban academic medical center from March to June 2016. We video-recorded 24 primary care encounters with patients and conducted post-visit interviews with both patients and physicians. The study setup used a second EHR monitor in the exam room on an articulating arm connected to the exam room's computer and two cameras at different angles to capture the interactions between the patients and clinicians (Figure 6.3). Video recording was used to quantify eye gaze behaviors of both physicians and patients. The interviews were transcribed and analyzed using inductive content analysis.

We identified four main themes from the patient interviews. Patients thought the mirrored EHR screen: (1) served as a catalyst for patient engagement through design; (2) augmented the clinical visit in a meaningful way; (3) improved the transparency of the care process; and (4) was a substantially different experience than sharing a single screen. The second screen encouraged patient involvement in the care process more so than a single screen. Patients described the design as "more inclusive," "more personal," and "more important." Patients also reported being more engaged

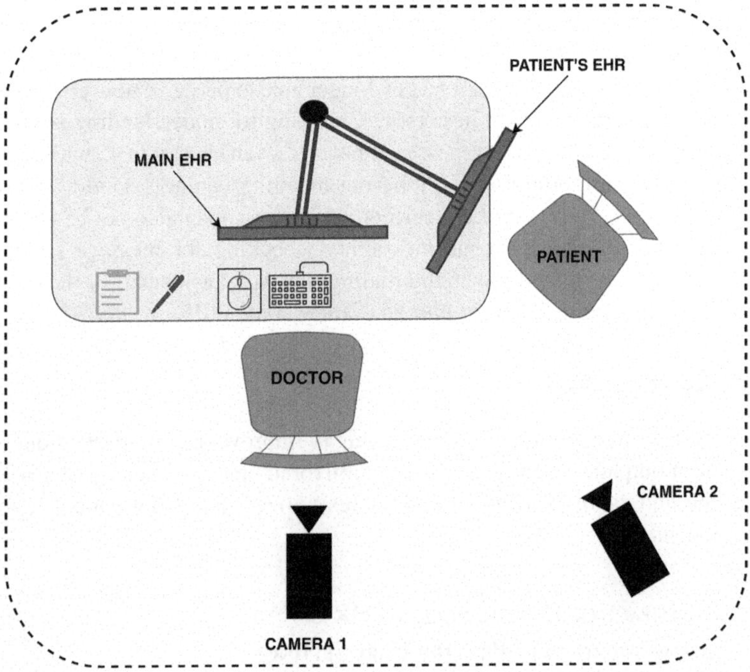

FIGURE 6.3 Double screen exam room setup.

in the clinical note-writing process, for example, clarifying a new diagnosis. Patients perceived the second screen improved their experience of care, including better patient education, better discussions, and an ability to clarify concepts, instructions, and orders, as voiced by one patient:

> I think it generates a more inclusive feeling. When you're there and not looking at it and just listening to the doctor explain to you what's happening, it – it's a little bit more like you're a bystander. When you're watching the – the secondary screen, it's more inclusive. And, I don't know. I don't—I don't wanna oversell it, but I—to me, it feels like you're taking more ownership in your own process and health and what should be happening next.

Double screen use encouraged both auditory and visual pathways for learning. The transparency provided by the second screen also allowed patients to see the process of clinical documentation, diagnoses, goals, and instructions and thus enhanced their confidence in the care process. The EHR interface was found to be confusing to several patients and some asked for a simplified interface design.

Physician interviews also identified that: (1) the second screen provided an opportunity to promote engagement and (2) documentation was transparent to patients, with related benefits and concerns. Similar to patients, clinicians acknowledged that the second screen could potentially enhance patient engagement and improve patient education. Most physicians thought the second screen made screen sharing substantially easier. However, some reported that improvement in patient engagement and education depends on patient interest. Physicians' level of comfort in sharing documentation with their patients emerged as a concern; some were completely at ease with sharing, whereas others were scared the notes might contain sensitive information, and unfinished notes might lead to misunderstanding.

The video analysis helped quantify the visit length and both physicians' and patients' interactions with the EHR. The average adjusted visit length, which excludes the physical exam period, was 23.6 min (SD 11.2). Physicians looked at the EHR (main screen) 39.1% (SD 14.4%) of adjusted visit length and typed/documented 8.2% (SD 6.3%) of the adjusted visit length on average. Physicians typed for documentation purposes in 19 encounters (79%). Patients also looked at the "patient display" 25% (SD 16.7%) of the adjusted visit length, notably longer than our previous studies, which reported patient gaze at EHR in single screen settings (Montague & Asan, 2014). The satisfaction score was a median of 5 [interquartile range (IQR) 4.5–5] out of 5 for patients and the median was 3.3 (IQR 2.3–3.9) out of 5 for physicians.

Overall, patients perceived more benefits than physicians and were more satisfied with the additional screen. The second screen was more inviting of patient participation in the care process, encouraged conversation to enhance comprehension, and increased trust in their physicians. However, physicians may also need to trust their patients, such that the distributed control will not interfere with or distract from tasks but rather make the visits even more useful for patients. A principal inference from this study is that the patients felt more in control of visits.

The use of the second EHR screen is similar to the concept of transparency and "nothing about me without me" that has been the mantra of the OpenNotes concept. OpenNotes is a movement to enable patients to read their clinicians' progress notes, which include information about the history, examination, and clinician's assessment and plan (Bell et al., 2015). The concept was initially tested on a wide scale in 2011. In the OpenNotes study, 13,000 patients received access to their notes across three hospitals. The findings showed that patients valued access to their notes far above the expectations of their clinicians, suggesting that clinicians may take data for granted but patients increasingly are seeing direct access as valuable. It is therefore unsurprising that patients also valued the ability to see clinicians create their notes in real time. Also similar to OpenNotes is the concept that clinicians are developing trust in their patients. The future of OpenNotes is shared note creation among patients and clinicians. The second screen may facilitate such future endeavors. However, patient literacy levels, including health and computer literacy specifically, likely will impact benefits experienced by patients (Asan et al., 2014, 2015).

EHR screens may also contain other sensitive information, such as other patients' names on a calendar or dashboard, which may be seen on the screen after a clinician logs in. A second EHR screen might foster a privacy threat in such situations. Another concern is that the interfaces or displays of modern EHRs are poorly laid out and cognitively challenging, limiting the potential to make use of the screen for patient interactions. EHR usability is a well-known challenge to effective clinician work, but poor usability also makes it challenging to present data to the patient. A patient-centered interface that focuses on single tasks may help facilitate discussions. Despite these challenges, however, patients were able to identify the benefits of viewing their records. Future iterations may also include dedicated "patient views" that simplify the tasks at hand while enabling clinicians to access advanced options when needed.

As it relates to our case (Box 6.5), Lee and Dr. Green shared a monitor that was large and positioned on the table in a neutral position as to not be necessarily 'owned' by any party, unlike how computers are commonly placed in examination rooms. The principle of 'sharing by design' applies, in that the physical environment of Lee and Dr. Green was capable of fostering a shared experience with the health record. Whether a single monitor or two screens, this is an example where conscious design and HFE factors can foster good communication.

6.5.2 EHR-enabled Patient Contextual Data and Patient-centered Communication

Our second case study relates to an intervention designed to improve contextual awareness of patients, leading to good patient–professional communication and experiences, by collecting, summarizing, and sharing patient contextual data (PCD) ahead of visits. A secure web application that interfaced with an EHR was used for the project in a primary care environment. The intervention invited patients to create a profile of their PCD, including needs, values, preferences, barriers, and visit agenda. This was done by sending an email invitation to patients, through which they could create or update their PCD.

The study qualitatively analyzed the impact of the PCD system on communication. The analysis consisted of focus groups and semi-structured interviews as a means of understanding patient ($n=26$) and clinician ($n=20$) perspectives. The

TABLE 6.2
Themes and Representative Quotes Regarding Communication and PCD among Care Team Members and Patients

Themes	Meaning
Enhancing Patient Voice	Providing opportunities for patients to express themselves, their beliefs, and their preferences
Space for Sensitive Topics	Offering a space to communicate emotionally charged topic that patients would otherwise feel uncomfortable sharing face-to-face
Rapport Building	Encouraging deeper relationships between patient and care team members
Alignment of Patient–Clinician Goals	Opportunity to establish mutual understanding around patient preferences for what to accomplish during the office visit
Potential Mistrust if Unattended	Potential for hindering effective communication and trust by ignoring PCD
Reflection	The process of developing PCD prompted reflection on their health and well-being they otherwise would not have
Humanizing	PCD facilitated the communication between patients and providers by making the patient feel more like a person, knowing their provider has more context to who they are as an individual

clinicians were purposefully selected with the help of clinical managers to ensure an inclusive representation by role, experience, and engagement with the program. All focus groups and interviews were audio-recorded with consent from the participants and then transcribed and anonymized to retain privacy. The study employed qualitative content analysis including immersion-crystallization and constant comparison techniques to analyze data and identify themes emerging from the data.

The study identified seven themes from patient and clinician interviews (see Table 6.2). Four out of seven themes occurred in both clinician and patient comments: (1) PCD improves the patient's voice in clinical visits; (2) the technology acts as a safe platform for sharing sensitive topics; (3) PCD encouraged efficient and effective rapport building; and (4) PCD aligned patient–clinician goals for the visit. It aided several clinicians' preparation for communication with their patients at the visit and better tailor care plans. Patients observed that the opportunity to share contextual data encouraged both asynchronous and synchronous communication with their care team.

The study identified two exclusive themes among patients: (1) the technology provides an opportunity for personal reflection before a visit and (2) PCD humanizes patients in the clinical context by providing information about their nature, values, and preferences. Completing the PCD tool before visits equipped patients to better communicate with clinicians. Patients benefited from the sharing of thoughts and feelings as a medium via the PCD tool to establish interaction and communication with the care team. The study acknowledged that collecting PCD may potentially undermine trust if not reviewed by clinical teams. It was observed that patients might feel that their voice is neglected and not validated if the PCD are not used. Finally, clinicians thought PCD facilitated team-based care for more efficient and effective communication, thus addressing the needs of the patient.

This study emphasized patients' and clinicians' perspectives on how an EHR-integrated PCD digital tool affects communication in everyday practice. The study showed that the PCD tool was initially conceived to facilitate patient–clinician communication and enabled stronger connections between care team members and patients. Patients felt more included in the care process and also participated in clinical decision-making using the PCD tool. Patients liked the opportunity of sharing their sensitive information via this tool. Past studies have shown that patients feel more comfortable sharing sensitive information via computers rather than in-person (Slack et al., 2012). The PCD tool also made the patients well prepared to communicate at their upcoming appointment. This study acknowledged that PCD has the potential to foster patient-centered communication, benefiting both patients and clinicians. Dual participation and usage of the PCD tool might result in better effectiveness and efficiency.

6.6 RECOMMENDATIONS

Patient–professional communication is a critical process in the healthcare system. Researchers need to use a systems approach, addressing all the potential elements of the work system to study process, context, and outcomes of patient–professional communication in the healthcare context. HFE approaches including the Patient Work System framework should be used to study patient–professional communication and design of physical environments and communication technology.

REFERENCES

Als, A. B. (1997). The desk-top computer as a magic box: Patterns of behavior connected with the desk-top computer; GPS' and patients' perceptions. *Family Practice, 14*, 17–23.

Arar, N., Wen, L., McGrath, J., Steinbach, R., & Pugh, J. (2005). Communicating about medications during primary care outpatient visits: The role of electronic medical records. *Journal of Innovation in Health Informatics, 13*, 13–21.

Arora, N. K. (2003). Interacting with cancer patients: The significance of physicians' communication behavior. *Social Science & Medicine, 57*, 791–806.

Asan, O. (2017). Providers' perceived facilitators and barriers to EHR screen sharing in outpatient settings. *Applied Ergonomics, 58*, 301–307.

Asan, O., Kim, S. C., Iglar, P., & Yan, A. (2018). Differences in verbal and nonverbal communication between depressed and non-depressed elderly patients. *Journal of Communication in Healthcare, 11*, 297–306.

Asan, O., & Montague, E. (2012). Physician interactions with electronic health records in primary care. *Health Systems, 1*, 96–103.

Asan, O., Smith, P. D., & Montague, E. (2014). More screen time, less face time–implications for EHR design. *Journal of Evaluation in Clinical Practice, 20*, 896–901.

Asan, O., Tyszka, J., & Crotty, B. (2018). The electronic health record as a patient engagement tool: Mirroring clinicians' screen to create a shared mental model. *Journal of the American Medical Informatics Association Open, 1*, 42–48.

Asan, O., Young, H. N., Chewning, B., & Montague, E. (2015). How physician electronic health record screen sharing affects patient and doctor nonverbal communication in primary care. *Patient Education and Counseling, 98*, 310–316.

Bari, A., Khan, R. A., & Rathore, A. W. (2016). Medical errors; Causes, consequences, emotional response and resulting behavioral change. *Pakistan Journal of Medical Sciences, 32*, 523.

Barry, C. A., Bradley, C. P., Britten, N., Stevenson, F. A., & Barber, N. (2000). Patients' unvoiced agendas in general practice consultations: Qualitative study. *The BMJ, 320*, 1246–1250.

Bell, S. K., Folcarelli, P. H., Anselmo, M. K., Crotty, B. H., Flier, L. A., & Walker, J. (2015). Connecting patients and clinicians: The anticipated effects of Open Notes on patient safety and quality of care. *Joint Commission Journal on Quality and Patient Safety, 41*, 378–384.

Benda, N. C., Higginbotham, J., Fairbanks, R. J., Lin, L., & Bisantz, A. M. (2017). Using cognitive work analysis to design communication support tools for patients with language barriers. In *Proceedings of the Human Factors and Ergonomics Society Annual Meeting, 61*, 120–124.

Bieber, C., Mueller, K. G., Nicolai, J., Hartmann, M., & Eich, W. (2010). How does your doctor talk with you? Preliminary validation of a brief patient self-report questionnaire on the quality of physician-patient interaction. *Journal of Clinical Psychology in Medical Settings, 17*, 125–136.

Burnett, J. S., Mitzner, T. L., Charness, N., & Rogers, W. A. (2011). Understanding predictors of computer communication technology use by older adults. In *Proceedings of the Human Factors and Ergonomics Society Annual Meeting, 55*, 172–176.

Burt, J., Abel, G., Elmore, N., Campbell, J., Roland, M., Benson, J., & Silverman, J. (2014). Assessing communication quality of consultations in primary care: Initial reliability of the global consultation rating scale, based on the Calgary-Cambridge guide to the medical interview. *The BMJ Open, 4*, e004339.

Bylund, C. L., Peterson, E. B., & Cameron, K. A. (2012). A practitioner's guide to interpersonal communication theory: An overview and exploration of selected theories. *Patient Education and Counseling, 87*, 261–267.

Carayon, P. (2012). *Handbook of human factors and ergonomics in health care and patient safety* (2nd ed.). CRC Press, Boca Raton, FL.

Clayton, M. F., Latimer, S., Dunn, T. W., & Haas, L. (2011). Assessing patient-centered communication in a family practice setting: How do we measure it, and whose opinion matters? *Patient Education and Counseling, 84*, 294–302.

Crotty, B. H., Mostaghimi, A., O'Brien, J., Bajracharya, A., Safran, C., & Landon, B. E. (2015). Prevalence and risk profile of unread messages to patients in a patient web portal. *Applied Clinical Informatics, 6*, 375–382.

D'Agostino, T. A., & Bylund, C. L. (2014). Nonverbal accommodation in health care communication. *Health Communication, 29*, 563–573.

deBronkart, D., & Sands, D. Z. (2018). Warner slack: "Patients are the most underused resource. *The BMJ, 362*, k3194.

Delbanco, T., Walker, J., Bell, S. K., Darer, J. D., Elmore, J. G., Farag, N., Feldman, H. J., Mejilla, R., Ngo, L., & Ralston, J. D. (2012). Inviting patients to read their doctors' notes: A quasi-experimental study and a look ahead. *Annals of Internal Medicine, 157*, 461–470.

Eveleigh, R. M., Muskens, E., van Ravesteijn, H., van Dijk, I., van Rijswijk, E., & Lucassen, P. (2012). An overview of 19 instruments assessing the doctor-patient relationship: Different models or concepts are used. *Journal of Clinical Epidemiology, 65*, 10–15.

Farzadnia, S., & Giles, H. (2015). Patient-provider interaction: A communication accommodation theory perspective. *International Journal of Society, Culture & Language, 3*, 17–34.

Frankel, R., Altschuler, A., George, S., Kinsman, J., Jimison, H., Robertson, N. R., & Hsu, J. (2005). Effects of exam-room computing on clinician-patient communication. *Journal of General Internal Medicine, 20*, 677–682.

Gasiorek, J., & Giles, H. (2012). Effects of inferred motive on evaluations of nonaccommodative communication. *Human Communication Research, 38*, 309–331.

Gerbner, G. (1985). "Field Definitions: Communication Theory." In 1984–85 U.S. Directory of Graduate Programs, 9th edition. Princeton, NJ: Educational Testing Service.

Gorawara-Bhat, R., & Cook, M. A. (2011). Eye contact in patient-centered communication. *Patient Education and Counseling, 82*, 442–447.

Gorawara-Bhat, R., Dethmers, D. L., & Cook, M. A. (2013). Physician eye contact and elder patient perceptions of understanding and adherence. *Patient Education and Counseling, 92*, 375–380.

Hannawa, A., Wu, A., & Juhasz, R. (2017). *New horizons in patient safety: Understanding communication: Case studies for physicians.* Walter de Gruyter GMBD & Co KG, Berlin, Germany.

Henry, S. G., Fuhrel-Forbis, A., Rogers, M. A., & Eggly, S. (2012). Association between nonverbal communication during clinical interactions and outcomes: A systematic review and meta-analysis. *Patient Education and Counseling, 86*, 297–315.

Hickman, J. M., Caine, K. E., Stronge, A. J., Pak, R., Rogers, W. A., & Fisk, A. D. (2006). Doctor-patient communication: Guidelines for improvements. In *Proceedings of the Human Factors and Ergonomics Society Annual Meeting , 50*(10), 1078–1082.

Hickman, J. M., Pak, R., Stronge, A. J., & Jones, W. B. (2004). Exploring communication between health care professionals and older adults. In *Proceedings of the Human Factors and Ergonomics Society Annual Meeting, 48*, 2156–2160.

Holden, R. J., Carayon, P., Gurses, A. P., Hoonakker, P., Hundt, A. S., Ozok, A. A., & Rivera-Rodriguez, A. J. (2013). SEIPS 2.0: A human factors framework for studying and improving the work of healthcare professionals and patients. *Ergonomics, 56*(11), 1669–1686.

Holden, R. J., Cornet, V. P., & Valdez, R. S. (2020). Patient ergonomics: 10-year mapping review of patient-centered human factors. *Applied Ergonomics, 82*, 102972.

Holden, R. J., Schubert, C. C., & Mickelson, R. S. (2015). The patient work system: An analysis of self-care performance barriers among elderly heart failure patients and their informal caregivers. *Applied Ergonomics, 47*, 133–150.

Holden, R. J., Valdez, R. S., Schubert, C. C., Thompson, M. J., & Hundt, A. S. (2017). Macroergonomic factors in the patient work system: Examining the context of patients with chronic illness. *Ergonomics, 60*, 26–43.

Hsu, J., Huang, J., Fung, V., Robertson, N., Jimison, H., & Frankel, R. (2005). Health information technology and physician-patient interactions: Impact of computers on communication during outpatient primary care visits. *Journal of the American Medical Informatics Association, 12*, 474–480.

Institute of Medicine Committee on Quality of Health Care in, A. (2001). In *Crossing the quality chasm: A new health system for the 21st century.* Washington, DC: National Academies Press (US).

Irani, J. S., Middleton, J. L., Marfatia, R., Omana, E. T., & D'Amico, F. (2009). The use of electronic health records in the exam room and patient satisfaction: A systematic review. *The Journal of the American Board of Family Medicine, 22*, 553–562.

Jensen, B. F., Gulbrandsen, P., Benth, J. S., Dahl, F. A., Krupat, E., & Finset, A. (2010). Interrater reliability for the four habits coding scheme as part of a randomized controlled trial. *Patient Education and Counseling, 80*, 405–409.

Karsh, B., Holden, R., Alper, S., & Or, C. (2006). A human factors engineering paradigm for patient safety: Designing to support the performance of the healthcare professional. *The BMJ Quality and Safety, 15*, i59–i65.

Krupat, E., Frankel, R., Stein, T., & Irish, J. (2006). The four habits coding scheme: Validation of an instrument to assess clinicians' communication behavior. *Patient Education and Counseling, 62*, 38–45.

Lau, D. T., McCaig, L. F., & Hing, E. (2016). Toward a more complete picture of outpatient, office-based health care in the US. *American Journal of Preventive Medicine, 51*, 403–409.
Lelievre, S., & Schultz, K. (2010). Does computer use in patient-physician encounters influence patient satisfaction? *Canadian Family Physician, 56*, e6–e12.
Maguire, P., & Pitceathly, C. (2002). Key communication skills and how to acquire them. *BMJ, 325*, 697–700.
Makoul, G., Curry, R. H., & Tang, P. C. (2001). The use of electronic medical records: Communication patterns in outpatient encounters. *Journal of the American Medical Informatics Association, 8*, 610–615.
Makoul, G., Krupat, E., & Chang, C.-H. (2007). Measuring patient views of physician communication skills: Development and testing of the communication assessment tool. *Patient Education and Counseling, 67*, 333–342.
Margalit, R. S., Roter, D., Dunevant, M. A., Larson, S., & Reis, S. (2006). Electronic medical record use and physician-patient communication: An observational study of Israeli primary care encounters. *Patient Education and Counseling, 61*, 134–141.
Mayberry, R. M., Nicewander, D. A., Qin, H., & Ballard, D. J. (2006). Improving quality and reducing inequities: A challenge in achieving best care. *In the Baylor University Medical Center Proceedings, 19*(2), 103–118.
McGuire, K., Carayon, P., Hoonakker, P., Khunlertkit, A., & Wiegmann, D. (2010). Communication in the tele-ICU. In *Proceedings of the Human Factors and Ergonomics Society Annual Meeting, 54*, 1586–1590.
Mehrabian, A. (1971). *Silent messages* (Vol. 8). Wadsworth Belmont, Belmont, California, USA.
Montague, E., & Asan, O. (2014). Dynamic modeling of patient and physician eye gaze to understand the effects of electronic health records on doctor-patient communication and attention. *International Journal of Medical Informatics, 83*, 225–234.
Montague, E., Xu, J., Chen, P. Y., Asan, O., Barrett, B. P., & Chewning, B. (2011). Modeling eye gaze patterns in clinician-patient interaction with lag sequential analysis. *Human Factors, 53*, 502–516.
Montori, V. M., Breslin, M., Maleska, M., & Weymiller, A. J. (2007). Creating a conversation: Insights from the development of a decision aid. *PLOS Medicine, 4*(8), e233.
Naughton, C. A. (2018). Patient-centered communication. *Pharmacy, 6*, 18.
Patel, M. R., Vichich, J., Lang, I., Lin, J., & Zheng, K. (2017). Developing an evidence base of best practices for integrating computerized systems into the exam room: A systematic review. *Journal of the American Medical Informatics Association, 24*, e207–e215.
Robinson, J. D., Tate, A., & Heritage, J. (2016). Agenda-setting revisited: When and how do primary-care physicians solicit patients' additional concerns? *Patient Education and Counseling, 99*, 718–723.
Rose, D., Richter, L. T., & Kapustin, J. (2014). Patient experiences with electronic medical records: Lessons learned. *Journal of the American Association of Nurse Practitioners, 26*, 674–680.
Roter, D. (1982). Physician/patient communication: Transmission of information and patient effects. *Maryland State Medical Journal, 32*(4), 260–265.
Roter, D., & Larson, S. (2002). The Roter interaction analysis system (rias): Utility and flexibility for analysis of medical interactions. *Patient Education and Counseling, 46*, 243–251.
Rouf, E., Whittle, J., Lu, N., & Schwartz, M. D. (2007). Computers in the exam room: Differences in physician-patient interaction may be due to physician experience. *Journal of General Internal Medicine, 22*, 43–48.
Ruusuvuori, J. (2001). Looking means listening: Coordinating displays of engagement in doctor-patient interaction. *Social Science & Medicine, 52*, 1093–1108.
Saleem, J. J., Russ, A. L., Sanderson, P., Johnson, T. R., Zhang, J., & Sittig, D. F. (2009). Current challenges and opportunities for better integration of human factors research

with development of clinical information systems. *Yearbook of Medical Informatics, 18*, 48–58.

Sawyer, S. M., & Aroni, R. A. (2003). Sticky issue of adherence. *Journal of Paediatrics and Child Health, 39*, 2–5.

Shannon, C. (1948). A mathematical theory of communication. *The Bell System Technical Journal, 27*, 379–423.

Shatz, M. (2006). Language in mind: Advances in the study of language and thought, Dedre Gentner, Susan Goldin-Meadow. *Language, 82*, 174–176.

Slack, W. V. (1997). Claude Shannon and communication theory. *MD Computing: Computers in Medical Practice, 14*, 262–264.

Slack, W. V., Kowaloff, H. B., Davis, R. B., Delbanco, T., Locke, S. E., Safran, C., & Bleich, H. L. (2012). Evaluation of computer-based medical histories taken by patients at home. *Journal of the American Medical Informatics Association, 19*, 545–548.

Stewart, R. F., Kroth, P. J., Schuyler, M., & Bailey, R. (2010). Do electronic health records affect the patient-psychiatrist relationship? A before & after study of psychiatric outpatients. *BioMed Central Psychiatry, 10*, 3.

Strowd, R. E. (2014). Right brain: The elephant in the room: One resident's challenge in transitioning to modern electronic medicine. *Neurology, 83*, e125–e127.

Travaline, J. M., Ruchinskas, R., & D'Alonzo, G. E., Jr. (2005). Patient-physician communication: Why and how. *The Journal of the American Osteopathic Association, 105*, 13–18.

Ventres, W. B., & Frankel, R. M. (2010). Patient-centered care and electronic health records: It's still about the relationship. *Family Medicine, 42*, 364–366.

Warshawsky, S. S., Pliskin, J. S., Urkin, J., Cohen, N., Sharon, A., Binztok, M., & Margolis, C. Z. (1994). Physician use of a computerized medical record system during the patient encounter: A descriptive study. *Computer Methods and Programs in Biomedicine, 43*, 269–273.

Weiner, S. J., Schwartz, A., Sharma, G., Binns-Calvey, A., Ashley, N., Kelly, B., Dayal, A., Patel, S., Weaver, F. M., & Harris, I. (2013). Patient-centered decision making and health care outcomes: An observational study. *Annals of Internal Medicine, 158*, 573–579.

White, A., & Danis, M. (2013). Enhancing patient-centered communication and collaboration by using the electronic health record in the examination room. *Journal of the American Medical Association, 309*, 2327–2328.

Wooldridge, A. R., Carayon, P., Hoonakker, P., Musa, A., & Bain, P. (2016). Technology-mediated communication between patients and primary care clinicians and staff: Ambiguity in secure messaging. In *Proceedings of the Human Factors and Ergonomics Society Annual Meeting, 60*, 556–560.

Xue, Q.-L. (2011). The frailty syndrome: Definition and natural history. *Clinics in Geriatric Medicine, 27*, 1–15.

Zachary, W., Maulitz, R. C., Rosen, M. A., Cannon-Bowers, J., & Salas, E. (2012). Clinical communications – Human factors for the hidden network in medicine. In *Proceedings of the Human Factors and Ergonomics Society Annual Meeting, 56*, 850–854.

Zandbelt, L. C., Smets, E. M., Oort, F. J., & de Haes, H. C. (2005). Coding patient-centered behavior in the medical encounter. *Social Science and Medicine, 61*, 661–671.

Zeballos-Palacios, C. L., Hargraves, I. G., Noseworthy, P. A., Branda, M. E., Kunneman, M., Burnett, B., Gionfriddo, M. R., McLeod, C. J., Gorr, H., & Brito, J. P. (2019). Developing a conversation aid to support shared decision making: Reflections on designing anticoagulation choice. In the *Mayo Clinic Proceedings, 94(4)*, 686–696.

Zhou, Y. Y., Kanter, M. H., Wang, J. J., & Garrido, T. (2010). Improved quality at kaiser permanente through e-mail between physicians and patients. *Health Affairs, 29*, 1370–1375.

Zill, J. M., Christalle, E., Muller, E., Harter, M., Dirmaier, J., & Scholl, I. (2014). Measurement of physician-patient communication—A systematic review. *PLOS One, 9*, e112637.

7 Human Factors and Patient Self-Care

Barrett S. Caldwell
Purdue University

Siobhan M. Heiden
Puyallup, WA

Michelle Jahn Holbrook
Bold Insight

CONTENTS

For most populations over history, patient self-care has been a primary approach to maintaining health without reliance on the medical profession (MacGill, 2015). Tensions between clinician-based vs. patient-based care in the 20th century included concerns regarding sources of and control of health information that can influence the patient's status. Although strongly grounded in the scientific method, the subject of medicine—the patient's health state and wellness status—is largely unavailable for direct observation. Such fundamental issues as the experience of pain, the organization of brain structures affecting physical and cognitive processes, the variability of biochemistry and physiological systems, and the effects of social and cultural influences on the awareness and recognition of physical symptoms of health and disease have been topics of debate in medicine, psychology, and public health throughout the 20th and 21st centuries (see, for example, Bertolero & Bassett, 2019; Higginson, 1931; Kinsey et al., 1948).

An increase in human factors research in health care has paid specific attention to "patient work," which comprises tasks patients (or their informal caregivers) complete to coordinate and administer health-related activities including self-care (Holden et al., 2015, 2017) (see Chapter 1 in this volume). Three human factors concepts apply broadly to task performance in a variety of healthcare settings, including patient self-care (see also Chapter 2 in this volume):

- *Signal detection*: the ability to discriminate important items of interest from the noise that is not relevant to the current task (Wickens et al., 2004);
- *Situation awareness*: "being aware of what is happening around you and understanding what that information means to you now and in the future" (Endsley et al., 2003, p. 13); and
- *Usability*: the ability to achieve important system goals with effectiveness, efficiency, and satisfaction (ISO DIS 9421-11, 2018).

Each of these concepts is very sensitive to the context of the person completing the task. The Oxford English Dictionary defines context as, "The circumstances that form the setting for an event, statement, or idea, and in terms of which it can be fully understood and assessed." It is important from a human factors standpoint that the context and demands of patient self-care tasks are emphasized and prioritized (NAE/IOM, 2005). However, actually knowing the current state of the patient includes multiple problems of signal detection (what are indicators of the patient's current condition?) and situation awareness (how do those indicators combine to allow understanding of patient health?), across a variety of time scales ranging from milliseconds to months. Some physiological states can be effectively detected by objective instruments or methods even without a recognizable physiological signal being detected by the patient (e.g., hemoglobin A1C; imaging to detect cancer tumors). On the other hand, no instrument or method is available to fully capture or correlate a patient's experience of pain, or mood swings associated with major depression: these subjective experiences strongly influence the context in which a patient manages self-care tasks.

As we explore the intersection of patient self-care tasks and chronic disease management, there is a need to consider a broader constellation of contextual system factors including time scales, data availability, and modeling of changes in health status. This expanded discussion of patient self-care helps define the needs of patients, informal caregivers, and clinicians over multiple time scales (Caldwell et al., 2010; Garrett & Caldwell, 2009). In this chapter, we will also consider patient involvement (how and which tasks the patient conducts), sources of health information (when and by whom health data are collected and shared), and health status dynamics (which changes are noticed and considered important health "signals" that influence care). For example, patients may spend only an average of 15–20 minutes with a physician during a formal healthcare appointment (Young et al., 2018), limiting the time for both patient and clinician to collect and share information about relevant health signals. In addition, the time that physicians interact with the electronic medical record (EMR) and laboratory test results, which approximate or exceed the total time spent with the patients (Tai-Seale et al., 2017), would not include signals generated by patients using home health devices or other technologies not routinely integrated into the EMR.

Perhaps more controversially, we also consider the patient as a potential source of *expertise* regarding their care and health status, especially in an era of increased information and communication technology (ICT) systems. This statement is not to suggest that the average patient is more skilled or knowledgeable in the subject matter domain than any of their clinicians (although we will address potential examples of this situation later in this chapter). Instead, expertise can be considered a multidimensional construct, including more facets of task context and use of task interfaces to achieve performance goals, rather than just subject matter domain knowledge (Garrett et al., 2009). Especially in the case of chronic disease care, the patient may develop considerable expertise in the use of particular medications, devices, and other care processes as interface tools, as well as recognition of situational contexts that require the use of those tools and processes. As a result, we will place considerable focus in this chapter on chronic care.

7.1 SELF-CARE, CHRONIC CARE, AND THE CHRONIC CARE MODEL

Self-care is described as the capability for a patient to become engaged and proactively involved in their own healthcare delivery (Forkner-Dunn, 2003; Lupton, 2013). Chronic care is defined as a systematic and coordinated approach to preventing and managing chronic conditions (Glasgow et al., 2001), especially as they become a larger fraction of healthcare considerations and expenses. As a result of the necessity of self-care in chronic care, the role (and task contexts) of patients and informal caregivers is even more prominent in chronic care than in acute care.

There are multiple approaches used to increase patient empowerment, feedback, motivation, and participation in their healthcare delivery. However, regardless of approach, the foundation of patient self-care is built on the patient's recognition of current health information (signal detection), knowledge of health status (situation awareness), and use of self-care tools (system usability) (Figure 7.1).

As Figure 7.1 suggests, patient-care activities involve the patient at the center of their care with the patient-facing tools also at the center. This figure shows how the three concepts discussed in this chapter overlap and work together to support patient self-care activities. Signal detection first affects the patient through recognition of an event, sometimes through a patient-facing tool or technology. The patient can share that information and provide alerts to members of their care team. Together, using tools and technology to treat the resulting healthcare event, the patient and their care team experience the three levels of situation awareness: perception; cognition; and projection. The usability of healthcare tools and technologies need to be designed to

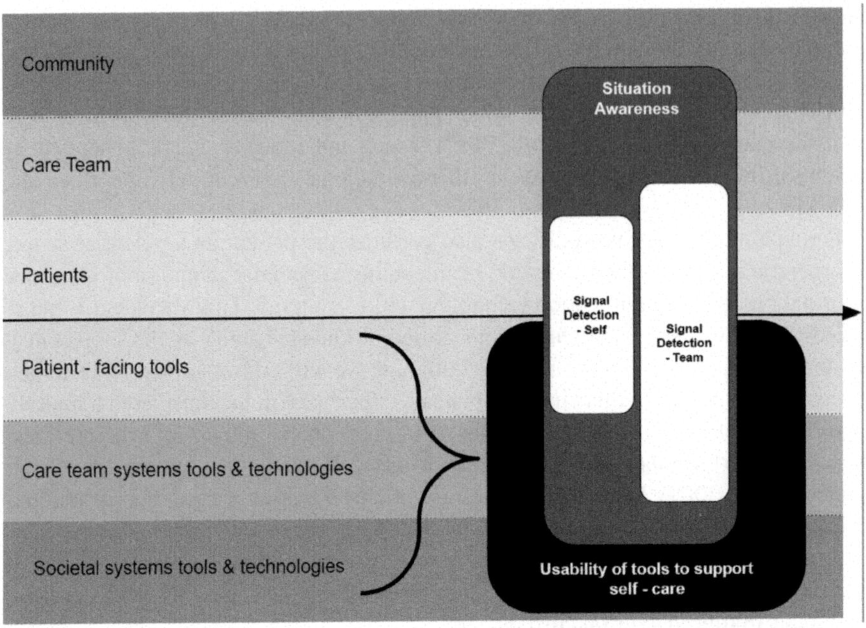

FIGURE 7.1 Diagram of human factors concepts as foundational tools for patient self-care within healthcare systems. The upper sections show patients, care teams, and community caregivers. The lower sections describe healthcare tools and technologies involved in supporting those individual patients, care teams, and communities.

support the signal detection and situation awareness needs for determining appropriate self-care activities and interventions.

7.1.1 Chronic Care Model and Self-care

The United States Institute of Medicine (IOM) estimated that more than 48 million Americans are living with one or more chronic conditions (IOM, 2012). Chronic care requires longitudinal care, much of which happens outside of clinical environments (i.e., self-care) (Glasgow et al., 2001; Sochalski et al., 2009). Examples of self-care in chronic conditions include medication management for addressing chronic pain, glucose monitoring for diabetes management, and applying external memory cues for coping with a cognitive deficit from acquired brain injury.

With the rise of chronic conditions (e.g., diabetes, heart disease, cancer, and arthritis), healthcare professionals and researchers have developed the Chronic Care Model (CCM) in attempts to redesign supporting healthcare system components to address deficiencies in chronic care processes (Wagner, 1998a). CCM is the most widely accepted and applied model for chronic care management. While not specifically focused on human factors considerations, the CCM is notable in its attempt to consider multiple contextual and systems dynamics factors in care management (Stuckey et al, 2011; Wagner et al., 2002).

There are six domains presented in the CCM: self-management support; delivery system design; decision-support; information systems; health systems; and community resources (Wagner, 1998a, 2000). Of the six domains presented in the CCM, the self-management support domain is the most intertwined with self-care activities. This self-management focus of the CCM model emphasizes patients' abilities to manage their own healthcare, whereas the other domains focus more on the technical aspects of the system and clinician interactions.

The CCM describes self-management as an integrated partnership between healthcare delivery systems and the community (Wagner et al., 2002). Most self-management support programs aim to increase patient participation in their health care, including self-care and patient–clinician interaction (Siminerio et al., 2004; Stuckey et al., 2011; USDHHS, 2010; Wagner, 1998b). Positive effects of such programs include improved patient self-efficacy (Lorig & Holman, 2003) and self-rated health (Grady & Gough, 2014; Lorig & Holman, 2003). Lorig and Holman (2003) suggest three classes of patient self-care *tasks* (medical management, role management, and emotional management) and six sets of self-management *skills* (problem-solving, decision-making, resource utilization, patient–clinician partnership forming, action planning, and self-tailoring).

Technology advances have contributed to newer iterations of the CCM for helping with supporting patient self-care task activities in the digital age (Gee et al., 2015). One version, the eHealth Enhanced Chronic Care Model (eCCM), incorporates additional task contexts (including social networking communities, telehealth, and mobile health interactions) within the scope of signal detection, situation awareness, and system usability considerations. Tools such as eHealth education are intended to change usability considerations to increase emphasis on facilitating communication and knowledge exchange between patients, clinicians, and eHealth communities (Gee et al., 2015).

7.2 COMMUNICATION, COORDINATION, HANDOFFS, AND TELEHEALTH

Patient safety aims are supported by effective care coordination and communication (Zachary et al., 2013) (see Chapters 6 and 8 in this volume). Substantial requirements for coordination and communication occur at care transitions, or handoffs, in which the participating members share information with the goal of shared situation awareness. Care transitions span a very wide range of time scales, coordination acts, and expertise sharing, ranging from seconds to months (see, for example, Garrett & Caldwell, 2009; Jahn Holbrook & Caldwell, 2019). In this chapter, we focus on three specific elements of distributed signal detection, situation awareness, and technology use applied to healthcare: expertise coordination, handoffs, and distributed electronic and ICT-based care delivery (also known as "telehealth").

7.2.1 Patient–Patient Expertise Coordination

The focus on the patient in many self-care tasks changes the emphasis of signal detection, situation awareness, and system usability for many care activities.

Depending on the setting, complexity of health condition, and care process, patients may interact more effectively with other clinicians, such as pharmacists, who may have greater insight into the drug interactions and limits on medication compliance and effectiveness because of their more frequent interactions with the patient (Benedict & Caldwell, 2011). Much research has been done in clinician–clinician and patient–clinician expertise coordination (e.g., Jahn Holbrook, 2019), but the focus of this chapter is on the more unique case of patient–patient expertise coordination.

Imagine a woman on a "cancer journey" experiencing surgery, chemotherapy, and radiation therapy after a diagnosis of breast cancer. While the surgeon, chemotherapy pharmacist, or radiation oncologist may be able to provide clinical information about the treatments she is experiencing, there are multiple other cognitive, emotional, social, and sociological considerations that she may only learn from other cancer survivors as ongoing situation awareness considerations. For instance, how and when does one procure wigs, hats, or other accessories to respond to hair loss? What skin care treatments best help address the sun- and heat-sensitivity resulting from cumulative radiation therapy fractions? Are there cognitive assists (or, more importantly, emotional support capabilities) to address the loss of mental acuity known as "chemo brain?" Regarding health status signal detection, what are important symptoms or side effects that she should report immediately to her cancer care team, and what are frequently experienced effects that can be addressed in a less urgent fashion? In these and other settings, the collected and shared experiences of multiple patients, from their own patient perspectives, represent a distinct form of signal detection, information assessment, and relevant situation awareness indicators to assist the woman's physical, emotional, and cognitive recovery.

With the increasing availability of consumer-level information and ICTs, patient and caregiver support groups have been able to share not only formal information and research studies about their conditions, but also informal (or even tacitly understood) experiences about their conditions, a form of shared understanding described in other communities of shared expertise (Guinery, 2011). Informal caregivers of patients with Alzheimer's disease as well as patients recovering from coronary artery bypass graft (CABG) surgery have used ICTs such as WebTV as far back as the 1990s (Brennan et al., 1998; Rogers et al., 1999). Clinical trials of such ICT use demonstrated that patients recovered faster to a stable health status, due in part to shared patient information about best practices, effective communication with nurses, and socioemotional support for patients and caregivers to give additional encouragement that "they are not alone" in their symptoms or experiences.

7.2.2 Handoffs

Human factors and ergonomics (HFE) work examining handoffs has primarily focused on clinician handoffs at the shift-change level (e.g., Furniss et al., 2016) and within-hospital department transfers (e.g., Apker et al., 2014; Fletcher et al., 2014). While the two handoff types described above are important, there are several other types of handoffs important to chronic care (and especially the involvement of patient self-care tasks) that have received little to no attention (Heiden & Caldwell, 2018).

An exploratory study using semi-structured ethnographic interviews ($n = 32$) with professional care providers in traumatic brain injury (TBI) rehabilitation helped to identify care handoffs directly involving patient tasks across various stages and levels of patient care and recovery (Heiden & Caldwell, 2018). Note that TBI is considered a chronic illness (Masel, 2009) and fits the chronic illness definition: "conditions that last a year or more and require ongoing medical attention and/or limit activities of daily living" (Warshaw, 2006, p. 5). The study identified the following handoffs: shift handoffs, session-to-session handoffs, department handoffs, facility handoffs, and health system handoffs. It is important to note the value of temporal handoffs supporting chronic care management of an individual patient over months and years, and not just handoffs of multiple patient status in settings such as emergency rooms or patient wards over time scales of minutes to hours (see Klein et al., 2005; Ye et al., 2007).

Successfully transferring all salient information at the handoff is rare. Heiden and Caldwell (2018) found that in handoffs when patients are referred from outside the health system of the receiving facility, it is more likely that information will be "lost." Patients play an important role in filling in "lost" information from incomplete handoffs (i.e., handoffs where not all salient information is transferred). Thus, receiving clinicians often re-assess the patient (in other words, engage in additional signal detection and situation awareness about health status) because the information provided is insufficient; patients may be expected to also support this repeated signal detection and situation awareness, even if they are not aware that the information was not provided from one clinician to another. Repeating information to multiple clinicians can be taxing to patients and their informal caregivers (Caldwell et al., 2010; Garrett & Caldwell, 2009) and can lead to more errors due to human memory recall limitations (see Proctor & Van Zandt, 2008).

The variety and disparity of TBI-related care handoffs demonstrate the complexity of navigating the current healthcare system as a patient with a chronic condition. Unsurprisingly, many patients or their informal caregivers find themselves devoting large amounts of time to coordinate and manage all the care points (e.g., where to go, whom to see, and when to see them) and integrate all the needed data (e.g., patient past procedures and symptoms, patient current health state).

7.2.3 Telehealth and Telemedicine

Although the use of electronic ICTs to support remote healthcare delivery (known as "telehealth" and "telemedicine") is usually associated with the advent of web-based communications and internet search technologies, the concept and potential benefits of telehealth were already known and discussed in the 1970s (Lehoux et al., 2000; Steele & Lo, 2013). Telemedicine was originally considered as a means to improve patient access and care in large, sparsely populated rural areas. As a result, the concept of telemedicine was specifically intended to increase the frequency and speed of detecting health status signals from patients to clinicians for whom extensive physical travel would be infeasible.

A variety of telehealth initiatives have been developed to help patients and their informal caregivers who cannot access formal healthcare settings on a regular basis;

these initiatives serve to emphasize the patients' self-care tasks and skills (Brennan et al., 1998; Powell et al., 2016) (see Chapter 5 in this volume). Telemedicine and telehealth have become increasingly used to assist patients who are isolated by physical disabilities, cognitive disabilities, and social challenges (Girard, 2007). It has also been shown that using it is not necessary to use the most advanced ICT tools available (Brennan et al., 1998): the goal of improved telehealth support is to enable improved signal detection and situation awareness during self-care tasks, not simply to use a particular technology.

Telehealth usability represents an important element of healthcare tool designs for patient and care team self-care tasks. Patients using early internet and social media tools such as WebTV were able to capitalize on a system development process that specifically addressed individual, care team, and community user needs. Resulting care tasks included development and information sharing and support between members of caregiver communities, as well as interactions between patients, their caregivers, and telehealth nurses, in settings where patients or caregivers would not have had opportunities to leave home to interact face-to-face (Brennan et al., 1998).

7.3 EXPERTISE IN PATIENT SELF-CARE

A broad, human factors conceptualization of expertise does not simply focus on factual knowledge in a specific domain, but the ability to successfully complete complex tasks in a variety of environments and contexts (Ericsson & Smith, 1991; Garrett et al., 2009). Effective recognition of environmental conditions (as signal detection), integration of those conditions into requirements for task performance (as situation awareness), and effective use of relevant tools to complete those tasks (as system usability) are crucial ways that experts demonstrate their expertise (Endsley, 1995; Endsley et al., 2003).

There are three key perspectives of expertise in patient self-care that are discussed in this section: (1) patients as the source of expertise of their condition, environment, interventions, and healthcare resources; (2) training programs to assist patients in becoming "experts" to manage their chronic disease(s) (i.e., self-management); and (3) the use of information resources (e.g., medical websites, social forums) to self-characterize and self-care for one's condition, particularly uncommon ones. Note that although we highlight the patient as an expert in this discussion, we also consider the role of informal caregivers (e.g., family members) as experts (Holden et al., 2015). These informal caregivers become particularly important as sources of critical health status signals and changes in the patient's capabilities when the patient's cognitive health is compromised, such as in dementia or TBI (Boustani et al., 2011; Heiden & Caldwell, 2017).

7.3.1 Patient as a Source of Expertise

Despite many technological advancements in medicine, patients are still the main source of data, particularly subjective data about themselves. These data include pain levels (Morone & Weiner, 2013; Mularski et al., 2006), depression scores such as PHQ-9 (Kroenke et al., 2001), and dementia symptom monitoring (Monahan et al., 2012, 2014).

These data are attempts to generate meaningful information to inform both clinicians and patients about signals regarding their health status. One challenge, of course, is that specific terms, references to medical literature, and even wording of probe questions may enhance differences in situation awareness and understanding between medical professionals and patients regarding the patient's condition and tasks (Douglas & Caldwell, 2011; Tattersall, 2002). Also, to effectively respond to and manage treatment, patients must know at least some facts about their illness from the clinician's medical standpoint, in order to effectively share information and situation awareness between clinician and patient (Heiden, 2018; Heiden & Caldwell, 2017).

Elements of situation awareness and performance expertise are not expected to all reside in a single individual, or in a declaration description of domain knowledge relevant to a particular disease. Based on work by Garrett and colleagues (2009), one can consider six distinct dimensions of expertise in a sociotechnical task context: (1) subject matter; (2) situational context; (3) interface tool; (4) expert identification; (5) communication; and (6) information flow path (Caldwell, 2005; Garrett et al., 2009). As this chapter highlights, the patient's context of self-care tasks, recognition and effective integration of multiple clinician and patient sources of knowledge, and appropriate design and usability of interfaces to support patient tasks are all important components in supporting distributed expertise for patient self-care.

For example, integration of patient self-care tasks and ICT tools, such as continuous blood glucose monitors, into care decisions demonstrates this multi-dimensional consideration of expertise at individual and team levels of performance (Pickup et al., 2011; Sharon, 2017).

7.3.2 The Expert Patient in Managing Chronic Illness

Patient self-management programs, such as those proposed by the CCM, strongly indicate a framework where the patient is seen as a proactive and knowledgeable participant in their own healthcare performance. With experience and learning, the patient is expected to become even more skilled and capable of performing physical tasks as well as maintaining situation awareness to enhance their own health status (Bodenheimer et al., 2002a, 2002b). For example, patients may use self-management tools to monitor their own status (e.g., blood pressure monitoring) and keep a diary log to track values over time to share with their care team. If a patient perceives their blood pressure values are consistently above a pre-determined range, it may prompt them to take an action toward caring for themselves (e.g., adjusting exercise, diet, or contacting their clinician).

As described above in the six-dimensional framework of expertise (Garrett et al., 2009), these types of skilled performances represent improvements in the patient's signal detection and situation awareness in the dimensions of situational context recognition, interface tool use, and domain knowledge of their health status. When the patient can share relevant status information with the proper specialist clinicians, the patient's self-care tasks can help to support shared situation awareness (Endsley, 1995) with members of the care team. Effective sharing of information and communication of relevant health status information by the patient would demonstrate effective expert identification and communication effectiveness; and access to relevant information shared between clinicians, patients, and ICT tool interfaces represents growing expertise in the use of appropriate information flow pathways.

This consideration of self-efficacy and proactive self-care in responding to chronic illness is acknowledged in the concept of the "expert patient" (Tattersall, 2002; Taylor & Bury, 2007). There are, within this concept, multiple elements of patient-contributed expertise and data integration associated with the process of living with chronic illness. At the very least, acknowledging that patients do have an active role in their own care and health status helps to engage and improve their health outcomes, including symptom management (Tattersall, 2002) and medication management (Mickelson et al., 2016). Especially when clinical care is fragmented or subject to multiple handoffs across diverse specialty clinicians, the patient or their informal caregiver(s) may bear the primary responsibility for maintaining information coordination and shared situation awareness across clinician groups (Heiden & Caldwell, 2018). This is not meant to suggest that healthcare professionals no longer have a role in healthcare delivery for chronic illness, as is sometimes suggested (Taylor & Bury, 2007). Patient self-advocacy in this sense represents a self-care skill of managing important health status signals with different meanings to different clinicians.

Presenting important information relevant to patient health status and self-care tasks in a patient-oriented format may help the patient's ability to successfully complete signal detection and situation awareness (understanding and engagement) tasks by improving health literacy (Douglas & Caldwell, 2011; Morrow et al., 2006). Furthermore, increases in patient-level (i.e., non-medical expert) access to medical information, shared patient experiences, and even research findings through internet resources can provide additional forms of expertise, and medically relevant signals, of use to both the patient and clinician. The next section addresses examples of how increased patient access and knowledge sharing (including improved situation awareness related to medical information) has played out in a range of self-care and expert patient settings.

7.3.3 Self-care Using Internet Resources

Although it is assumed that patients visit clinicians to obtain medical expertise, there are situations where the patient's health condition can change this flow of communication and situation awareness sharing. The availability of online medical information and analysis (such as electronic journals, specialized searches such as Google Scholar, or research communities such as ResearchGate) can further enrich the patient's capability of developing specialized expertise in their condition (see Chapter 13 in this volume). For example, the concept of prosopagnosia, or face-blindness, was long considered a rare medical anomaly, with only a few cases reported in the medical literature, and nearly all of those due to acute trauma (Leibach, 2016). There was little or no medical consideration of prosopagnosia as a common, chronic medical condition until Oliver Sacks (himself a profoundly face-blind and place-blind individual) wrote a medical case in 1985 popularized as "The Man Who Mistook His Wife for a

Hat" (Sacks, 2010). Thanks to online social network communications among persons with face-blindness, its incidence is now considered over 2% of the adult population (Leibach, 2016).

An even more instructive example is shown in the history of fibromyalgia, a disease syndrome only recognized in the 1980s. Fibromyalgia was often confused with other systemic syndromes such as Epstein–Barr or Lyme diseases and even assumed to be a psychiatric condition (Goldenberg, 1999). Fibromyalgia can be extremely debilitating but does not have well-defined objective or instrumental data "signals"; instead clinicians must rely on subjective patient reports of perceived pain. The primary signal detection and situation awareness processes were online information exchanges among patient self-help communities. Both face-blindness and fibromyalgia demonstrate interactions between content, context, and process of signal detection and shared situation awareness that support increased support for patients with these chronic conditions.

7.4 FUTURE OF SELF-CARE AND CHRONIC CARE

As described above, challenges of managing chronic illness include appropriately detecting relevant health signals, performing situation awareness tasks to make sense of those signals, and having access to well-designed devices (e.g., ICT) that increase usability to achieve desired health goals.

The potential for substantial increases in the ability to detect and collect health information via self-care devices is significant, but it also comes at a significant risk. Mobile smartphone devices (such as the iPhone) and purpose-built health devices (such as Fitbit) have enabled significant advances in the ability of patients to collect and track their own medical and health data (see Chapter 12 in this volume). These devices drastically increase the availability of self-care data for both signal detection and situation awareness regarding patients' conditions. The movement to leverage those data has been described as "the quantified self" (Gimpel et al., 2013; Sharon, 2017; Swan, 2009). However, concerns have been raised about the usability of these devices, as well as the gains in situation awareness achieved if collected health signal data are not presented in a way that patients can understand or use (Evans et al., 2016; Martinez et al., 2018; Sun et al., 2018).

Advances in patient self-care include systems to help patients with Type 1 and Type 2 diabetes, collect and monitor their glucose levels, increasing situation awareness and signal detection for both patients and clinicians (see, for example, Pickup et al., 2011; Wu et al., 2017). The "self-management" tasks identified by Wu et al. (2017) regarding these signal detection and situation awareness considerations were significantly affected by system usability; the authors suggested that these issues be more strongly considered in the design of future mobile health apps.

The availability of usable devices and software applications that can provide healthcare signals compatible with a patient's situation awareness needs for self-care suggests that such patient-focused data can be an increasingly important component of the patient's health record. As some research has shown, signal detection through patient self-care tasks can be an effective form of chronic illness management (e.g., Sun et al., 2018; Wu et al., 2017). Precision medicine is an emerging field that may assist with the effective use of the variety of available patient data to enable customized care that incorporates each individual's unique combination of environmental, experiential, genetic, nutritional, and physiological expressions (Jahn et al., 2018). Future work will be needed at this intersection of precision medicine and self-care as nearly continuous signals from data sources ranging from genomic inheritance to individual responses to environmental conditions can greatly exceed the signal detection and sensemaking capability of healthcare professionals and patients.

7.4.1 Patient-centered Medical Home Model

Research into the design and implementation of chronic care management and healthcare delivery has begun to focus on a Patient-Centered Medical Home (PCMH) model. Shared situation awareness and expertise coordination between patients and clinicians, however, can be affected by differing contexts, goals, and priorities that will affect the perceived usability of PCHM tools and health status data. Ideally, both patients and clinicians would recognize that each is contributing complementary signals for improved understanding, and sharing patient-focused and understood data across multiple clinician types work to improve health status and healthcare cost-effectiveness (Pourat et al., 2015; Rosenthal et al., 2016; Yoon et al., 2015).

7.4.2 Medical and Personal Health Records

Situation awareness requires that the users have a clear understanding of the data providing signals about the state of the world, and how those signals can be used to make valid decisions and perform appropriate tasks (Endsley et al., 2003). Accurate health records should, as a result, be a high priority for both patients and clinicians. However, patients and clinicians disagree over the rights of access, ownership, and review of clinician-maintained EMRs (Vallette & Caldwell, 2013). In addition, there is a situation awareness and usability challenge as many patients do not have sufficient health literacy to understand the health data provided in clinician-maintained EMRs (Douglas & Caldwell, 2011).

By contrast, active patient self-care tasks can be used to generate a rich set of personal health record (PHR) data, including through the use of mobile health devices and software applications (see Chapter 5 in this volume). Some studies have specifically looked at clinicians "prescribing" the use of mobile device applications to help patients monitor and manage chronic conditions through self-care tasks (Lopez et al., 2018; Wu et al., 2017). Successful use in such settings has been linked not only to patient usability of resulting PHR data, but more consistent availability and situation awareness of health status signals between visits to primary care, specialist, or emergency health services. Improved patient understanding and situation awareness can also enhance the patient's ability to coordinate and share relevant PHR signals among multiple clinicians in a chronic care management team. This is likely to be an important area of improved care coordination and patient self-care task usability in the future.

7.5 A SELF-CARE EXEMPLAR

The following exemplar illustrates how signal detection, situation awareness, and usability affect self-care.

> Standing at the breakroom table filled with sweets, Jan sighs and mumbles "Diabetes". Jordan notices Jan's discomfort, and shares, "I hope I'm not being too forward—desserts are hard for my family too. My youngest, Sam, was diagnosed with Type 1 diabetes a few years ago. Now he's using the newest continuous glucose monitoring system and software." Jordan pulls out his smartphone and opens an app from the home page. "See? These are his blood glucose levels for the past week, and they are much easier to read in real-time and use colors to indicate how often he has been in his target ranges. The school nurse and I are both on the app as 'trusted followers,' and if his glucose goes too low or high, we get alerts. It's already saved us from a trip to the ER, and our pharmacist helped us set up the right glucose levels for alerts."
>
> "Oh, health alerts have been a lifesaver," Jo adds to the conversation. "I'm now a year cancer-free, but I really didn't understand how much the 'chemo brain' was going to affect my short-term memory. My peer cancer support group helped me put together some reminders on my smartphone as well. The doctor tried to show me, but it's hard when you're in the middle of chemo to remember instructions. The reminders helped me realize when I wasn't taking some of my meds, and I could add notes to share with the oncologist during my next visit. Those reminders helped us figure out that I was having a bad reaction and my doctor lowered the dose."

Human factors concepts in the above exemplar include the following.

- Signal detection: Sam's blood glucose level alerts; Jo's medication reminder alerts;
- Situation awareness: Remote alerts sent to Jordan and the school nurse; Jo's reminder alerts; Jan may lack situation awareness about whether cookies were okay to eat;
- Usability: Smartphone app design for Sam and Jordan to read displays and set personal alarms; capacity for Jo to use reminders although impaired by chemotherapy.

7.6 SUMMARY AND CONCLUDING RECOMMENDATIONS

The role of the patient as an active participant in their own health care and important performer of self-care tasks has changed substantially over the past 100 years. HFE considerations of signal detection and situation awareness and HFE methods to enhance system usability (see Chapter 10 in this volume) have contributed to improvements in healthcare tools and processes. Patient self-care tasks are a suitable and important focus of HFE work, particularly as a growing number of consumer devices are intended to assist in generating and monitoring healthcare data to be used as important signals describing the patient's health status.

Limited patient health literacy and poor shared situation awareness between patients and clinicians can degrade the patient's ability to effectively participate in tasks to help improve and sustain health status (DeWalt et al., 2004; Osborn et al., 2011). New medical care approaches, such as the CCM and PCMH, represent opportunities for integrating patient and clinician situation awareness over multiple care visits spanning months or years (NIH, 2011). As a result, one area of recommended future work is to examine the complexity of healthcare handoffs to understand the patient roles and responsibilities for managing data that serve as signals across multiple clinician areas of expertise, as well as the signals generated by personal health devices and applications and managed by the patients themselves.

Health-related situation awareness is not only the processing and understanding of the patient's formal test results entered into a clinician-maintained EMR. Successful exchange of meaningful information between patients and clinicians includes integration of subjective signals (such as a subjective pain report) and tools that enhance the usability of critical patient–clinician communication and patient self-care tasks. This integration of formal EMR and PHR data is another valuable area of future work. Processes for integrating signals (with meaning to both patients and clinicians) and usability of resulting systems should be determined to help patients manage and improve their health status. Both individual-level patient self-care tasks and knowledge sharing among patient communities are important in developing patient situation awareness that is supported by easily accessed and interpreted health status signals available from all data sources.

ACKNOWLEDGMENTS

Portions of this research were supported by the Purdue University School of Industrial Engineering, the Purdue University Doctoral Fellowship (second and third authors), and the National Science Foundation Graduate Research Fellowship Program (Grant #DGE-1333468: third author).

REFERENCES

Apker, J., Beach, C., O'Leary, K., Ptacek, J., Cheung, D., & Wears, R. (2014). Handoff communication and electronic health records. *Proceedings of the International Symposium on Human Factors and Ergonomics in Health Care, 3*(1), 162–169.

Benedict, A. J., & Caldwell, B. S. (2011). Media usage for feedback communication in an outpatient prescribing setting. In *Proceedings of the Human Factors and Ergonomics Society Annual Meeting* (Vol. 55, No. 1, pp. 768–772). Sage, CA: Los Angeles, CA: SAGE Publications.

Bertolero, M., & Bassett, D., S. (2019, July). How the mind emerges from the brain's complex networks. *Scientific American, 174*. Retrieved from https://www.scientificamerican.com/article/how-the-mind-emerges-from-the-brains-complex-networks/

Bodenheimer, T., Wagner, E. H., & Grumbach, K. (2002a). Improving primary care for patients with chronic illness. *Journal of the American Medical Association, 288*(14), 1775–1779.

Bodenheimer, T., Wagner, E. H., & Grumbach, K. (2002b). Improving primary care for patients with chronic illness: The chronic care model, Part 2. *Journal of the American Medical Association, 288*(15), 1909–1914.

Boustani, M. A., Sachs, G. A., Alder, C. A., Munger, S., Schubert, C. C., Guerriero Austrom, M., . . . Callahan, C. M. (2011). Implementing innovative models of dementia care: The Healthy Aging Brain Center. *Aging & Mental Health, 15*(1), 13–22.

Brennan, P. F., Caldwell, B. S., Moore, S. M., Sreenath, N., & Jones, J. (1998). Designing HeartCare: Custom computerized home care for patients recovering from CABG Surgery. *Proceedings of the AMIA Symposium (Nov 7-11, Lake Buena Vista, FL, pp 381–385). PMID: 9929246.*

Caldwell, B. S. (2005). Analysis and modeling of information flow and distributed expertise in space-related operations. *Acta Astronautica, 56*(9–12), 996–1004.

Caldwell, B. S., Garrett, S. K., & Boustany, K. (2010). Healthcare team performance in time critical environments: Coordinating events, foraging, and system processes. *Journal of Healthcare Engineering, 1*(2), 255–276.

DeWalt, D. A., Berkman, N. D., Sheridan, S., Lohr, K. N., & Pignone, M. P. (2004). Literacy and health outcomes: A systematic review of the literature. *Journal of General Internal Medicine, 19*(12), 1228–1239.

Douglas, S. E., & Caldwell, B. S. (2011). Design and validation of an individual health report (IHR). *International Journal of Industrial Ergonomics, 41*, 352–359.

Endsley, M. R. (1995). Toward a theory of situation awareness in dynamic systems. *Human Factors: The Journal of the Human Factors and Ergonomics Society, 37*(1), 32–64.

Endsley, M. R., Bolté, B., & Jones, D. G. (2003). *Designing for situation awareness: An approach to user-centered design*. New York: Taylor & Francis.

Ericsson, K. A., & Smith, J. (Eds.). (1991). *Toward a general theory of expertise*. Cambridge: Cambridge University Press.

Evans, J., Papadopoulos, A., Silvers, C. T., Charness, N., Boot, W. R., Schlachta-Fairchild, L., . . . Ent, C. B. (2016). Remote health monitoring for older adults and those with heart failure: Adherence and system usability. *Telemedicine and e-Health, 22*(6), 480–488.

Fletcher, K. A., Bedwell, W. L., Rosen, M., Catchople, K., & Lazzara, E. (2014). Medical team handoffs. *Proceedings of the Human Factors and Ergonomics Society Annual Meeting, 58*(1), 654–658.

Forkner-Dunn, J. (2003). Internet-based patient self-care: The next generation of health care delivery. *Journal of Medical Internet Research, 5*(2), e8.

Furniss, S. K., Burton, M. M., Larson, D. W., & Kaufman, D. R. (2016). Modeling patient-centered cognitive work for high-value care goals. *Proceedings of the International Symposium on Human Factors and Ergonomics in Health Care, 5*(1), 112–119.

Garrett, S. K., & Caldwell, B. S. (2009). Healthcare systems engineering. In M. Kutz (ed.), *Standard handbook of biomedical engineering and design* (2nd Ed., Vol. 2, Ch 27, pp 731–747). New York: McGraw-Hill Professional.

Garrett, S. K., Caldwell, B. S., Harris, E. C., & Gonzalez, M. C. (2009). Six dimensions of expertise: A more comprehensive definition of cognitive expertise for team coordination. *Theoretical Issues in Ergonomics Science, 10*(2), 93–105.

Gee, P. M., Greenwood, D. A., Paterniti, D. A., Ward, D., & Miller, L. M. S. (2015). The eHealth enhanced chronic care model: A theory derivation approach. *Journal of Medical Internet Research,* 17(4): e86. doi:10.2196/jmir.4067.

Gimpel, H., Nißen, M., & Görlitz, R. (2013). Quantifying the quantified self: A study on the motivations of patients to track their own health. *Thirty Fourth International Conference on Information Systems*, Milan. Available at https://pdfs.semanticscholar.org/7ccb/e2e99078317a8657a2d362cdeb755b323cf4.pdf

Girard, P. (2007). Military and VA telemedicine systems for patients with traumatic brain injury. *Journal of Rehabilitation Research & Development, 44*(7), 1017–1026.

Glasgow, R. E., Tracy Orleans, C., Wagner, E. H., Curry, S. J., & Solberg, L. I. (2001). Does the chronic care model serve also as a template for improving prevention? *The Milbank Quarterly, 79*(4), 579–612.

Goldenberg, D. L. (1999). Fibromyalgia syndrome a decade later: What have we learned? *Archives of Internal Medicine, 159*(8), 777–785.

Grady, P. A., & Gough, L. L. (2014). Self-Management: A comprehensive approach to management of chronic conditions. *American Journal of Public Health, 104*(8), e25–31.

Guinery, J. (2011). Capturing decision input content to support work system design. Paper Presented at *Tenth International Symposium on Human Factors in Organizational Design and Management*, April 4-6, Grahamstown, South Africa, IEA Press.

Heiden, S. M. (2018). *Information and coordination for tracking traumatic brain injury recovery: A functional needs assessment.* Doctoral Dissertation. West Lafayette, IN: Purdue University.

Heiden, S. M., & Caldwell, B. S. (2018). Considerations for developing chronic care system for traumatic brain injury based on comparisons of cancer survivorship and diabetes management care. *Ergonomics, 61*(1): 134–137. doi:10.1080/00140139.2017.1349932.

Heiden, S. M., & Caldwell, B. S. (2018). *Multi-level, multi-discipline, and temporally-diverse handoffs in traumatic brain injury rehabilitation.* Paper presented at the 2018 International Symposium on Human Factors and Ergonomics in Health Care, Boston, MA.

Higginson, G. D. (1931). *Fields of psychology: A study of man and his environment.* New York: Henry Holt and Company.

Holden, R., Schubert, C. C., & Mickelson, R. S. (2015). The patient work system: An analysis of self-care performance barriers among elderly heart failure patients and their informal caregivers. *Applied Ergonomics, 47*, 133–150.

Holden, R., Valdez, R., Schubert, C. C., Thompson, M., & Hundt, A. S. (2017). Macroergonomic factors in the patient work system: Examining the context of patients with chronic illness. *Ergonomics, 60*(1), 26–43.

IOM. (2012). *Living well with chronic illness: A call for public health action.* Washington, DC: Institute of Medicine.

ISO DIS 9421-11. (2018). *Ergonomics of human-system interaction—Part 11: Usability: Definitions and concepts.* Retrieved January 7, 2020, from https://www.iso.org/obp/ui/#iso:std:iso:9241:-11:ed-2:v1:en

Jahn Holbrook, M. (2019). *A systems engineering analysis of opportunities for pharmacists on diabetes care teams.* Doctoral Dissertation. West Lafayette, IN: Purdue University.

Jahn Holbrook, M., & Caldwell, B. S. (2019). Development of a multi-layer systems engineering visualization for diabetes team coordination. *Proceedings of the Human Factors and Ergonomics Society 2019 International Annual Meeting*, Oct 31–Nov 4, Seattle Los Angeles, CA: Sage Publications, pp 643-647. doi: 10.1177/1071181319631195.

Jahn, M. A., Heiden, S. M., & Caldwell, B. S. (2018). *Identifying improvements in healthcare systems engineering models for chronic care and precision medicine applications.* Paper presented at the 2018 International Symposium on Human Factors and Ergonomics in Health Care, Boston, MA.

Kinsey, A. C., Pomeroy, W. B., & Martin, C. E. (1948). *Sexual behavior in the human male.* Philadelphia: W. B. Saunders Company.

Klein, K. R., Brandenburg, D. C., Atlas, J. G., & Maher, A. (2005). The use of trained observers as an evaluation tool for a multi-hospital bioterrorism exercise. *Prehospital and Disaster Medicine, 20*, 159–163.

Kroenke, K., Spitzer, R. L., & Williams, J. B. (2001). The PHQ-9: validity of a brief depression severity measure. *Journal of general internal medicine, 16*(9), 606-613. Accessed Jan 9, 2020, doi: 10.1046/j.1525-1497.2001.016009606.x

Lehoux, P., Battista, R. N., & Lance, J. M. (2000). Telehealth: Passing fad or lasting benefits? *Canadian Journal of Public Health, 91*(4), 277–280.

Leibach, J. (2016). "What is face blindness?" *Ask Science Friday.* Retrieved from https://www.sciencefriday.com/articles/what-is-face-blindness/

Lopez Segui, F., Bufill, C. P., Gimenez, N. A., Roldan, J. M., & Cuyas, F. G. (2018). The prescription of mobile apps by primary care teams: A pilot project in Catalonia. *Journal of Medical Internet Research mHealth and uHealth, 6*(6), e10701.

Lorig, K. R., & Holman, H. R. (2003). Self-management education: History, definition, outcomes, and mechanisms. *Annals of Behavioral Medicine, 26*(1), 1–7.

Lupton, D. (2013). The digitally engaged patient: Self-monitoring and self-care in the digital health era. *Social Theory & Health, 11*(3), 256–270.

MacGill, M. (2015). Does an apple a day really keep the doctor away? *Medical News Today.* Accessed July 17, 2019 Retrieved from https://www.medicalnewstoday.com/articles/291683.

Martinez, W., Threatt, A. L., Rosenbloom, S. T., Wallston, K. A., Hickson, G. B., & Elasy, T. A. (2018). A patient-facing diabetes dashboard embedded in a patient web portal: Design sprint and usability testing. *Journal of Medical Internet Research Human Factors, 5*(3), e26.

Masel, B. (2009). *Conceptualizing brain injury as a chronic disease.* Vienna, VA: Brain Injury Association of America.

Mickelson, R. S., Unertl, K., & Holden, R. (2016). Medication management: The macrocognitive workflow of older adults with heart failure. *Journal of Medical Internet Research Human Factors, 3*(2): e27. doi: 10.2196/humanfactors.6338.

Monahan, P. O., Alder, C. A., Khan, B. A., Stump, T., & Boustani, M. A. (2014). The healthy aging brain care (HABC) monitor: Validation of the patient self-report version of the clinical tool designed to measure and monitor cognitive, functional, and psychological health. *Clinical Interventions in Aging, 9*, 2123–2132.

Monahan, P. O., Boustani, M. A., Alder, C., Galvin, J. E., Perkins, A. J., Healey, P., . . . Callahan, C. (2012). Practical clinical tool to monitor dementia symptoms: The HABC-Monitor. *Clinical Interventions in Aging, 7*, 143–157.

Morone, N. E., & Weiner, D. K. (2013). Pain as the fifth vital sign: Exposing the vital need for pain education. *Clinical therapeutics, 35*(11), 1728–1732.

Morrow, D., Clark, D., Tu, W., Wu, J., Weiner, M., Steinley, D., & Murray, M. D. (2006). Correlates of health literacy in patients with chronic heart failure. *The Gerontologist, 46*(5), 669–676.

Mularski, R. A., White-Chu, F., Overbay, D., Miller, L., Asch, S. M., & Ganzini, L. (2006). Measuring pain as the 5th vital sign does not improve quality of pain management. *Journal of General Internal Medicine, 21*(6), 607–612.

NAE / IOM. (2005). *Building a better delivery system: A new engineering / health care partnership.* Washington, DC: National Academy of Engineering and Institute of Medicine.

NIH. (2011). Redesigning the health care team: Diabetes prevention and lifelong management. Retrieved July 26, 2017, from https://www.niddk.nih.gov/health-information/health-communication-programs/ndep/health-care-professionals/team-care/Documents/-NDEP-37_508-2.pdf

Osborn, C., Cavanaugh, K., Wallston, K., Kripalani, S., Elasy, T., Rothman, R., & White, R. (2011). Health literacy explains racial disparities in diabetes medication adherence. *Journal of Health Communication, 16*(Suppl 3), 268–278.

Pickup, J. C., Freeman, S. C., & Sutton, A. J. (2011). Glycaemic control in type 1 diabetes during real time continuous glucose monitoring compared with self monitoring of blood glucose: meta-analysis of randomised controlled trials using individual patient data. *The The BMJ, 343*, d3805.

Pourat, N., Davis, A. C., Chen, X., Vrungos, S., & Kominski, G. F. (2015). In California, primary care continuity was associated with reduced emergency department use and fewer hospitalizations. *Health Affairs, 34*(7), 1113–1120.

Powell, J., Fraser, R. T., Brockway, J. A., Temkin, N. R., & Bell, K. R. (2016). A telehealth approach to caregiver self-management following traumatic brain injury: A randomized controlled trial. *Journal of Head Trauma Rehabilitation, 31*, 180–190.

Proctor, R. W., & Van Zandt, T. (2008). *Human factors in simple and complex systems* (2nd ed.). Boca Raton, FL: CRC Press, Taylor & Francis Group.

Rogers, M., Caldwell, B., Marmet, G. J., & Brennan, P. F. (1999). Usability design considerations in providing computer support for health informatics. *Proceedings of the Human Factors and Ergonomics Society 43rd Annual Meeting*, Sep 27 -- Oct 1, Houston, Santa Monica, CA: HFE, pp 855–858.

Rosenthal, M. B., Alidina, S., Friedberg, M. W., Singer, S. J., Eastman, D., Li, Z., & Schneider, E. C. (2016). A difference-in-difference analysis of changes in quality, utilization and cost following the Colorado multi-payer patient-centered medical home pilot. *Journal of General Internal Medicine, 31*(3), 289–296.

Sacks, O. (2010). Face-Blind: Why are some of us terrible at recognizing faces?. *New Yorker (New York, NY: 1925)*, August 30 issue, pp. 36-43. Accessed July 17, 2019 from https://www.newyorker.com/magazine/2010/08/30/face-blind.

Sharon, T. (2017). Self-tracking for health and the quantified self: Re-articulating autonomy, solidarity, and authenticity in an age of personalized healthcare. *Philosophy & Technology, 30*(1), 93–121.

Siminerio, L. M., Zgibor, J. C., & Solano, F. X., Jr. (2004). Implementing the chronic care model for improvements in diabetes practice and outcomes in primary care: The University of Pittsburgh medical center experience. *Clinical Diabetes, 22*(2), 54–58.

Sochalski, J., Jaarsma, T., Krumholz, H. M., Laramee, A., McMurray, J. J., Naylor, M. D., . . . Stewart, S. (2009). What works in chronic care management: The case of heart failure. *Health Affairs, 28*(1), 179–189.

Steele, R., & Lo, A. (2013). Telehealth and ubiquitous computing for bandwidth-constrained rural and remote areas. *Personal and Ubiquitous Computing, 17*(3), 533–543.

Stuckey, H. L., Adelman, A. M., & Gabbay, R. A. (2011). Improving care by delivering the chronic care model for diabetes. *Diabetes Management, 1*, 37+.

Sun, R., Korytkowski, M. T., Sereika, S. M., Saul, M. I., Li, D., & Burke, L. E. (2018). Patient portal use in diabetes management: Literature review. *Journal of Medical Internet Resesrch Diabetes, 3*(4), e11199.

Swan, M. (2009). Emerging patient-driven health care models: An examination of health social networks, consumer personalized medicine and quantified self-tracking. *International Journal of Environmental Research and Public Health, 6*(2), 492–525.

Tai-Seale, M., Olson, C. W., Li, J., Chan, A. S., Morikawa, C., Durbin, M., . . . Luft, H. S. (2017). Electronic health record logs indicate that physicians split time evenly between seeing patients and desktop medicine. *Health Affairs, 36*(4), 655–662.

Tattersall, R. (2002). The expert patient: A new approach to chronic disease management for the twenty-first century. *Clinical Medicine, 2*(3), 227–229.

Taylor, D., & Bury, M. (2007). Chronic illness, expert patients and care transition. *Sociology of Health & Illness, 29*(1), 27–45.

USDHHS. (2010). *Multiple chronic conditions—A strategic framework: Optimum health and quality of life for individuals with multiple chronic conditions.* Washington, DC: U.S. Department of Health and Human Services.

Vallette, M. A., & Caldwell, B. S. (2013). *Patient and provider perspectives on electronic health record (EHR) information access and rights. In Proceedings of the International Symposium on Human Factors and Ergonomics in Health Care* (Vol. 2, No. 1, pp. 64–68), March 10-13, Baltimore. Los Angeles, CA: SAGE Publications.

Wagner, E. H. (1998a). Chronic disease management: What will it take to improve care for chronic illness? *Effective Clinical Practice: ECP, 1*(1), 2–4.

Wagner, E. H. (1998b). Chronic disease management: What will it take to improve care for chronic illness? *Effective Clinical Practice, 1*(1), 1–4.

Wagner, E. H. (2000). The role of patient care teams in chronic disease management. *The BMJ, 320*(7234), 569.

Wagner, E. H., Davis, C., Schaefer, J., Von Korff, M., & Austin, B. (2002). A survey of leading chronic disease management programs: Are they consistent with the literature? *Journal of Nursing Care Quality, 16*(2), 67–80.

Warshaw, G. (2006). Introduction: Advances and challenges in care of older people with chronic illness. *Generation, 30*(3), 5–10.

Wickens, C. D., Lee, J. D., Liu, Y., & Gordon Becker, S. E. (2004). *An introduction to human factors engineering* (2nd ed.). Upper Saddle River, NJ: Pearson Prentice Hall.

Wu, Y., Yao, X., Vespasiani, G., Nicolucci, A., Dong, Y., Kwong, J., . . . Li, S. (2017). Mobile app-based interventions to support diabetes self-management: A systematic review of randomized controlled trials to identify functions associated with glycemic efficacy. *Journal of Medical Internet Research mHealth and uHealth*, *5*(3), e35.

Ye, K., Taylor, D., Knott, J. C., Dent, A., & MacBean, C. E. (2007). Handover in the emergency department: Deficiencies and adverse effects. *Emergency Medicine Australasia, 19*, 433–441.

Yoon, J., Liu, C.-F., Lo, J., Schectman, G., Stark, R., Rubenstein, L. V., & Yano, E. M. (2015). Early changes in VA medical home components and utilization. *The American Journal of Managed Care, 21*(3), 197–204.

Young, R. A., Burge, S. K., Kumar, K. A., Wilson, J. M., & Ortiz, D. F. (2018). A time-motion study of primary care physicians' work in the electronic health record era. *Family Medicine, 50*(2), 91–99.

Zachary, W., Maulitz, R., Iverson, E., Onyekwelu, C., Risler, Z., & Zenel, L. (2013). A data collection framework for care coordination and clinical communications about patients (CAPs). *Proceedings of the International Symposium on Human Factors and Ergonomics in Health Care, 2*(1), 27–29.

8 Patient Engagement in Safety

Patients Are the Ultimate Stakeholders

Elizabeth Lerner Papautsky
University of Illinois at Chicago

CONTENTS

Patient engagement in their own care is now considered one of the many safety interventions for addressing medical errors, the third leading cause of death in the United States (Makary & Daniel, 2016). Leape (2004) stated that "as eventual recipients of that care, we [the public] all have a stake in this enterprise" (p. 11). Indeed, patients are the ultimate stakeholders in health care because they have the most to lose. As part of their 2007 *National Patient Safety Goals*, the Joint Commission mandated that healthcare organizations "encourage patients' active involvement in their own care as a patient safety strategy" (The Joint Commission, 2008). This mandate propelled research on patient engagement in safety. Now the idea that patients are in a position to affect and monitor their safety has accrued enough research evidence to no longer be a question, but rather be part of the answer (Berger et al., 2014).

Active involvement of the patient (and potentially, their informal caregivers such as family and friends) and the public in their health and health care, rather than being passive recipients, is a commonly accepted definition of patient engagement (Coulter, 2011). or a more detailed description of patient engagement and comparison of related terms such as patient- and family-centered care and patient activation, the reader can refer to Carman et al.'s (2013) patient engagement framework. Patient safety refers to "freedom from accidental (or preventable) injury" (AHRQ, 2019; Kohn et al., 1999) and includes not just outcomes but also processes and structures of care. Consistent with recent publications on this topic (Boggan, 2019; Sharma et al., 2018), I will use the term *patient engagement in safety.*

Although clinicians have deep knowledge of the science of diagnosis and treatment, patients are the ones who have experience with their bodies and disease, privileged knowledge about themselves, and the continuity across the care continuum (Papautsky, 2019) (see also Chapter 7 in this volume). Given they are the only constant across the care continuum (Wright, 2019), "patients and people who have spent time with them, are in the best position to gather data and keep watch" (The World Health Organization, 2007).

8.1 PATIENTS' ROLE IN REPORTING ERRORS

Patients and informal caregivers can engage in safety by directly reporting care deviations, adverse events, and harm that have already occurred. Reporting is primarily discussed in studies of adult inpatient settings (Armitage et al., 2018; Weingart et al., 2005; Weissman et al., 2008) and the emergency department (Friedman et al., 2008). Khan et al. (2017, 2016) also did work in pediatric inpatient settings. For instance, they described a parent-reported care deviation of a 12-hour delay in treatment of a patient with fluid accumulation in the lungs despite parent reports of respiratory symptoms (Khan et al., 2017). Other reports described missed diagnoses and medication errors. Studies suggest that not only are the patients' and families' safety concerns valid, they are also unique. For example, their 2017 study found that families reliably reported an additional 16% of errors and 10% of adverse events above the clinicians (Khan et al., 2017). Extending this work to an ongoing multi-site patient- and family-centered program to improve communication between patients/families and clinicians during and after rounds, an intervention called I-PASS was coproduced by families and other stakeholders. I-PASS was implemented in pediatric units in seven US hospitals, and the findings suggest a 38% reduction in harmful medical errors through systematically engaging families in communication (Haskell, 2018; Khan, 2018; Khan et al., 2018).

Multidisciplinary research teams that include human factors researchers and medical professionals at Vanderbilt University Medical Center have been examining and characterizing patient- and family-reported nonroutine events. Nonroutine events are anything that happens during a care episode that is a deviation from optimal care. These studies have captured reporting across multiple clinical settings including adult and pediatric oncology, adult outpatient surgery, pediatric cardiac surgery, and cardiac catheterization (Cleary et al., 2017; Tippey et al., 2018; Troy et al., 2016). Their findings also suggest that patients and families are a valuable and potentially unique source of information regarding deviations in care that affect not just psychological distress and satisfaction but also safety and risks of harm. In addition, the information reported by patients provides almost no overlap with the information reported by the healthcare team caring for them (Anders, 2016).

8.2 HUMAN FACTORS AS A LENS FOR PATIENT ENGAGEMENT IN SAFETY

Theories of naturalistic decision-making and distributed cognition have provided human factors science with perspectives and approaches to study cognitive work (see Chapter 2 in this volume). This research has traditionally focused on professional experts conducting purposeful activities in naturalistic contexts. Patient cognitive work also takes place in naturalistic contexts. Not only do patient contexts vary across the care continuum, but there is also variability in patient factors including health literacy, culture, socioeconomic status, technology access, work and life context, goals, preferences, attitudes, beliefs, and information needs. To characterize patient opportunities for and barriers to engaging in safety and inform tailored interventions, these factors need to be examined and considered.

Patient ergonomics highlights the application of human factors science to supporting patient work across contexts of health within and beyond clinical settings (to home, work, and community)—or "contexts of daily living"—and with consideration for patient factors. To summarize recent efforts in patient ergonomics, Holden et al. (2020) conducted a mapping review of conference publications during 2007–2017. The review included studies examining care processes associated with safety, knowledge, and self-care. Several studies addressed medication safety for those with chronic illnesses such as diabetes (Lippa et al., 2008; Lippa & Klein, 2008) and heart failure (Mickelson & Holden, 2013). Notably, authors mentioned that studies on patient decision-making are few. Decision-making involves cognitive work of managing information of meaningful and functional significance, which may have safety implications. In the below discussion, we propose a framework of patient cognitive work during engagement in safety. The framework positions the patient as central to the information space, through possessing, acquiring, seeking, applying, and sharing patient-held information.

To illustrate, we present a case study of Mrs. Dunbar, a surgical patient whose care takes place across (a) inpatient, (b) home, and (c) outpatient settings (Box 8.1). As the contexts of care shift outside of clinical settings, more of the information remains patient-held, highlighting the patient role as central to the information space.

8.3 PATIENTS ARE CENTRAL TO THE INFORMATION SPACE

Conceptualized by Hutchins (1995) using the example of an airplane cockpit, the theory of distributed cognition states that cognitive tasks are undertaken by a network

BOX 8.1 CASE STUDY OF A SURGICAL PATIENT, MRS. DUNBAR

Inpatient setting: Mrs. Dunbar has surgery and remains in the hospital for three nights for postoperative care. Her husband sleeps on the pullout sofa that is provided in the room. While in the hospital, clinicians monitor, assess, and observe her, administer pain management medications, and provide wound care, assistance with toileting and moving around, as well as physical and occupational therapy. Mrs. Dunbar is only asked to report on pain level (on a scale of 1–10), as well as on her perception of physical and psychological state (e.g., how do you feel?). During her stay, Mrs. Dunbar and her husband are in a position to monitor medication safety, falls, emergent conditions (respiratory and cardiovascular), as well as potential changes from baseline across shifts (clinicians and time). However, most of the clinically relevant information is held by clinicians.

Home setting: Based on the discharge instructions and with her husband's help, Mrs. Dunbar recovers at home by self-administering multiple medications based on a schedule (pain relief, blood thinner, stool softener, and others), conducting wound care (cleaning and bandaging), toileting, showering, and taking short walks around her bedroom. By noting her energy levels and wound healing, Mrs. Dunbar assesses trends in her recovery. In addition, Mrs. Dunbar has been told to monitor for blood clots and signs of infection. To monitor for infection, she was given a list of signs associated with the appearance and temperature of the surgical site, as well as elevated body temperature. She makes decisions on whether and when to report concerns to the surgeon by calling, emailing, and sending photographs of surgical sites (with clinician missing vital perceptual cues such as temperature and "feel"). At this time, much of the information space is patient-held, including home context, medication adherence, proper and sanitary wound care practices, and unreported signs and symptoms. While Mrs. Dunbar convalesces at home, she manages uncertainty and makes decisions about when to take action-all with no medical training.

Outpatient setting: After a week at home, Mrs. Dunbar, accompanied by her husband, returns to her surgeon for a postoperative appointment. The nurse takes vital signs and asks for a pain rating; the surgeon examines the surgical sites. The surgeon asks Mrs. Dunbar how she is feeling and whether she has any questions or concerns, as well as her perception of physical and psychological state. In this example, the surgeon is privy to perceptual cues to identify infection or blood clots, mitigating immediate safety concerns. However, much of the information may not be evident upon examination by a surgeon, and remains patient-held (e.g., regarding home context, medication adherence, and signs and symptoms).

of actors with varied roles and expertise. More specifically, the cognitive system of structures and processes is distributed across a group of individual minds, space, and time, including internal (in the head) and external (in the world) representations and artifacts (Zhang & Patel, 2006). It has long been accepted that patient information is distributed (and fragmented) across clinicians, technologies, and artifacts (e.g., clinician roles and teams, and within and across technologies such as electronic health records [EHRs]). In their human factors research on shared decision-making in patient–clinician interactions in management of multiple sclerosis, Lippa et al. (2015, 2016a, b, 2017) extended the construct of distributed cognition to the patient. The authors suggest that information such as disease history and symptoms is held by patients and thus, sometimes only the patient is privileged to it. These studies challenge the traditional conception of medical cognition as clinician-centric and demonstrate mutual dependence in the patient–clinician dyad (see also Chapters 6 and 7 in this volume). Distributed cognition points to a unit of analysis that includes the patient in goal-directed and coordinated activities and interactions with their clinician(s) (Hazlehurst et al., 2008). Thus, given that patients manage privileged information, they are not just central to the work system, but to the information space.

Patient-held information may exist as a function of experience with oneself, but also sought, gathered, shared, and applied to inform decision-making across time and space. This cognitive work may have safety implications. Figure 8.1 represents the patient as central to the information space that is fragmented across multiple interprofessional clinicians. In this central role, the patient contributes to their safety by managing safety-relevant information.

8.3.1 Patients Possess and Acquire Information

Patients (and informal caregivers) possess and acquire information by being observers of themselves, their condition, and their care across space and time of the care continuum. These observations comprise historical information including experience with their body (Pierret, 2003) and the trajectory of their condition (Holman & Lorig, 2004). The term *patient expertise*, particularly in Great Britain, is used to refer to a patient who is not just a consumer of health care, but a producer of health through possession of deep knowledge (Tattersall, 2002). Lippa et al. (2008) suggest that patients' problem detection skills and strategies acquired through practice may be indicative of proficient or expert performance. For example, patients with chronic illness are able to identify problems and take appropriate actions to address them, as well as adapt to changing circumstances, as a function of their experience with their illness (Bodenheimer et al., 2002).

Examples of patient-held information include explicit knowledge such as medication and treatment adherence status (e.g., *I do not take my blood pressure medication every day*); procedural knowledge such as self-care and information management processes and strategies (e.g., *this is how I keep track of all my medications*); tacit knowledge including physical and psychological baselines and deviations from them (e.g., *I feel more fatigued in the last week than usual; the surgical site looks more irritated than it did yesterday*); and symptom phenomenology (e.g., *this is how I experience a migraine*).

FIGURE 8.1 Patient (and caregiver) is central to the information space that is fragmented across multiple interprofessional clinicians. (Created by Danielle Robinson, MS in Biomedical Visualization, University of Illinois at Chicago).

8.3.2 Patients Actively Seek Information

Patients may decide to actively seek out and gather information about their health on the internet and in the library. A body of literature has examined patient information-seeking behaviors particularly around complex and serious illnesses such as cancer, diabetes, and cardiovascular disease. Information seeking may be motivated by the need to make decisions about treatment options, as well as informing oneself about disease mechanisms, progression, as well as management. On the one hand, misinformation and vulnerability to confirmation bias may compromise a patient's safety (Meppelink et al., 2019). On the other hand, through actively seeking out information, patients can become more informed about their condition, enabling them to engage in their safety (Househ et al., 2014).

8.3.2.1 Engaging with Information on the Internet and Online Patient Communities

Useful and relevant information that informs safe and appropriate care provision can and does exist outside of clinical settings and traditional medical literature, such as on the internet. There are numerous examples of patient-found information making a difference in care provision, outcomes, and safety. For example, after a bleak diagnosis of stage IV renal cell carcinoma, deBronkart (2013) accessed the Association of Cancer Online Resources (ACOR) website to discover an experimental treatment that was not initially offered to him. Throughout the treatment, he continued to reach out to disease-specific online patient communities to learn about the treatment side effects, not covered in traditional medical literature. Thirteen years following his diagnosis, deBronkart is alive and well. His experience served as an impetus for a patient advocacy movement, empowering patients to engage with their health information through being e-patients—patients who are equipped, engaged, empowered, and enabled (Riggare, 2018). Similarly, Kushniruk (2019), a health informatics and human factors specialist describes how patient and caregiver information seeking on the internet saved his life. He recounts his journey of finding effective treatment through internet sources, followed by navigating recovery for a cancer he was initially told had a poor prognosis (Kushniruk, 2019). Such stories are not unique. However, they are rarely captured in scholarly literature. In an essay penned for the *BMJ*, another e-patient, Riggare (2018), argues that patients strategize to improve their well-being by acquiring information. Riggare conducted her own research at the library about her symptoms and used it to actively seek a diagnosis of Parkinson's disease and subsequently strategize to manage her disease (Riggare, 2018). These examples highlight information seeking as a way of acquiring expertise in one's illness. Although these examples are not directly related to safety outcomes, they highlight the idea that patient-found information positions the patient to engage in safety by finding appropriate health professionals, asking the right questions, selecting treatments understanding and managing risk, and managing one's illness more effectively.

As of 2019, it is estimated that approximately 72% of Americans use social media (Pew Research Center, 2019). Online patient communities on social media platforms such as Facebook and Twitter are a source of support and a space for information sharing; examples include Brain Tumor Social Media (#BTSM), Breast Cancer Social Media (#BCSM), Young Survival Coalition, Breast Cancer Straight Talk, Patients Like Me, and Mayo Clinic Connect. Geographically distributed patients may share lived experiences, treatment options, recommendations for self-care management, and tips for recovery. Participation in these communities may range from passive observation to active inquiry and response. Young (2013) shares effective strategies for building and sustaining online health communities. Although many online patient communities are run by patients, integrating clinicians (and other experts) is critical to safety by seeding the communities with evidence-based information and monitoring for errors.

8.3.2.2 Engaging with Personal Health Information

Opportunities are growing for patients to engage with their personal health information through patient-facing health information technologies such as patient portals (also known as personal health record and patient-accessible health records) (see Chapter 5 in this volume). The patient portal is the patient-facing side of the EHR, and it is intended to aid remote communication through messaging, deliver test results, and track appointments, among other features. As inpatient care settings are vulnerable to a higher risk of medical errors, Kelly et al. conducted a systematic review of inpatient portal use (2018). Authors found that patients, caregivers, and clinicians perceive patient portals as an opportunity to improve patient safety through patient and caregiver identification and interception of medical errors, particularly for pediatric populations. In addition, clinicians see patient portals as a medium to provide patients with information regarding medication safety, as well as other safety and precaution topics. A recent study was implemented in six acute care units of an academic medical center—a patient portal with safety features including safety education and reminders, fall prevention, and safety concern reporting (Schnock et al., 2019). They found patient portal users had higher levels of patient activation. As of 2020, patient portals at University of California San Francisco Hospitals provide access to radiographic study images (X-rays, magnetic resonance images, computed tomography scans, and some ultrasounds).

OpenNotes, an innovation implemented at three sites in 2010 and now more widespread, invites patients to read their clinical visit notes (see also Chapter 6 in this volume). Initially, some physicians reported a perceived increase in patient safety (along with patient engagement) (Delbanco et al., 2012). In a survey study with 99 physicians and >4,500 patients, about one-quarter of physicians expressed concerns about the accuracy of their notes (Bell et al., 2017). Although only 7% of patients contacted the physician's office after reading their notes, about one-quarter of them reported a potential safety concern, such as this family caregiver's report:

> "[I] just wanted to confirm accuracy. My husband's note says he has a 40-year [history of] back pain, but it was actually only a 4-year [history of] back pain. When clinicians copy and paste, the errors just keep propagating and never get corrected unless we see our notes."
>
> *(Bell et al., 2017, p. 266)*

With over 40 million US patients now having access, OpenNotes is an opportunity for patients and caregivers to engage with the information space at an unprecedented level and play a role in safety by monitoring the accuracy of their records.

8.3.3 Patients Apply Information to Decision-Making

Human factors is well versed in studying the cognitive work of decision-makers whose performance has safety implications (see Chapters 2 and 7 in this volume). Like operators in other high stakes work domains (e.g., fire fighters, soldiers, and pilots), patients make decisions in uncertain and complex naturalistic settings. Unlike such operators, patients may do so with no training or experience, but with a strong implicit objective to maximize safety and effectiveness. Naturalistic decision-making (NDM) provides both a theoretical perspective and research tools to examine how people, primarily experts, make decisions and perform cognitively and perceptually complex work in domains characterized by uncertainty, time limitations, and high cost of error (Klein, 2008). NDM approaches have focused on capturing and characterizing acquisition, content, and application of knowledge to decision-making, particularly focused on macrocognitive processes of problem detection, sensemaking, uncertainty and risk management, planning, and common grounding, among others (Klein et al., 2003). Some research has applied NDM perspective to studying individual patient decision-making (Lippa et al., 2008). Lippa et al. interviewed type II diabetes patients to characterize decision-making processes and knowledge associated with diabetes self-management. They found that diabetes self-management "draws on the same cognitive skills found in experts from diverse professional domains" (Lippa et al., 2008, p. 112). Patients' decision-making skills can affect adherence and self-care and thus, downstream outcomes such as patient safety.

As opposed to chronic conditions where patients have the opportunity to acquire deep knowledge through experience, acute situations call upon patients to make decisions with likely no experience. For example, monitoring for infection following surgery requires attention to complex perceptual cues (e.g., surgical site is red, warm, or tender) without medical training. Furthermore, by choosing whether, when, and how to initiate requests for care, patients make judgments about urgency that can be the difference between life and death (e.g., sepsis diagnosis and management).

8.3.4 Patients Share Information

Effective communication is at the heart of safety. Traditionally, research has focused on information transfer among clinicians, but patient–clinician communication is also important (Chapter 6 in this volume). Patients may share privileged information through face-to-face or technology-mediated (e.g., patient portal, email, text, and phone) interactions with clinicians, within and outside of clinical encounters.

8.3.4.1 Clinician-Elicited Information

In service of effective decision-making and safer care, clinicians consider patients to be a source of clinically relevant information. How can clinicians *partnering with patients make a difference in the safety of care provision?* We asked this question to an emergency physician, family physician, and nurse practitioner in a panel on *The Patient in Patient Safety: Clinicians' Experiences Engaging Patients as Partners in Safety* (Papautsky et al., 2019). Dr. Gruss (an advanced practice nurse) discussed a case study where she elicited a history of significant radiation exposure by a patient seeking fertility counseling. The elicited information may not have been revealed by traditional history taking but led her to consult with other professionals to develop a safer and more tailored care plan. Panelists discussed other examples of patient-held information such as life history, baselines, and previously shared complaints, as necessary to the provision of safer care (Papautsky et al., 2019). There is a need to characterize areas in which clinicians most benefit from partnering with patients through information elicitation and to prioritize these areas in research.

In their research on patient–clinician primary care encounters with unannounced, standardized patient actors, Weiner and Schwartz highlight the need for physicians to elicit individual patient contextual factors (life circumstances and preferences) often missing from the EHR, to prevent contextual errors (Weiner et al., 2010, 2016). Contextual errors are defined as inappropriate care plans in light of the patient's situation or life context, which may result in adverse consequences. Related work has studied the barriers and facilitators of using consumer health technology to deliver patient-generated contextual factors to inform individualized care (Holt et al., 2019). Even in critical care, clinicians seek information from patients' families such as new and unusual patient behaviors, allergies, and preferences, in a manner that may be ad hoc or informal (Papautsky et al., 2017). Patient contributions of clinically relevant information yield a more comprehensive clinical picture, informing more appropriate treatment plans and improving patient safety and health outcomes (Mackay, 2015).

8.3.4.2 Patient-Generated Health Data

Patient information sharing may be mediated by technology including, but not limited to, self-monitoring technologies that may be integrated with the EHR. Patient-generated health data (PGHD) are "created, recorded, or gathered by or from patients" outside of clinical encounters (*HealthIT.gov*, 2018) are now being applied to clinical practice to inform a more complete clinical picture (Estabrooks et al., 2012). PGHD range from biometric data (e.g., heart rate, blood glucose, and blood pressure) to lifestyle behaviors (e.g., exercise and diet) to subjective interpretations of health status (e.g., symptoms indicative of a migraine). PGHD are distinct from other clinical data in that they are patient driven—recorded, captured, and shared by patients. Several entities are developing policies for systematic sharing of PGHD through patient portal capture or entry by patients and delivery to clinicians through the EHR (e.g., *HealthIT.gov*, 2018). Nevertheless, research is lacking on user needs (patients' and clinicians') for PGHD technology.

8.3.4.3 Patient (and Caregiver) Engagement in Information Exchange

Patient and caregiver engagement in information sharing can benefit formation of a common ground and facilitate the raising of concerns and asking of questions. Family-centered rounds, pervasive in pediatric clinical settings, are an example of patient collaboration integral to patient safety (Xie et al., 2015). Indeed, rounds serve as a unique opportunity for families to concurrently engage with multiple care professionals. Nursing change of shift reports at the bedside or family-centered nursing hand-offs in the neonatal intensive care unit can provide similar benefits (Griffin, 2010). Despite evidence of coordination and safety benefits, family-centered rounds and nursing hand-offs are not currently standard practice in adult care across (or even within) hospital systems. Especially for long hospital stays for complicated patients whose care is particularly fragmented, family-centered rounds and hand-offs can provide an opportunity to ensure relevant patient information does not fall through the cracks (a common cause of medical errors).

8.4 PATIENT–CLINICIAN TEAMING

Patient engagement in safety requires a culture shift. Given the mutual dependence between the patient and their clinician(s), the patient should be considered a partner or team member on their care team. Framing of the patient as a team member yields opportunities to extend human factors teamwork models and approaches to patient–clinician teams. Particular focus should be paid to processes associated with effective teamwork such as coordination, communication, common grounding, and goal setting (Marks et al., 2001). Moreover, team research is focused on formal (e.g., surgical teams and profession-specific teams) rather than ad hoc teams. However, individual patient-specific care team membership varies from patient to patient (in terms of professions, roles, and individuals). Even if roles remain constant, specific individuals may change over time (e.g., as a function of changes in shifts or assignments). Thus, although referred to as teams, patient-specific care teams may not demonstrate teamwork as typically defined.

Complex care, such as cancer care, is comprised of multiple interprofessional teams (e.g., oncology, radiology, surgery, social work, physical therapy, and dentistry). A recent article on the delivery of coordinated cancer care called the patient a "unifying member of the team of teams" (Henry et al., 2016, p. 992). The authors explicitly recognize that there is not one individual who possesses the full clinical picture. However, effective interprofessional collaborative practice (let alone including the patient) remains an objective rather than a reality. There is a need to develop solutions that support the needs of both patients and interprofessional clinicians

TABLE 8.1
Opportunities for Patient Engagement in Safety

	Description	Examples	Intervention Types	Patient Ergonomics Considerations
Prevention	Infection control	Carrying out safe, effective, sanitary wound care and medication adherence following injury or surgery	Patient education and continued support with user-centered framing and placement	Message framing and placement
	Fall risk management	Engaging in fall prevention practices	Patient education on risk factors and mitigation strategies	Design of tools and technologies for fall prevention and notification
Detection	Self-monitoring for abnormalities	Breast cancer, testicular cancer, skin cancer, unusual signs and symptoms	Patient education with focus on perceptual cues; patient decision aids	Identification and judgment of perceptual cues
	Self-monitoring for infection	Relevant to immunocompromised populations; patients with wounds, surgical sites, etc.	Safety reminders such as text messages or email signatures with signs of infection; nurse check-ins	Identification and judgment of perceptual cues
	Self-monitoring for dehydration	Patients with nausea and vomiting due to illness, pregnancy, or chemotherapy	Safety reminders; clear procedures on who/how to contact	Identification and judgment of physical cues

(Continued)

TABLE 8.1 (*Continued*)
Opportunities for Patient Engagement in Safety

	Description	Examples	Intervention Types	Patient Ergonomics Considerations
Self-Management	Wound and surgical site care	Hand washing, sterile procedures	Patient education and knowledge assessment using training principles at discharge; decision support for assessment	Application of training principles to patient education including knowledge assessment, hands-on practice of perceptual tasks
Self-data collection	Observing, gathering, documenting, & understanding data on self	Blood glucose levels, blood pressure, trends and patters	Manual data collection strategies; consumer mobile health technologies	Characterizing acquisition and state of patient knowledge and mental models
Communication	Asking questions; delivering information	Receiving adequate information regarding cancer treatment options	Patient-tailored support for patient-clinician encounters (e.g., setting agenda, what to tell and ask clinician, speaking up regarding concerns)	Common grounding; mental models
Reporting	Safety event reporting	Medication error in inpatient setting	User-centered processes and technologies for reporting for inpatient and outpatient settings	Design of user-centered tools
Engaging in research	Engaging patients as citizen scientists in patient safety research	Identifying research gaps and questions, serving as patient co-investigators, analyzing data, co-producing interventions	N/A	N/A
	Engaging patients in co-production of safety interventions	Interventions co-produced with patients and caregivers	N/A	N/A

managing and integrating data, information, and knowledge in the service of safe treatment delivery.

8.5 RECOMMENDATIONS

How can patients be engaged and supported in their engagement in safety? Patients should be informed that they are in a position to engage in safety and that there is both a need and opportunity to do so. Patients should be educated on risks, self-assessment, and self-monitoring strategies with evidence-based information. Safety interventions, including technologies, must serve to enhance and augment the patient–clinician relationship, not replace it. Table 8.1 provides a further list of opportunities for patient engagement in safety through prevention, detection, self-management, and other activities, along with examples, intervention types, and considerations for the science of patient ergonomics.

Traditional models of medical cognition do not recognize the mutual dependency between patients and clinicians. However, the provision of appropriate and safe care depends upon the patient cognitive work associated with managing information. If a patient delays a workup on a suspicious lump, misses an infection, or discontinues a treatment regimen, there is an implication for their safety. Key to empowering and engaging patients and supporting patient–clinician teaming is ensuring that patients and their informal caregivers understand how and when they can support safety and providing them with opportunities and tools to do so. After all, the patient is the one that has the most to lose—they are the ultimate stakeholder in patient safety.

REFERENCES

AHRQ. (2019). *Measurement of patient safety | AHRQ patient safety network*. Retrieved from https://psnet.ahrq.gov/primers/primer/35/Measurement-of-Patient-Safety

Anders, S. (2016). *Nonroutine events in ambulatory surgery: Comparison of analysis by clinicians and patients*. International Symposium on Human Factors and Ergonomics in Health Care, San Diego, CA.

Armitage, G., Moore, S., Reynolds, C., Laloë, P.-A., Coulson, C., McEachan, R., . . . O'Hara, J. (2018). Patient-reported safety incidents as a new source of patient safety data: An exploratory comparative study in an acute hospital in England. *Journal of Health Services Research & Policy*, *23*(1), 36–43.

Bell, S. K., Mejilla, R., Anselmo, M., Darer, J. D., Elmore, J. G., Leveille, S., . . . Walker, J. (2017). When doctors share visit notes with patients: A study of patient and doctor perceptions of documentation errors, safety opportunities and the patient–doctor relationship. *The BMJ Quality & Safety*, *26*(4), 262–270.

Berger, Z., Flickinger, T. E., Pfoh, E., Martinez, K. A., & Dy, S. M. (2014). Promoting engagement by patients and families to reduce adverse events in acute care settings: A systematic review. *The BMJ Quality & Safety*, *23*(7), 548–555.

Bodenheimer, T., Lorig, K., Holman, H., & Grumbach, K. (2002). Patient self-management of chronic disease in primary care. *Journal of the American Medical Association, 288*(19), 2469–2475.

Boggan, J. (2019). Top 2018 articles on patient and family engagement – BMJ Quality & Safety. *Blog: The BMJ Quality & Safety*. https://blogs.bmj.com/qualitysafety/2019/05/08/top-2018-articles-on-patient-and-family-engagement/

Carman, K. L., Dardess, P., Maurer, M., Sofaer, S., Adams, K., Bechtel, C., & Sweeney, J. (2013). Patient and family engagement: A framework for understanding the elements and developing interventions and policies. *Health Affairs*, *32*(2), 223–231.

Cleary, R. K., Moroz, S., Tippey, K. G., Xu, J., Slagle, J., Weinger, M. B., & Kachnic, L. A. (2017). Evaluating the use of a novel patient-reported outcomes measure in cancer care: A pilot study in patients receiving radiation therapy. *International Journal of Radiation Oncology*, *99*(2), E549–E550.

Coulter, A. (2011). *Engaging patients in healthcare*. McGraw-Hill Education (UK).

deBronkart, D. (2013). How the e-patient community helped save my life: An essay by Dave deBronkart. *The BMJ*, *346*.

Delbanco, T., Walker, J., Bell, S. K., Darer, J. D., Elmore, J. G., Farag, N., . . . Leveille, S. G. (2012). Inviting patients to read their doctors' notes: A quasi-experimental study and a look ahead. *Annals of Internal Medicine*, *157*(7), 461–470.

Estabrooks, P. A., Boyle, M., Emmons, K. M., Glasgow, R. E., Hesse, B. W., Kaplan, R. M., . . . Taylor, M. V. (2012). Harmonized patient-reported data elements in the electronic health record: Supporting meaningful use by primary care action on health behaviors and key psychosocial factors. *Journal of the American Medical Informatics Association*, *19*(4), 575–582.

Friedman, S. M., Provan, D., Moore, S., & Hanneman, K. (2008). Errors, near misses and adverse events in the emergency department: What can patients tell us? *Cjem*, *10*(05), 421–427.

Griffin, T. (2010). Bringing change-of-shift report to the bedside: A patient-and family-centered approach. *The Journal of Perinatal & Neonatal Nursing*, *24*(4), 348–353.

Haskell, H. (2018). Improving patient safety? Ask the patient. *The BMJ*. Retrieved from https://blogs.bmj.com/bmj/2018/12/06/helen-haskell-improving-patient-safety-ask-the-patient/

Hazlehurst, B., Gorman, P. N., & McMullen, C. K. (2008). Distributed cognition: An alternative model of cognition for medical informatics. *International Journal of Medical Informatics*, *77*(4), 226–234.

HealthIT.gov. (2018). HealthIT.Gov | the official site for health IT information. Retrieved from https://www.healthit.gov/

Henry, E., Silva, A., Tarlov, E., Czerlanis, C., Bernard, M., Chauhan, C., . . . Stewart, G. (2016). Delivering coordinated cancer care by building transactive memory in a team of teams. *Journal of Oncology Practice*, *12*(11), 992–999.

Holden, R. J., Cornet, V. P., & Valdez, R. S. (2020). Patient ergonomics: 10-year mapping review of patient-centered human factors. *Applied Ergonomics*, *82*, 102972.

Holman, H., & Lorig, K. (2004). Patient self-management: A key to effectiveness and efficiency in care of chronic disease. *Public Health Reports*, *119*(3), 239–243.

Holt, J. M., Cusatis, R., Asan, O., Williams, J., Nukuna, S., Flynn, K. E., . . . Crotty, B. H. (2019). Incorporating patient-generated contextual data into care: Clinician perspectives using the consolidated framework for implementation science. *Healthcare*, 100369.

Househ, M., Borycki, E., & Kushniruk, A. (2014). Empowering patients through social media: The benefits and challenges. *Health Informatics Journal*, *20*(1), 50–58.

Hutchins, E. (1995). How a cockpit remembers its speeds. *Cognitive Science*, *19*(3), 265–288.

Kelly, M. M., Coller, R. J., & Hoonakker, P. L. T. (2018). Inpatient portals for hospitalized patients and caregivers: A systematic review. *Journal of Hospital Medicine*, *13*(6), 405–412.

Khan, A. (2018). Ensure that the family's voice is heard first and last, and in their own words. *The BMJ*. Retrieved from https://blogs.bmj.com/bmj/2018/12/06/in-clinical-practice-i-ensure-that-the-familys-voice-is-heard-first-and-last-and-in-their-own-words/

Khan, A., Coffey, M., Litterer, K. P., Baird, J. D., Furtak, S. L., Garcia, B. M., . . . Yu, C. E. (2017). Families as partners in hospital error and adverse event surveillance. *Journal of the American Medical Association Pediatrics*.

Khan, A., Furtak, S. L., Melvin, P., Rogers, J. E., Schuster, M. A., & Landrigan, C. P. (2016). Parent-reported errors and adverse events in hospitalized children. *Journal of the American Medical Association Pediatrics, 170*(4), e154608–e154608.

Khan, A., Spector, N. D., Baird, J. D., Ashland, M., Starmer, A. J., Rosenbluth, G., . . . Landrigan, C. P. (2018). Patient safety after implementation of a coproduced family centered communication programme: Multicenter before and after intervention study. *The BMJ, 363*, k4764.

Klein, G. (2008). Naturalistic decision making. *Human Factors: The Journal of the Human Factors and Ergonomics Society, 50*(3), 456–460.

Klein, G., Ross, K. G., Moon, B. M., Klein, D. E., Hoffman, R. R., & Hollnagel, E. (2003). Macrocognition. *Institute of Electrical and Electronics Engineers Intelligent Systems, 18*(3), 81–85.

Kohn, L. T., Corrigan, J. M., & Donaldson, M. S. (1999). *To err is human: Building a safer health system*. National Academies Press.

Kushniruk, A. (2019). The importance of health information on the internet: How it saved my life and how it can save yours. *Journal of Medical Internet Research, 21*(10), e16690.

Leape, L. L. (2004). Human factors meets health care: The ultimate challenge. *Ergonomics in Design, 12*(3), 6–12.

Lippa, K. D., Feufel, M. A., Robinson, F. E., & Shalin, V. L. (2017). Navigating the decision space: Shared medical decision making as distributed cognition. *Qualitative Health Research, 27*(7), 1035–1048.

Lippa, K. D., & Klein, H. A. (2008). Portraits of patient cognition: How patients understand diabetes self-care. *Canadian Journal of Nursing Research Archive, 40*(3).

Lippa, K. D., Klein, H. A., & Shalin, V. L. (2008). Everyday expertise: Cognitive demands in diabetes self-management. *Human Factors, 50*(1), 112–120.

Lippa, K. D., & Shalin, V. L. (2015). Stepping up to the blackboard: Distributed cognition in doctor-patient interactions. *Cognitive Science*.

Lippa, K. D., & Shalin, V. L. (2016a). Creating a common trajectory: Shared decision making and distributed cognition in medical consultations. *Patient Experience Journal, 3*(2), 73.

Lippa, K. D., & Shalin, V. L. (2016b). Distributed cognition in the past progressive: Narratives as representational tools for clinical reasoning. *Cognitive Science*.

Mackay, E. A. (2015). Patients, consumers, and caregivers: The original data stewards. *Journal for Electronic Health Data and Methods, 3*(1).

Makary, M. A., & Daniel, M. (2016). Medical error—The third leading cause of death in the US. *The BMJ, 353*, i2139.

Marks, M. A., Mathieu, J. E., & Zaccaro, S. J. (2001). A temporally based framework and taxonomy of team processes. *Academy of Management Review, 26*(3), 356–376.

Meppelink, C. S., Smit, E. G., Fransen, M. L., & Diviani, N. (2019). "I was right about vaccination": Confirmation bias and health literacy in online health information seeking. *Journal of Health Communication, 24*(2), 129–140.

Mickelson, R., & Holden, R. (2013). Assessing the distributed nature of home-based heart failure medication management in older adults. *Proceedings of the Human Factors and Ergonomics Society Annual Meeting, 57*, 753–757.

Papautsky, E. L. (2019). Piecing the patient story back together: Why the patient and caregiver contribution matters. *Journal of Patient Experience, 7*, 151–154.

Papautsky, E. L., Abdulbaseer, U., & Faiola, A. (2017). What is the role of patient families? An exploratory study in a medical intensive care unit. *Proceedings of the Human Factors and Ergonomics Society Annual Meeting, 61*(1), 550–554.

Papautsky, E. L., Holden, R. J., Valdez, R. S., Gruss, V., Panzer, J., & Perry, S. J. (2019). The patient in patient safety: Clinicians' experiences engaging patients as partners in safety. *Proceedings of the International Symposium on Human Factors and Ergonomics in Health Care, 8*, 265–269.

Pew Research Center. (2019). *Demographics of social media users and adoption in the United States*. Pew Research Center: Internet, Science & Tech. Retrieved from https://www.pewresearch.org/internet/fact-sheet/social-media/

Pierret, J. (2003). The illness experience: State of knowledge and perspectives for research. *Sociology of Health & Illness*, *25*(3), 4–22.

Riggare, S. (2018). E-patients hold key to the future of healthcare. *The BMJ*, *360*, k846.

Schnock, K. O., Snyder, J. E., Fuller, T. E., Duckworth, M., Grant, M., Yoon, C., . . . Dykes, P. C. (2019). Acute care patient portal intervention: Portal use and patient activation. *Journal of Medical Internet Research*, *21*(7), e13336.

Sharma, A. E., Rivadeneira, N. A., Barr-Walker, J., Stern, R. J., Johnson, A. K., & Sarkar, U. (2018). Patient engagement in health care safety: An overview of mixed-quality evidence. *Health Affairs*, *37*(11), 1813–1820.

Tattersall, R. (2002). The expert patient: A new approach to chronic disease management for the twenty-first century. *Clinical Medicine*, *2*(3), 227–229.

The Joint Commission. (2008). *National Patient Safety Goals*. Retrieved from http://psnet.ahrq.gov/issue/national-patient-safety-goals

The World Health Organization. (2007). Communication during patient hand-overs. *The Nine Patient Safety Solutions*. Retrieved from http://www.who.int/patientsafety/events/07/02_05_2007/en/

Tippey, K. G., Slagle, J. M., Cleary, R. K., Friedman, D. L., Kachnic, L. A., Shotwell, M. S., . . . Weinger, M. B. (2018). Evaluating the use of patient-reported non-routine events in pediatric and radiation oncology: A pilot study. *Proceedings of the Human Factors and Ergonomics Society Annual Meeting*, *62*(1), 538–542.

Troy, L., Pete, F., Jason, S., Anders, S., & Gagandeep, J. (2016). Patient and clinician reported non-routine events during periprocedural cardiac catheterization. *Circulation: Cardiovascular Quality and Outcomes*, *9*(suppl_2), A226–A226.

Weiner, S. J., & Schwartz, A. (2016). Contextual errors in medical decision making: Overlooked and understudied. *Academic Medicine*, *91*(5), 657–662.

Weiner, S. J., Schwartz, A., Weaver, F., Goldberg, J., Yudkowsky, R., Sharma, G., . . . Persell, S. D. (2010). Contextual errors and failures in individualizing patient care: A multicenter study. *Annals of Internal Medicine*, *153*(2), 69–75.

Weingart, S. N., Pagovich, O., Sands, D. Z., Li, J. M., Aronson, M. D., Davis, R. B., . . . Phillips, R. S. (2005). What can hospitalized patients tell us about adverse events? Learning from patient-reported incidents. *Journal of General Internal Medicine*, *20*(9), 830–836.

Weissman, J. S., Schneider, E. C., Weingart, S. N., Epstein, A. M., David-Kasdan, J., Feibelmann, S., . . . Gatsonis, C. (2008). Comparing patient-reported hospital adverse events with medical record review: Do patients know something that hospitals do not? *Annals of Internal Medicine*, *149*(2), 100–108.

Wright, M. (2019). *TEDxBoise: Partnering to beat sepsis*. Retrieved from https://www.youtube.com/watch?v=J_e77rsuqWU&t=64s.

Xie, A., Carayon, P., Cartmill, R., Li, Y., Cox, E. D., Plotkin, J. A., & Kelly, M. M. (2015). Multi-stakeholder collaboration in the redesign of family-centered rounds process. *Applied Ergonomics*, *46PA*, 115–123.

Young, C. (2013). Community management that works: How to build and sustain a thriving online health community. *Journal of Medical Internet Research*, *15*(6), e119.

Zhang, J., & Patel, V. L. (2006). Distributed cognition, representation, and affordance. *Pragmatics & Cognition*, *14*(2), 333–34.

Section IV

Patient Ergonomics Methods

9 Field Methods for Patient Ergonomics

Interviews, Focus Groups, Surveys, and Observations

Kathleen Yin and Annie Y. S. Lau
Australian Institute of Health Innovation,
Macquarie University

CONTENTS

As the name patient ergonomics suggests, the understanding of health-related work as it is conducted by the patient is the cornerstone of this discipline. Field methods are used to answer research questions that revolve around patients' subjective experiences, their health-related behaviors, and contextual factors influencing these behaviors. Field methods are primarily used to comprehend patient attitudes, perspectives, experiences, and values. Multi-method research is also becoming more popular in projects that seek to holistically understand complicated and multilayered health phenomena (Holden et al., 2015; Kelly, 2010). In this chapter, we examine how field methods—namely interviews, focus groups, surveys, and observations—have been utilized in patient ergonomics, their appropriateness and implications for different situations, and how they can be used to triangulate results to enrich the findings of a study.

9.1 INTERVIEWS

Interviewing is one of the principal field methods, ideal for studying past experiences, beliefs, and thoughts of participants that are difficult to observe. Within patient ergonomics, interviews have been used in many situations, ranging from documenting how patients form new routines (Hammarlund et al., 2017) to how patients harbor complex thoughts and feelings toward self-care (Holden et al., 2015). Interviews should at least be audio recorded, with video recordings providing further context such as body language, facial gestures, and how people behave and interact with contextual barriers (for more on nonverbal data, see also Chapter 6 in this volume).

Depending on the research question, interviews are generally divided into the following categories:

- **Structured interviews**: Researchers ask every participant identical questions in the same sequence. This method is good for projects with a clear goal, trying to understand one specific experience from participants (e.g., using a specific website, visiting a specific clinic). The answers from structured interviews are precise (such as "the belt was not wide enough" in the study by Ehmen et al. [2012]) and contain actionable feedback, which also makes data easy to analyze. As answers for these interviews often lack depth, they are not ideal for understanding complicated phenomena (e.g., reasons behind noncompliance).
- **Semi-structured interviews:** Researchers create a list of questions, but the exact wording and order of the questions can change depending on the flow of the interview and the participant's previous answers. A popular choice in patient ergonomics, this method is good for projects that have a theoretical background but are also looking to explore new concepts (e.g., understanding the contextual factors influencing the quality of care received by patients from one clinic). This method provides a level of flexibility in its questions and gives participants room to elaborate on their answers and has therefore been adopted for many patient ergonomics domains, such as studying how patients create self-care routines amid contextual barriers and facilitators (Bukhave & Huniche, 2014) and identifying contextual barriers that reduced self-management efficiency and increased risk (Dehghanzadeh et al., 2017). This method, however, needs good sampling techniques to provide range in the participant population, and the researcher needs to remain consistent in their questioning style between participants.
- **Open-ended interviews**: Researchers pose very few (typically 1–3) questions, and the interview flows as a natural conversation instead of a question-and-answer format (Murphy et al., 1998). This methodology is good for projects that are exploratory (Catlin et al., 2016) or highly individual-focused (e.g., exploring the lived experience of patients managing an inherited rare disease). These interviews give participants maximum freedom in their answers, can assess how a participant interprets the question, and provide rich personal data that explore the individual's perception of experience. Since answers can vary significantly between participants, making data difficult to analyze and the outcomes of the research highly dependent on the interpersonal skill of the researcher, this is a rather uncommon method.

For best practices, interview questions should be planned in consultation with existing models (Holden et al., 2015) and piloted in a few participants before being finalized. The appropriate sampling strategy (often purposive sampling, selecting participants according to predefined criteria to represent the desired population) should be used and sufficient time and resources budgeted for recruitment, conducting interviews, traveling, recording, audio transcription, and data analysis. For open-ended interviews, researchers also need to establish rapport with participants in the early part of (or prior to) the interview and be able to ask the appropriate follow-up questions at the right time during the conversation. Researchers are recommended to take field notes during and after each interview to provide a contextual understanding of the participant and the conversation. While interviews provide rich personal data, the information is purely self-reported and often recited from memory. Triangulating interviews with objective data sources such as observations can validate findings and improve relevance, benefits that we illustrate in our multi-method case study later in the chapter.

Thematic analysis, where information in the interview transcript is sorted into themes that emerge through repeated reading, is the most common analysis method for interview data in patient ergonomics. The themes could be derived either deductively or inductively, with the former analyzing data according to themes within existing models and the latter creating themes from the ground up as the data are being analyzed. A combined deductive–inductive approach is a common hybrid in patient ergonomics, where the researcher deductively fits the data into themes within an existing patient self-management framework and inductively creates new themes from data that did not fit into the framework. Other research approaches used in patient ergonomics include grounded theory, which studies and constructs new theories on a novel issue based on the gathered field data (Holden et al., 2017), and phenomenology, which investigates how participants subjectively experience and perceive their reality (Hammarlund et al., 2017). Another approach to interview data analysis is using hermeneutic analysis (i.e., analyzing the wording, grammar, and interpretation of the transcript) to understand the participant's underpinning values and beliefs. All the above qualitative analysis methods can be applied to interviews as well as to focus groups, which is the topic of the next section.

9.2 FOCUS GROUPS

Focus groups are a subset of interviews, best for studying the experience or behaviors of a specific group (e.g., patients currently using a home dialysis machine). A focus group session is facilitated by at least one researcher with a small group of participants who share a common trait of interest. The researcher would produce questions encouraging group discussion, record the resultant conversation between participants, and analyze the discussion content as well as the group interaction and dynamics.

Focus groups are more time-efficient than interviews and allow better access to the shared meaning, experiences, vocabulary, and values in a collective (Murphy et al., 1998). In comparison, interviews offer better opportunities to hear individual stories, personal viewpoints, and provide privacy for participants. In a focus group setting, researchers can also study how participants reach consensus (or not!) on the discussion topic, with the participants "bouncing ideas off each other," revealing different views within a group that may appear homogenous to the outside (Pauling et al., 2018). Focus groups may also be used in exploratory studies (Bowling et al., 2017), or examine the differences between participants at various study sites across multiple cities or countries (Czuber-Dochan et al., 2012). This method, however, could be difficult to use for sensitive health topics (e.g., sexual health), and the quality of the data is highly dependent on the facilitator's skills in managing a group and driving a discussion. Recruitment for focus groups can often be assisted by patient associations, many of which have existing patient support group meetings that can accommodate a few sessions of researcher-facilitated focus groups. Such practices can lead to convenience sampling (Au et al., 2014; Czuber-Dochan et al., 2012), where participants are recruited simply on the basis that they were convenient to reach, potentially introducing bias.

Best practices for using the focus group methodology include all the recommendations given for interviews and some extra considerations with recruitment. Researchers need to decide the composition of participants in each focus group and the total number of groups. Attending the focus groups could also be an issue for some participants, and researchers may need to over-recruit if there are potential issues with no-shows, especially for participants whose symptoms may suddenly deteriorate (e.g., those with mental health conditions, motor neuron diseases). To improve data quality and the flow of the conversation, it is vital for the group to dedicate time at the beginning of the session to build rapport among the participants and with the facilitator. There should also be more than one facilitator per group, with one driving discussion and the other assisting, observing behavior, and taking notes.

Questions used in focus groups deserve special mention, as these questions should encourage group discussion and allow participants to respond with different opinions. The questions should be developed with care, primarily using a small number of open-ended questions that allow detailed discussion between participants. For this reason, focus groups need to be kept small, with each group no bigger than 10 people—otherwise the group will become too large to ensure everyone has an equal chance to speak. After a round of discussions, the researcher could also probe participants using short questions such as "why would you say that?" "could you explain that further?" to elicit richer responses relevant to the research aims. Other question formats can also be used to increase participation, such as asking each participant to introduce themselves and their viewpoints, using rankings or multiple-choice questions, or asking participants to draw their experiences or perceptions on paper. Due to the large variation between the questions asked in different projects, studies conducted via interviews and focus groups are recommended to be reported using standard guidelines, such as the Consolidated Criteria for Reporting Qualitative Research (Tong et al., 2007).

9.3 QUALITATIVE AND QUANTITATIVE SURVEYS

Surveys have been used in patient ergonomics to gather self-reported data such as perceptions of self-management, effect of chronic illness on daily life, or feedback on mobile devices (Ehmen et al., 2012; Holden et al., 2017; Kawi, 2014). Paper surveys are completed face-to-face or mailed out, whereas digital surveys are now deployed online, supporting fast data collection. Checklists such as CHERRIES (Checklist for Reporting Results of Internet E-Surveys) are good starting points for constructing and reporting online surveys (Eysenbach, 2004) and should be consulted when the researcher is planning such studies for the first time.

Surveys are flexible because both qualitative and quantitative data may be obtained from one survey. Qualitative, open-ended survey questions allow participants to write free text as answers (Kawi, 2014), whereas quantitative questions translate behaviors and feelings into numeric measures, using scales, rankings, or ratings (see Chapter 11 in this volume). For qualitative questions, descriptive statistics may be generated using content analysis (e.g., counting the prevalence of a characteristic) to show the features of the sampled population. Quantitative questions, on the other hand, may enable inferential statistics, unveiling statistical differences between groups (Roberts et al., 2017) or finding associations between clinical outcomes and specific behavior patterns.

Compared to interviews, surveys are less expensive to implement, the results could be analyzed faster (especially quantitative ones), and there are fewer logistics (especially for online surveys). Surveys can also be completely anonymous, making it a confidential option in health research. While surveys can yield similar information to interviews, they are more useful for obtaining specific details and examples and concrete feedback, such as what medications people took and what kind of exercise they conducted (Kawi, 2014), instead of the rich qualitative data offered by interviews. Surveys have therefore been used as the preliminary step to test the general experience of the population, before researchers formulate detailed interview questions. Alternatively, interviews can be used prior to surveys, gathering insight which can be used to develop and direct a later survey.

One of the biggest considerations when planning surveys is the variable response rate between different recruitment methods, such as incentives versus no incentives, advance notice versus no advance notice, long questions versus short questions (Asch et al., 1998; Jepson et al., 2005; Kaplowitz et al., 2004). For mailed surveys, the response rate may be much lower than expected (Asch et al., 1998). For best practices, a realistic response rate should be planned to allow for appropriate recruitment.

9.4 OBSERVATIONS

Observations have the capacity to record behaviors and processes in context, able to examine the setting and meaning of actions as they are performed in their native environment—for example, within the patient's home and private lives. Observations can record patient work as conducted in the patient's daily life, within the context of home, employment, and other locations patients move through in life. Patient ergonomics researchers can also play a more active role during the observation period, asking participants why they acted in a certain way at the time of the action, gathering psychological as well as physical context to holistically understand work as done. Ward et al. (2010) observed patients in their homes for how medication packaging affected patient safety, recording how patients opened the packaging and whether they needed any assistance. This method is best used to assess health activities carried out and to investigate the processes of patient work. Observations can also note actions, habits, or contexts that participants may not be able to recall in an interview setting.

Data gathered from observations can include participants' body language, facial expression, and other nonverbal indications of thoughts and feelings. Similarly, it is always beneficial to have a dedicated researcher acting as the observer when conducting other methods (e.g., interviews or focus groups), who can observe and note participant behavior during the session. If interviews occurred in the participant's home, the observer should also note the physical environment the participant lives in and clues to patient work tasks the patient may be involved in but did not mention verbally (e.g., inferred from the presence of an asthma inhaler or eye-drop bottle in the house).

Researchers could carry out observations physically, shadowing participants for extended periods of time. Technical advances, such as video recordings and wearable cameras, have enabled longer and more unobtrusive observations (see Chapter 12 in this volume). Observational data are exceedingly rich, providing a wealth of information about the environment and work as it is done. A significant barrier for this methodology is participant acceptance of the observational method—whereas some participants might be happier wearing a camera for hours, some others are more comfortable with a shorter duration of a human observer shadowing them. The benefit to the participant should be carefully weighed against the logistics of the observation. The consent of other people who may be observed with the participant, such as family and friends, may also be required before the observation can be undertaken. The presence of a researcher may also trigger the observer effect (McCambridge et al., 2014), wherein participants behave differently compared to normal due to the knowledge that they are being observed.

Observational data (e.g., recorded footage) could undergo quantitative content analysis or qualitative inductive analysis. For example, the frequency of distractions or interruptions in health behavior could be counted and descriptive statistics derived from one reading of the footage (Ward et al., 2010), whereas a second reading would analyze the contextual influences, barriers, and facilitators to such interruptions. Observations yield highly holistic, interpretive, and rich data unrivaled by other field methods, especially with ethnographic observational studies where the researcher participates and observes a community for a long period of time to understand how the group members act and why they act in such a way.

9.5 CASE STUDY

9.5.1 A Mixed-method, Multi-method Study

Our ongoing mixed-method study (Figure 9.1) provides an example of integrating various field methods to investigate the patient ergonomics of community-dwelling individuals living with Type 2 diabetes, having at least one chronic comorbidity, and conducting self-care. The study aimed to explore the work tasks conducted by the participants and analyze the contextual factors influencing the tasks. More details on this methodology are available in Table 9.1 and elsewhere (Yin et al., 2018).

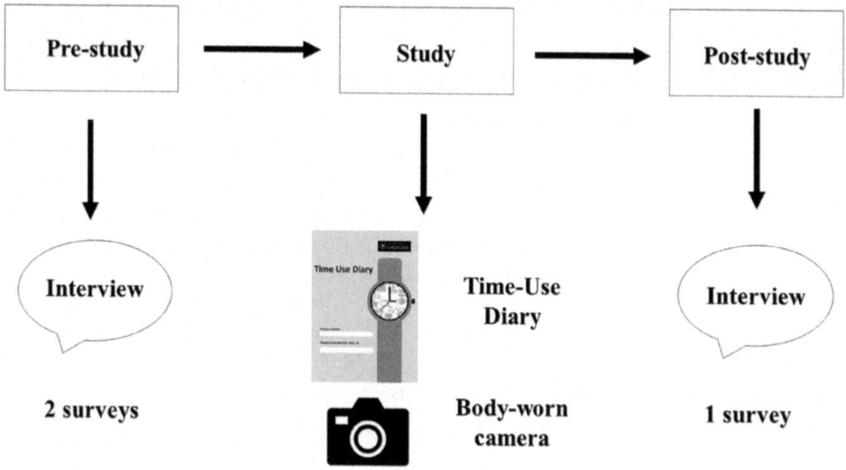

FIGURE 9.1 The study employed a combination of field methods, such as semi-structure interviews, observations via body-worn cameras, surveys, and time-use diaries.

The rationale for a multi-method methodology was the desire to explore how multiple layers of patient ergonomics could be investigated from one information collection session. We wanted to capture how patient ergonomics functions in the real world, with chronic illness and comorbidities. In particular, Type 2 diabetes is an incurable chronic disease that can cause various complications such as retinopathy and often copresents with cardiovascular diseases and dyslipidemia (Khunti et al., 2018), making Type 2 diabetes self-care full of interconnected self-care tasks that serve a variety of different purposes (e.g., taking medication and blood glucose monitoring). We also wanted a way to observe the patient without adding to the patient's stress, as well as reducing the person-hours associated with having researchers physically shadowing participants. Lastly, we were interested in the potential discrepancy between self-reported activity and activity recorded by a body-worn camera, and whether this could be resolved using patient feedback during interviews.

The Time-Use Diary used in this study is a part of the Harmonised European Time Use Surveys, developed by the European Statistical Office (Eurostat), a European Union organization that harmonizes methods used in member states and provides statistics regarding residents with the European Union. The Time-Use Diary (Eurostat, 2009) is a diary divided into 10-minute blocks over 24 hours, and participants can record a maximum of 2 activities conducted concurrently. The Time-Use Diary functions as a self-reported activity survey over one day, which we used to triangulate with activity data recorded by the body-worn camera.

Twenty-six participants were recruited from endocrinology specialist clinics. The selection criteria were that the participants should have been diagnosed with Type 2 diabetes, have at least one comorbidity, speak fluent English, and be able to give consent. Participants in health studies can regard video recording in research as acceptable and worthwhile, as long as the associated risk and purpose are explained fully and specific concerns raised by the participants are well managed (Parry et al., 2016).

TABLE 9.1
Methods Used in a Mixed-method Study to Assess Patient Work

Method	Aspect of Patient Work Studied	Why Method Was Chosen
Face-to-face semi-structured interview	• Health history, experiences with health conditions, attitude and beliefs about health and self-care, barriers and facilitators of self-care.	• To understand participants' self-reported routines, health history, experiences and contextual influences that may not be captured in observations. • To understand the motivation behind participant's self-care behavior.
Quantitative survey (Multimorbidity Illness Perceptions Scale [Gibbons et al., 2013])	• Demographic data. • Attitudes and perceptions of their illnesses.	• To provide statistics regarding subjective reporting of treatment burden.
Digital observation (body-worn camera, model Edesix VB-300)	• Self-care behavior, daily routines, contextual factors influencing patient work.	• To silently record the participant's waking hours during the study day.
Time-Use Diary (Eurostat, 2009)	• Self-care behavior and processes. • Contextual factors influencing patient work.	• To triangulate the finding of the body-worn camera. • To compare differences between self-reported routine and observed routine. • Backup data capture if patient turned off camera.
Qualitative survey	• Feedback and study experience.	• To obtain feedback on the study experience.
Field notes (prepared by researchers postinterview)	• Physical contextual factors. • Patient body language during interviews.	• To capture researchers' insights and impressions of the participant, physical surroundings and the interview.

To further address ethical concerns regarding privacy, patients were advised to speak to family members before agreeing to participate, all recorded human faces and identifying information were blurred, and the body-worn camera recorded silently. Participants were explicitly told the camera could be removed or turned off at any time without having to provide an explanation and were given a laminated card explaining the purpose of the study that could be shown to others. The camera footage was downloaded and converted to images at the end of the study, and all images were viewed by the participant (where they could delete any image) before researchers had access. Our protocol paper provides more detailed information on the ethical concerns and how we addressed them (Yin et al., 2018). For a very helpful list of recommendations of what to do when conducting video recording in healthcare research, please see the article by Parry et al. (2016).

All semi-structured interviews were audio recorded, transcribed, and then analyzed via thematic analysis. The body-worn camera screenshots were analyzed to provide a detailed portrait of the patient's journey through time, the amount of time they spent doing each task, and contextual factors that acted as barriers or facilitators.

9.5.2 The Interplay of Different Field Methods

The different parts of the methodology triangulated and complemented each other successfully, painting a comprehensive and holistic profile of each participant's patient work. The prestudy interview established the participant's health history, past health experiences, health beliefs, and contextual barriers and facilitators acting on the participant's self-care. The cameras provided objective observation with minimal disruptions. All but one participant in our cohort of 26 reported they soon forgot the camera was even there. Participants recorded most of their activities on the paper Time-Use Diary and noted information for time periods when they removed the camera or when images were unclear. Only 3 of the 26 participants did not complete the entirety of the Time-Use Diary during the study day and completed it at the poststudy interview. Conversely, the body-worn camera screenshots provided context to the actions written in the Time-Use Diary, which were sometimes brief or abbreviated.

For data analysis, we were able to directly compare the activities recorded by the camera to those self-reported in the Time-Use Diary. The interview transcripts were thematically analyzed to uncover contextual influencers of self-care, and participant perceptions of their comorbidities were assessed via the quantitative survey. All data were collected from only one information collection session per participant, minimizing the logistic burdens of the researchers and the participants.

9.5.3 Challenges of the Methodology

Combining a multitude of different technologies presented its own challenges, and there were many things we wish we had known before. The most unexpected was that our methodology formed a barrier to recruitment. More females declined to participate than males, with many women citing privacy and security concerns. Two male potential participants also declined on their wives' recommendations. This led to a skewed gender representation: of the 26 participants, only 10 were female.

Our methodology also resulted in recruiting a specific population of patients with Type 2 diabetes—those who were not too healthy nor too sick, not having to go to work every weekday, and not looking after someone else as a caregiver. Patients who were too sick, or acting as a caregiver to others, were unable to participate due to their frailty or the amount of health-related work they had to conduct. Conversely, patients who were very healthy were more likely to hold down a full-time job, thus not willing to wear the body-worn camera due to privacy concerns. Consequently, our population were mostly older, retired diabetics who have their conditions under moderate control.

The complicated instructions for participants also posed a challenge. Numerous participants had difficulties understanding the operation of the camera, which was made worse by the discreet appearance of the camera and the limited digital literacy of the older participants. Some participants also struggled to fill in the Time-Use Diary. The diary recorded a prospective 24-hour period. However, the printed

Time-Use Diary started at 4 am and participants felt compelled to start recording at 4 am. In response, the researchers took pictures of the camera at different stages of operation and made a visual aid, and made a large sign large sign on the Time-Use Diary to remind the participant they should only start recording in the diary from the end of the prestudy interview.

Technically, we had concerns with how to present up to 16 hours of video footage during the poststudy interview. We overcame this difficulty by making an in-house software, which transformed the video into a series of screenshots taken every 10 seconds. This improved the viewing experience of the footage immensely, allowing the researchers and participants to view all the footage within an hour.

BOX 9.1 RECOMMENDATIONS

Multi-method research: Allows projects to examine multiple facets of patient ergonomics at once. Researchers need to decide whether they wish for the methods to occur one after another (with data from the first method helping the development of the later methods), or for the methods to occur during the same session (where each method would need to examine a different aspect of patient ergonomics). All methods used need to have questions that closely adhere to the research aims, especially for interviews and surveys, where each method should provide complementary information. Once experimental design has been finalized, it is imperative that the entire multi-method design is pilot tested on a few participants (usually no more than 3) to assess feasibility. Any feedback from the research team and participants should be taken seriously at the pilot stage, and the experimental design changed accordingly. Data from the pilot stage could be either included or excluded from the final analysis (this is dependent on how much the experiment changed after pilot testing).

Interviews: Use to obtain individual experiences and beliefs. Offers rich subjective data. Plan thoroughly for logistics (e.g., transport), pilot test interview questions, keep field notes, and develop the questions based on the type of analysis the transcript will undergo.

Focus groups: Use to obtain collective experiences and beliefs. Offers rich subjective data. Plan in a similar way to interviews; however, be wary of convenience or snowball sampling and make notes on the group dynamics during the group discussion.

Surveys: Use to obtain quantitative or qualitative data. Enables analytical statistics. Plan for a low response rate and understand the qualitative data would not be as rich as that yielded in interviews.

Observations: Use to obtain quantitative or qualitative data that is subjective and visible to others. Observation recording and notes reanalyzed via content analysis may yield different kinds of data. Plan extensively for logistics and privacy concerns, consider the influence of the observer's effect, and explore novel technologies that allow remote and discrete observations.

Despite such challenges in our methodology, the data analysis was a success in triangulating camera images and activities recorded in the Time-Use Diary, with the camera's time stamp feature playing a critical role in ensuring the two data sources matched. Moreover, the team transposed all camera data and Time-Use Diary data into Excel spreadsheets, enabling "clean" data that could be statistically analyzed further.

9.6 RECOMMENDATIONS

In Box 9.1 we compile a list of recommendations for the use of field methods in patient ergonomics. Each method has its own benefits and uncertainties and is appropriate to answer different kinds of research questions. Interviews are ideal for obtaining individual experiences and options. Focus groups allow participants to discuss issues between each other and explore differences in opinions. Surveys provide a quick method to reach many participants. Observations provide rich data that investigate the contexts of the participant's behavior. For the future researcher, a thorough understanding and innovative combination of field methods would be beneficial to unravel the multifaceted field of patient ergonomics.

REFERENCES

Asch, D. A., Jedrziewski, M. K., & Christakis, N. A. (1998). Response rates to mail surveys published in medical journals. *Journal of Clinical Epidemiology, 50*(10), 1129–1136.

Au, T. S. Y., Wong, M. C. M., McMillan, A. S., Bridges, S., & McGrath, C. (2014). Treatment seeking behaviour in southern Chinese elders with chronic orofacial pain: A qualitative study. *BioMed Central Oral Health, 14*, 8.

Bowling, C. B., Vandenberg, A. E., Phillips, L. S., McClellan, W. M., Johnson, T. M., & Echt, K. V. (2017). Older patients' perspectives on managing complexity in CKD self-management. *Clinical Journal of the American Society of Nephrology, 12*, 635–643.

Bukhave, E. B., & Huniche, L. (2014). Activity problems in everyday life – Patients' perspectives of hand osteoarthritis: "Try imagining what it would be like having no hands." *Disability and Rehabilitation, 36*(19), 1636–1643.

Catlin, A., Ford, M., & Maloney, C. (2016). Determining family needs on an oncology hospital unit using interview, art, and survey. *Clinical Nursing Research, 25*(2), 209–231.

Czuber-Dochan, W., Dibley, L. B., Terry, H., Ream, E., & Norton, C. (2012). The experience of fatigue in people with inflammatory bowel disease: An exploratory study. *Journal of Advanced Nursing, 69*(9), 1987–1999.

Dehghanzadeh, S., Nayeri, N. D., Varaei, S., & Kheirkhah, J. (2017). Living with cardiac resynchronization therapy: Challenges for people with heart failure. *Nursing and Health Sciences, 19*, 112–118.

Ehmen, H., Haesner, M., Steinke, I., Dorn, M., Gövercin, M., & Steinhagen-Thiessen, E. (2012). Comparison of four different mobile devices for measuring heart rate and ECG with respect to aspects of usability and acceptance by older people. *Applied Ergonomics, 43*, 582–587.

Eurostat. (2009). Harmonised European time use surveys 2008 guidelines [Press release]. Retrieved from https://ec.europa.eu/eurostat/ramon/statmanuals/files/KS-RA-08-014-EN.pdf

Eysenbach, G. (2004). Improving the quality of web surveys: The checklist for reporting results of internet e-surveys (CHERRIES). *Journal of Medical Internet Research, 6*(3), e34.

Gibbons, C. J., Kenning, C., Coventry, P. A., Bee, P., Bundy, C., Fisher, L., & Bower, P. (2013). Development of a multimorbidity illness perceptions scale (MULTIPleS). *Plos One, 8*, e81852.

Hammarlund, C. S., Lexell, J., & Brogårdh, C. (2017). Perceived consequences of ageing with late effects of polio and strategies for managing daily life: A qualitative study. *BioMed Central Geriatrics, 17*, 179.

Holden, R. J., Schubert, C. C., & Mickelson, R. S. (2015). The patient work system: An analysis of self-care performance barriers among elderly heart failure patients and their informal caregivers. *Applied Ergonomics, 47*, 133–150.

Holden, R. J., Valdez, R. S., Schubert, C. C., Thompson, M. J., & Hundt, A. S. (2017). Macroergonomic factors in the patient work system: Examining the context of patients with chronic illness. *Ergonomics, 60*(1), 26–43.

Jepson, C., Asch, D., Hershey, J. C., & Ubel, P. A. (2005). In a mailed physician survey, questionnaire length had a threshold effect on response rate. *Journal of Clinical Epidemiology, 58*(1), 103–105.

Kaplowitz, M. D., Hadlock, T. D., & Levine, R. (2004). A comparison of web and mail survey response rates. *Public Opinion Quarterly, 68*(1), 94–101.

Kawi, J. (2014). Chronic low back pain patients' perceptions on self-management, self-management support, and functional ability. *Pain Management Nursing, 15*(1), 258–264.

Kelly, M. (2010). The role of theory in qualitative health research. *Family Practice, 27*(3), 285–290.

Khunti, K., Kosiborod, M., & Ray, K. K. (2018). Legacy benefits of blood glucose, blood pressure and lipid control in individuals with diabetes and cardiovascular disease: Time to overcome multifactorial therapeutic inertia? *Diabetes, Obesity & Metabolism, 20*, 1337–1341.

McCambridge, J., Witton, J., & Elbourne, D. R. (2014). Systematic review of the Hawthorne effect: New concepts are needed to study research participation effects. *Journal of Clinical Epidemiology, 67*(3), 267–277.

Murphy, E., Dingwall, R., Greatbatch, D., Parker, S., & Watson, P. (1998). Qualitative research methods in health technology assessment: A review of the literature. *Health Technology Assessment Reports, 2*(16), 1–278.

Parry, R., Pino, M., Faull, C., & Feathers, L. (2016). Acceptability and design of video-based research on healthcare communication: Evidence and recommendations. *Patient Education and Counseling, 99*, 1271–1284.

Pauling, J. D., Domsic, R. T., Saketkoo, L. A., Almeida, C., Withey, J., Jay, H., . . . Hewlett, S. (2018). Multinational qualitative research study exploring the patient experience of Raynaud's phenomenon in systemic sclerosis. *Arthritis Care & Research, 70*(9), 1373–1384.

Roberts, A. R., Adams, K. B., & Warner, C. B. (2017). Effects of chronic illness on daily life and barriers to self-care for older women: A mixed-methods exploration. *Journal of Women & Aging, 29*(2), 126–136.

Tong, A., Sainsbury, P., & Craig, J. (2007). Consolidated criteria for reporting qualitative research (COREQ): A 32-item checklist for interviews and focus groups. *International Journal for Quality in Health Care, 19*(6), 349–357.

Ward, J., Buckle, P., & Clarkson, J. (2010). Designing packaging to support the safe use of medicines at home. *Applied Ergonomics, 41*, 682–694.

Yin, K., Harms, T., Ho, K., Rapport, F., Vagholkar, S., Laranjo, L., . . . Lau, A. Y. S. (2018). Patient work from a context and time use perspective: A mixed-methods study protocol. *The BMJ Open, 8*(12), e002163.

10 Design and Usability Methods

Agile Innovation and Evaluation of Interventions for Patients and Families

Richard J. Holden and Malaz A. Boustani
Indiana University School of Medicine

CONTENTS

Given that patients, family members, and other nonprofessionals perform effortful work in the service of health-related goals, it follows that they, like other workers, can benefit from better tools, processes, policies, and physical spaces. In some cases, an intervention already exists, and the challenge is to implement, diffuse, or sustain it. However, in most cases, either there is no suitable evidence-based solution or existing solutions require redesign to become implementation-ready. In such cases, the challenge is one of innovation, or designing "an idea, practice, or object that is perceived as new" (Rogers, 2003, p. 12). User-centered design, simply put, is innovation with user input.

User-centered design gathers data from and about users, considers user needs during design or involves users as codesigners, and evaluates the product with users during and after design. In the context of patient ergonomics, the primary user is often a patient, family member, or community member involved in the patient's care. Healthcare professionals may also be involved as secondary or tertiary users. Being a "user" could refer to using technology, interacting with the physical environment, or performing a process. User-centered evaluation often means assessing the innovation's usability and acceptability for users, to determine whether the innovation should be terminated, redesigned, or advanced to additional testing, for example, for efficacy, effectivenes, and scalability. Testing for usability also informs specific redesign decisions and therefore should be interleaved with design activities, rather than "saved for the end."

The tools and processes of user-centered design and evaluation of patient- or family-facing innovations largely resemble those extensively described for other areas of design and testing, for example, of regulated medical devices (Weinger et al., 2011), nonmedical consumer products (Beyer & Holtzblatt, 2017), services (Lockwood, 2009), and physical spaces such as hospitals (Joseph et al., 2018). Those designing for patients and families should also be responsive to the many international standards for user-centered design and other resources amassed by cross-disciplinary scholars and practitioners. We, therefore, present herein a general user-centered innovation process built on such prior work but ostensibly suitable for patient ergonomics. At the same time, patients, families, and nonprofessionals have distinctive needs and characteristics. Design and evaluation methods may require adjustment to suit these populations or subpopulations thereof, for example, older adults, caregivers of pediatric patients, residents of rural areas, or low-income families. Minimally, one should recognize the advantages and limitations of applying general methods with specific populations that have specific needs. We, therefore, present important lessons learned from multiple design and evaluation studies performed with patients and family caregivers, drawing primarily on our work with older adult patients and their family caregivers. We conclude with a specific case example of one of those projects, Power to the Patient (P2P), showing some of the ways we adapted general methods for the specific needs of patient ergonomics in one particular patient population.

10.1 AGILE INNOVATION: A USER-CENTERED DESIGN PROCESS

The Agile Innovation process (Figure 10.1) was developed based on user-centered design best practices (International Organization for Standardization, 2010) and its "older sister," the Agile Implementation process. Agile Implementation is a set of principles and processes developed to ensure the viability of evidence-based solutions in actual healthcare delivery organizations (Boustani et al., 2019). Both Agile Innovation and Implementation draw heavily on complex adaptive system and behavioral economics theories. A complex adaptive system is an evolving sociotechnical network of competing and collaborating individuals who exchange energy and information with each other and with their social and physical environment in nonlinear ways for the purpose of accomplishing complex tasks that increase the fitness of each member within their surrounding environment (Boustani et al., 2010; Carayon, 2006; Young et al., 2017) (see also Chapter 4 in this volume). Behavioral economics is a set of social-cognitive theories built on the assumption that human behaviors are shaped by the surrounding social and physical environment within the restraints of the biological processes governing human thoughts (Kahneman, 2003). Such human behaviors do not fit with traditional assumptions about rational choice (Simon & March, 1958). Thus, emotions play a major role in shaping human decision-making and behavior within sociotechnical systems (Pasmore, 1988). These theories produced important steps in Agile Implementation and Agile Innovation such as assessing demand for

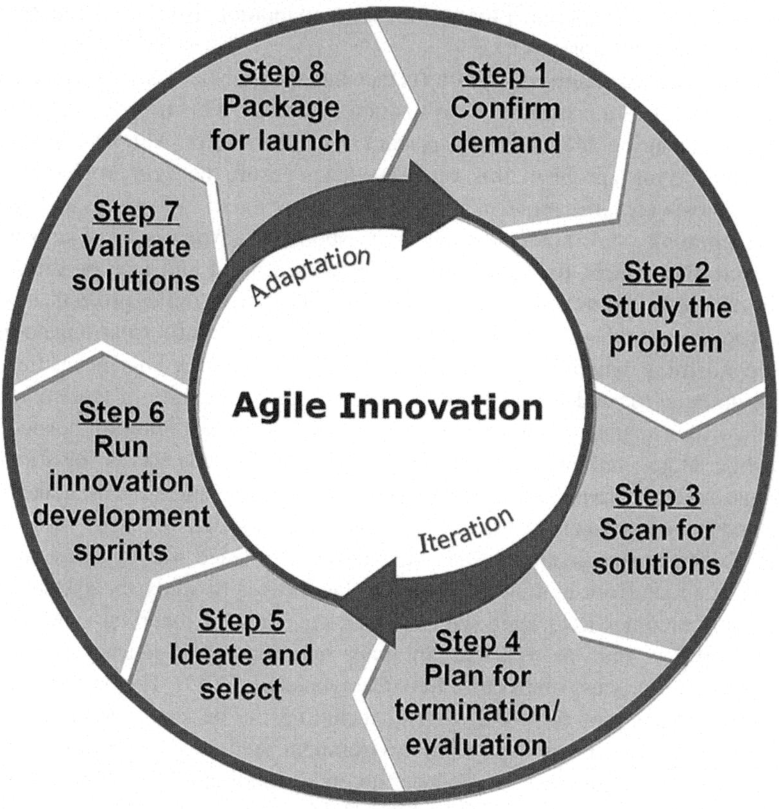

FIGURE 10.1 The eight-step Agile Innovation process.

the solution and creating a termination plan, two steps meant to avoid poor investments of resources. By complex adaptive system and behavioral economics theories, Agile Innovation differs from more reductionist and lab-based approaches that are predicated on rational-choice models, or view innovation as a controllable and linear process (for discussion of such approaches in human factors and ergonomics, see, e.g., Meister & Enderwick, 2001; Whitefield et al., 1991).

Agile Innovation is a process that can be performed by teams of various compositions. For example, the team may consist of any combination of patients, families, community members, researchers, consultants, sponsors, healthcare professionals or leaders, design professionals, regulators, subject matter experts, or clients. When a patient ergonomics project uses Agile Innovation to create something that will affect patients, families, or other stakeholders, it is advisable for these stakeholders to be invited to be on the project team. This practice—called participatory design, codesign, or more broadly coproduction—can improve the innovation and increase its likelihood of acceptance.

The first four Agile Innovation steps are considered planning activities, whereas the final four steps are execution activities. We call them steps to encourage forward motion, but recognize the need for nonlinearity, adaptation as needed, and iteration (i.e., carrying out multiple cycles).

Step 1: Confirm demand. Before further investing resources, designers should identify that there is a problem whose solution is desired. This requires defining the problem and ensuring adequate support from key stakeholders. Although it is tempting to begin solving a problem, this step is needed to ensure the *right* problem is being solved. In most cases, the problem is a social construction of individuals and groups, such as community stakeholders/beneficiaries, sponsors, advocates, or organizational leaders; in other words, the problem is in someone's mind and may be differently perceived by different stakeholders or groups. In patient ergonomics projects, it is best when stakeholders who are or else represent patients and families participate in the step of confirming demand (see also Chapter 13 in this volume). This means the project was initiated (or codeveloped) by the stakeholder community or that stakeholders or their representatives reviewed the project and deemed it relevant, needed, and worthwhile. Major challenges for engaging stakeholders include access; logistics (e.g., payments, meetings, availability); equitable treatment and attention to stakeholder needs; and representativeness, especially among diverse groups.

Step 2: Study the problem. Once demand is confirmed, the designer employs standard methods from human factors and related fields to study the systems, processes, and outcomes associated with the problem. Process mapping in its various forms—workflow analysis, value stream mapping, user stories, journey maps—can be done, or prior process maps consulted (Carayon et al., 2012). The designer should perform an analysis of the work system, including the person(s), tools and technologies, tasks, organization, physical environment, and external environment. In patient ergonomics, this analysis produces an understanding of the patient work system (Holden, Schubert, Eiland, et al., 2015; Holden, Schubert, & Mickelson, 2015) (see Chapter 4 in this volume). To create such an understanding, one may conduct interview or observations in the field (see Chapter 9 in this volume) or laboratory experiments (Chapter 11 in this volume), identify user needs, and create personas representing groups of users (Holden et al., 2017; Holden, Daley, et al., 2020). The pitfall of this step is spending too much or too little time on it. To avoid delays, it is advisable to (1) when possible, rely on preexisting research and data; (2) use methods designed for speed and simplicity such as rapid ethnography (Millen, 2000), personas (Adlin & Pruitt, 2010), surveys (Chapter 9 in this volume), and sensor-based passive data collection (Cornet & Holden, 2018) (see also Chapter 12 in this volume); (3) decide how data will be analyzed before beginning analysis, to avoid scope creep; (4) designate one or more Problem Masters to know the problem and related data as deeply as possible; and (5) reserve time throughout the design process to reassess the data and consult the Problem Master, who can also be the "voice" of the user. The Problem Master is someone who understands the research about the problem, was directly involved in the research, has relevant direct personal experiences (e.g., is a patient themselves), or has some combination of these and other insights into the problem.

Step 3: Scan for solutions. Innovations are often the result of combining two well-known concepts or applying something well known to a new problem or in a novel way (Dyer et al., 2011; Kelley, 2001). Therefore, before creating something new, one should scout and analyze existing solutions or platforms. Platforms are existing infrastructures or practices that expand the realm of possibilities (the "adjacent possible") (Johnson, 2010). For example, one might use an existing solution such as Facebook Groups to solve the problem of a safe space for information sharing among family caregivers of children with cancer. One might draw inspiration from existing "gig" or "sharing" platforms such as Care.com, Airbnb, or Lyft to create a novel solution for matching patients with available caregivers. Sources for solutions and platforms include the literature, competitors, conferences and experts from unfamiliar fields, store catalogs, and even history books. Existing solutions may also be observed in the data gathered during step 2, because a study of current systems and processes reveals clever ways systems were adapted by professionals (Holden et al., 2013) or patients (Mickelson & Holden, 2018) to accomplish their goals. For example, a small study of homemade tools patients invent to keep track of medications (Mickelson & Holden, 2017) can inspire solutions that benefit thousands or millions of others.

Step 4: Plan for termination and evaluation. The final planning step is to determine how the innovation will be evaluated and, if necessary, terminated. Specific termination criteria and timelines should be documented and a review should be formally scheduled, to avoid continuing to invest resources into something that failed. Failure, when soundly identified, should be a welcomed finding that results in a brief study of what went wrong; then, a decision should be made to completely stop versus restart the project. In our experience, termination is often a difficult, emotional process that we tend to avoid, preferring to rationalize continued investments into failures. Therefore, we recommend using clear, specific criteria and viewing the termination decision through an entrepreneurial lens—as an opportunity for a fresh start, one more likely to result in eventual success. Termination and evaluation planning and decisions should be made by the project team, which may include patient and family representatives. If not, those stakeholders should be consulted when possible and their input into termination and evaluation criteria sought.

Step 5: Ideate and select. This step is the first of four related to execution and is well described in writing on design thinking and related approaches (Lockwood, 2009; Shah et al., 2003). Ideation is a creative process but must be accompanied by more mundane tasks: recruiting the idea-generators, seeking as much diversity as possible; facilitating judgment-free, constructive, out-of-the-box ideation; motivating as many ideas as possible from as many people as possible; and documenting each idea. Diversity refers here to demographics (e.g., age, race, ethnicity, and gender), experience, discipline, role, and other factors. A formal group-based ideation method is the *Innovation Forum*, developed by the Indiana University Center for Health Innovation and Implementation Science (Boustani et al., 2012). The Innovation Forum recruits a diverse group of stakeholders who perform work in the targeted work system or are impacted by it. It is an example of participatory design or codesign that is fast, engaging, and promotes the sharing of "time and space" between stakeholders and local experts (see Chapter 13 in this volume); yet, it is not as deep, engaging, or time- and resource-intensive as other recently published examples of participatory design with patients (Ahmed et al., 2019; Reddy et al., 2019). Box 10.1 describes the team and procedures of an Innovation Forum.

BOX 10.1 TEAM AND PROCEDURES RECOMMENDED FOR THE INNOVATION FORUM

INNOVATION FORUM TEAM (DISTINCT ROLES CAN SOMETIMES BE PLAYED BY THE SAME INDIVIDUAL)

- **Presenter**. Owns the design problem and is responsible for identifying a small group of individuals to whom a personal invitation will be sent. May be a designer or user; may be a patient, caregiver, community member, leader, or healthcare professional.
- **Forum coordinator**. The primary organizer of the event; responsible for ongoing monitoring and evaluation of the event and maintaining communication with the presenter on any forum-related needs or preferences.
- **Invited participants**. During the Innovation Forum, participants request clarification about the problem and generation solutions. Network with the presenter during and after the Forum. Can be users, local system experts, or other types of stakeholders.
- **Administrative coordinator**. Provides logistical and administrative support during the planning process and at the event.
- **Solution tracker**. Records and distributes notes during and after the Innovation Forum. Records solutions during the event.
- **Facilitator**. Conducts the Innovation Forum, ensures smooth knowledge transfer between presenter and audience, and profiles and engages the audience while clarifying meaning during discussion. The facilitator is not a content expert but rather promotes conversation and understanding and is selected for empathy and listening skills.

INNOVATION FORUM PROCEDURE

- *Opening networking (30 minutes).* Allows time and space for attendants to network.
- *Presenting the problem (10 minutes).* Reserved for the presenter to describe their problem. The presenter may use any visual aids or handouts. The presenter receives guidance from the Forum coordinator on how to present to the specific audience.
- *Clarifying questions from the audience (5 minutes).* Done to clarify anything from the problem presentation. The facilitator must enforce use of question form and that no solutions are generated during this time.
- *Generating solutions and discussion (45 minutes).* Facilitator calls on audience members to generate solutions through brainstorming or question storming. Ideas are not discarded or criticized. The facilitator ensures diversity in participation.
- *Closing networking (30 minutes).* Intended to provide closure to the discussion in a more informal environment, as people are encouraged to move around the room.

Once ideas are generated, clear criteria for selecting top candidates should be applied, gradually converging on fewer and fewer viable candidates that will be more fully developed. A formal selection process may be useful, applying a set of criteria to each idea in a matrix. Example criteria might be support of leadership for the idea, cost or effort to produce it, likelihood it will solve the problem, scalability, and feasibility or viability. Given that those in power create the selection criteria, patient ergonomics projects must either include patient, family, or community stakeholders in this step or invoke their needs and preferences when setting the criteria. For example, when selecting from virtual reality solutions differing in out-of-pocket costs, a consumer or patient advocacy group can set cost thresholds for specific patient populations (e.g., no more than $5/month and $75/year).

Step 6: Run innovation sprints. Ideas become prototypes when they are ready to be tested. This is not to diminish the importance of or time required for prototyping, but to remind designers of the importance of iterative and rapid prototyping (Beevis & St Denis, 1992), as a strategy toward generating external user feedback as soon and as often as possible. In our experience, a lot of time spent prototyping and self-critiquing could be converted to time spent arranging for tests, including establishing a pool or panel of testers. Prototype testing occurs in sprints: quick, iterative tests that provide feedback for revising the prototype in advance of the next sprint. Generally, tested prototypes progress from lower to higher fidelity, but major adjustments or new hypotheses may require reducing the fidelity temporarily. This holds for prototypes of software (e.g., wireframe screens to field-ready software), physical products (e.g., drawings to 3D printed artifacts), physical space (e.g., cardboard to stable architecture), and processes (e.g., verbal vignettes to dress rehearsals). The various methods of user testing described at length elsewhere, and remarkably similar across innovation types, should be used to learn from these sprints (Charlton & O'Brien, 2010; Jacko, 2012; Nielsen, 1993; Weinger et al., 2011).

Step 7: Validate solutions. A set of sprints will produce what agile software development calls the minimum viable product (MVP), meaning it does not contain all possible functionality but that all included functions are properly working. The concept applies beyond software, for example, in our notion of the minimally viable evidence-based service (Boustani et al., 2019). The MVP can be subjected to formative testing on established evaluation criteria, for example, ability to complete specific tasks, usability for a range of users, and desirability to potential consumers. During validation, it is necessary to monitor for and address unintended consequences or unexpected benefits. For example, a project to better inform families in a pediatric hospital might select a solution such as an in-room large-screen display of the patient's electronic health record (Asan et al., 2016; Holden, Asan, et al., 2016); testing the display might reveal unintended positive consequences such as families feeling more empowered and negative consequences such as concerns about the privacy of displayed patient information (Asan et al., 2018; Asan et al., 2019).

Step 8: Package for launch. The final step is an important one, because it bridges innovation development with actual implementation. At this step, designers create the hand-off package, which includes a business plan, MVP, and clear specifications for use. The package can undergo additional validation if needed or be customized

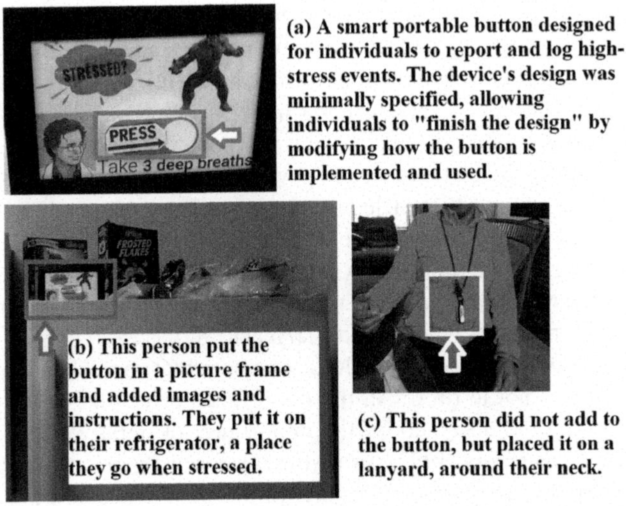

STRESSED?
PRESS
Take 3 deep breaths
(a) A smart portable button designed for individuals to report and log high-stress events. The device's design was minimally specified, allowing individuals to "finish the design" by modifying how the button is implemented and used.
FROSTED FLAKES
(b) This person put the button in a picture frame and added images and instructions. They put it on their refrigerator, a place they go when stressed.
(c) This person did not add to the button, but placed it on a lanyard, around their neck.

FIGURE 10.2 Example of users "finishing the design" of a smart portable button (panel a), by adding design components and specifying how it is implemented and used (panels b and c).

for local deployment and implemented. It may be tempting to continue refining and testing the solution, but one should consider that much of an innovation's success is related to how it is localized to its implementation setting (Boustani et al., 2019); therefore, the designer must allow for local users to "finish the design," as Rasmussen observed (Rasmussen & Goodstein, 1987). For instance, we used a commercially available smart button as a portable device for individuals to use to report and log high-stress events (Figure 10.2). As the figure shows, individuals finished the minimally specified design of this solution by adding design components such as personal instructions and choosing its implementation and use.

A hand-off package is important so that future implementers do not need to reconstruct essential elements of the innovation or abandon the innovation because it is difficult to adopt.

10.2 LESSONS LEARNED FROM DESIGN AND EVALUATION FOR PATIENTS AND FAMILIES

Across multiple projects, we have learned and documented the challenges (and joys) of user-centered design and evaluation. Table 10.1 lists ten types of challenges associated with user research with patients and families (Holden, McDougald Scott, et al., 2015). These challenges are most salient when working with vulnerable individuals, including those who are older, disabled, poorer, less-educated, or rural-dwelling (Holden, Toscos, et al., 2020). From another project, we identified 12 practical considerations for applying user-centered design and evaluation methods in health and health care (Cornet, Toscos, et al., 2020), summarized in Table 10.2. Among the challenges is to "juggle between patient and clinician perspectives" to avoid two errors (Cornet et al., 2019):

TABLE 10.1
Challenges Conducting User Research with Patients and Families

Challenge	Examples
Differences in priorities	Researchers and participants may have different goals and expectations (e.g., learning versus doing)
Mistrust and misunderstanding	Participants do not always want strangers in their homes or asking about sensitive topics
Differences in language, perspective, and norms	The jargons, terms, or assumptions of researchers and participants may differ and may not be mutually understood or welcomed
Competing life and health demands	Participants and researchers alike may have life and health situations and responsibilities that conflict with research (e.g., making it hard to make time or to travel)
Psychosocial, cognitive, and perceptual limitations	Instructions may be difficult to read or understand; other communication may be hard (e.g., quiet researcher and hearing-impaired participant speaking in a noisy room)
Participant identification and recruitment challenges	Mistrust born of previous experiences leads to a gatekeeper (e.g., a family member or clinician) preventing contact with target individuals; calls or emails are unanswered from fear of telemarketers and scammers
Logistical and transportation issues	Participants or research staff may feel unsafe or inconvenienced attending face-to-face meetings
Maintaining privacy and confidentiality	Data collection may encroach on cultural or personal boundaries, for example, collection of photos or videos; participants may grant broad consent to access their medical data containing sensitive information they would not have shared if specifically asked about it
Conflicts regarding compensation and risk of coercion	Authorities may prohibit cash payment, despite participant preference for cash over gift cards
Questions about scientific validity, interpretation, or integration of data	Mistrust or social desirability may lead participants to "sugarcoat" the truth or over- or underreport something; patient and family perspectives may differ

Adapted from (Holden, McDougald Scott, et al., 2015; Holden, Toscos, et al., 2020; Valdez & Holden, 2016).

- **Type 1 Design Error or "User-Reality Error":** when designers do not accommodate user characteristics, tasks, contexts of use, needs, or preferences. An example of this error is when a physician's opinions of what is best for patients, and not the patients' needs, drive the design of a mobile app for patients.
- **Type 2 Design Error or "Clinical-Reality Error":** when designers do not accommodate the clinical reality, including biomedical knowledge, clinical workflows, and organizational requirements. An example of this error is when a designer creates a patient–clinician communication platform without adequate consideration of federal and organizational privacy rules or the extra burden it will impose on clinicians to use it.

TABLE 10.2
Considerations for Applying User-Centered Design with Patients and Families

Consideration	Description
Deciding the number of iterations	The number of design-test cycles is often specified at or before project commencement, but depends on how the project unfolds
Managing user-centered design logistics	Logistic factors such as recruitment, privacy, confidentiality, compensation, and communication influence project success
Collaborating as a multidisciplinary team	Disciplinary diversity is important but creates communication and decision-making challenges, regardless of the disciplines involved. One solution is to have one or more individuals be a discipline-spanning "multilinguistic (symphonic) conductor"
When and how much to involve stakeholders	Designers must decide how to budget limited resources, access to stakeholders, and tolerance for delays as they determine how many stakeholders to involve, when, and how—e.g., as advisors versus codesigners; a further trade-off is in the acceptability and sustainability of the intervention, which may depend on the depth, duration, and nature of stakeholder involvement
Choosing stakeholder representatives	Stakeholder selection should attend to issues such as diversity, representativeness, commitment, and ability to work together
Interactions between stakeholders and designers	There should be some representation of stakeholders or users on the design team, whether directly, via access to data, or a third party who speaks with the "voice" of the stakeholders
Overcoming designer assumptions	Designers' assumptions may persist despite evidence to the contrary and can influence data interpretation and decisions
Managing project scope and complexity	Ideas are easier to generate than to dismiss, requiring prioritization, vigilance for scope creep, and discipline
Maintaining innovation equilibrium	Designers balance between what is immediately practical and conventional versus disruptive, innovative, or futuristic
Laboratory versus in-the-wild testing	Testing in a well-equipped laboratory generates key findings but cannot always simulate realistic conditions
Adapting methods to users	Methods for general populations or developed with young, healthy individuals require adaptation, for example, for older users
Number of evaluation methods	Number of methods affects depth of data, participant burden, staffing, and the pace of progress (e.g., due to analysis time)

Adapted from (Cornet et al., 2019; Cornet, Toscos, et al., 2020; Holden, Binkheder, et al., 2018).

We are also aware of challenges to being "agile" in an academic environment, including differences of opinion between academic researchers and clinical staff partners, equipment purchasing delays, and the competing time demands and commitments of students and faculty (Holden, Bodke, et al., 2016). To address these and other conflicts between the constraints of academia and the desire to be agile and innovative, we recommended to:

- Partner with clinical staff to create shared responsibility and shared project goals;

- Have clinical staff implement components of research (e.g., usability testing), by training people in certain roles (e.g., community health workers) to perform design and evaluation;
- Establish "living laboratories" with community partners to increase quick access to participants and advisors;
- Hire full-time design staff or faculty.

These suggestions are only realistic in a hybrid model that blends research, clinical practice, and innovation activities. Because each activity contributes to success, there is funding and support for each in the hybrid model and incentives to staff them with cross-trained individuals. Such hybrids are not often encountered due to narrowly focused reimbursement models and incentive structures but may grow through emerging initiatives such as innovation districts (Katz & Wagner, 2014).

We provide other recommendations for managing the practical challenges of user-centered design and evaluation elsewhere (Cornet et al., 2019; Cornet, Toscos, et al., 2020; Holden, Bodke, et al., 2016; Holden, McDougald Scott, et al., 2015). In particular, we have stressed the importance of having a "multilinguistic (symphonic) conductor" (Holden, Binkheder, et al., 2018, p. 142), a team leader with sufficient knowledge of multiple disciplines to ably bridge them (Schall & McAlister, 2019).

10.3 APPLYING USER-CENTERED DESIGN AND EVALUATION IN THE POWER TO THE PATIENT PROJECT

We illustrate the application of user-centered design and evaluation to create and test innovative products for patient and family users by describing work done as part of the P2P project (R21 HS025232). Our goal was to create and test P2P, a technology intervention to improve self-care management among older adults with chronic heart failure (CHF). An innovative aspect of P2P was its use of data from patients' cardiovascular implantable electronic devices (e.g., pacemakers and defibrillators) to personalize and inform self-care management decisions. Additional case studies of user-centered design and evaluation of technology for CHF self-care are available in two other publications (Cornet, Daley, et al., 2020; Srinivas et al., 2017).

Consistent with the Agile Innovation process, the study included phases of studying the problem, examining the landscape of existing innovations, ideating, and sprints of design-test cycles, with early cycles focusing on detecting correctable usability problems and later cycles on user acceptance and likelihood of future use. Other Agile Innovation steps were not as evident in the project, for example, we ran out of time and funds before we could create a hand-off package. (Such discrepancies between actual projects and ideal processes are often seen when resources are limited.) Even before beginning the project, we confirmed demand and readiness for a solution. Leading work on patients with CHF established both the importance of and breakdowns in self-care management, which includes understanding, recognizing, and acting on symptoms in a timely manner (Daley et al., 2019; Riegel et al., 2011). Our clinical research partner, Parkview Health and its cardiology group, desired improvements in CHF self-care. More importantly, Parkview Health and the Parkview Research Center had at the time of project conceptualization recently pioneered the transmission of implantable device data directly to patients and were advocating for developments in this area (Daley et al., 2017; Mirro et al., 2018). The multidisciplinary team of human factors, human–computer interaction, medical informatics, and cardiology experts had aligned goals and recent experience working on a similar project for delivering device data to patients with CHF (Ahmed et al., 2019; Ghahari et al., 2018).

To further study the problem of self-care management decisions made by patients with CHF, with and without data from implantable devices, we found it necessary to adapt traditional human factors methods to our older adult patient population. With the aid of six pilot tests, we developed the patient-centered cognitive task analysis (P-CTA) method (Holden, Daley, et al., 2020), modified from the Critical Incident Technique and Scenario-based Interviewing methods. Description of the method and its use in data collection sessions with 24 patients and 14 support persons (informal caregivers) are provided elsewhere (Holden, Daley, et al., 2020).

During ideation and design, we assembled a design team of students and faculty. They drew on collected data and occasional meetings with clinical experts (cardiologists, device clinic staff, telehealth leadership) and two patient advisors who were adults with CHF. One of the design team members had participated in the prior data collection with patients and caregivers, was dually employed by the university and its clinical partner, and was an ideal Problem Master (i.e., speaking with the voices of our stakeholders). Design was informed by several important design artifacts, including empirically derived use-case scenarios, decision-making personas, a naturalistic decision-making model of self-care management, and a requirements document (see e.g., Cornet et al., 2019; Daley et al., 2018; Holden, Joshi, et al., 2018). Ten interactive prototypes were created iteratively, and at three intervals, we conducted sprints of user testing sessions.

Testing sprints used progressively advanced prototypes and increasing sample sizes: 4 users in round 1, 8 in round 2, and 12 in round 3. Testing was performed in a laboratory setting but in the final round, to enhance realism, we created scenarios to simulate longitudinal use of P2P over a role-played 14-day period, as we had done in a prior study (Cornet et al., 2017). During each testing sprint, we collected (1) video-recording and written observation notes as users performed preset tasks, with think aloud, (2) self-reported usability, (3) the NASA Task Load Index measure of workload, (4) a 33-item survey of technology acceptance measures, and (5) demographic and medical data. Our measures were chosen and adapted to suit older adult patient participants. For example, we used the simplified System Usability Scale (SUS) for cognitively impaired and older adults (Holden, 2020), presented in Figure 10.3, which we adapted in 2016. Lastly, three experts external to the team conducted independent heuristic evaluations of P2P using a framework modified for the project, based on eight consolidated principles for good design (Holden, Voida, et al., 2016).

10.4 CONCLUSIONS

As seen elsewhere in this volume (Chapters 9 and 11), patient ergonomics benefits from using general methods for data collection, design, and evaluation

Thinking about using (system)...	Strongly disagree	(CIRCLE ONE ANSWER PER QUESTION)			Strongly agree
1. I would use (system) frequently	1	2	3	4	5
2. (system) is too complex for me	1	2	3	4	5
3. (system) was easy to use	1	2	3	4	5
4. I really need help from someone to use (system)	1	2	3	4	5
5. The various parts of (system) were well integrated	1	2	3	4	5
6. (system) was confusing for me	1	2	3	4	5
7. Learning to use (system) was quick for me	1	2	3	4	5
8. (system) was hard to use	1	2	3	4	5
9. I felt confident using (system)	1	2	3	4	5
10. I will need to learn a lot before using (system)	1	2	3	4	5

FIGURE 10.3 Simplified system usability scale for cognitively impaired and older adults. (Adapted from the System Usability Scale, Bangor et al., 2008.)

(Holden, Cornet, et al., 2020). At the same time, we should apply emerging new methods (e.g., Novak et al., 2016) (see Chapter 12 in this volume) and adapt standard methods to the unique characteristics of patient work (Holden & Mickelson, 2013). In light of the need for both standard and adapted methods, we presented here a general Agile Innovation process and tailored methods such as the simplified SUS and P-CTA. In addition to the methods themselves, there are also unique considerations for implementing them in the context of patient work.

We propose several future directions in the design and evaluation of innovations for patients and families. First, we encourage the pursuit of partnerships between academic and nonacademic groups, to accelerate the innovation process while maintaining appropriate evidence-based design and evaluation. Second, we promote the participation of patient, family, and community stakeholders as codesigners of innovation rather than simply "informants" or "testers" (see Chapter 13 in this volume). Third, we believe too often the scope of projects in human factors and related disciplines is restricted to design and formative evaluation. We recommend expanding the scope to add packaging, implementing, and diffusing the intervention. In the P2P case study, we could have worked with health system informatics and clinical leads, patient advisory boards, and a device vendor to create and implement a commercial product. Such steps require human factors professionals to learn to work with entrepreneurs, insurers, and other atypical partners.

ACKNOWLEDGMENTS

The smart button example (Figure 10.1) was one of several in the Happy Medium project supported by a Regenstrief Innovation award to Drs. Preethi Srinivas and Richard Holden. The Power to the Patient (P2P) project was supported by an Agency for Healthcare Research & Quality (AHRQ) award to Dr. Holden (R21 HS025232). We thank the P2P team, patient advisors, and study participants for their contributions to the project. The content is solely the responsibility of the authors and does not necessarily represent the official views of AHRQ.

REFERENCES

Adlin, T., & Pruitt, J. (2010). *The persona lifecycle: Keeping people in mind throughout product design.* San Francisco, CA: Morgan Kaufmann.

Ahmed, R., Toscos, T., Ghahari, R. R., Holden, R. J., Martin, E., Wagner, S., . . . Mirro, M. (2019). Visualization of cardiac implantable electronic device data for older adults using participatory design. *Applied Clinical Informatics, 10*(4), 707–718.

Asan, O., Holden, R. J., Flynn, K. E., Murkowski, K., & Scanlon, M. C. (2018). Providers' assessment of a novel interactive health information technology in a pediatric intensive care unit. *Journal of the American Medical Informatics Association Open, 1*(1), 32–41.

Asan, O., Holden, R. J., Flynn, K. E., Yang, Y., Azam, L., & Scanlon, M. C. (2016). Provider use of a novel EHR display in the pediatric intensive care unit: Large customizable interactive monitor (LCIM). *Applied Clinical Informatics, 7*(3), 682–692.

Asan, O., Scanlon, M. C., Crotty, B., Holden, R. J., & Flynn, K. E. (2019). Parental perceptions of displayed patient data in a pediatric intensive care unit: An example of unintentional empowerment. *Pediatric Critical Care Medicine, 20*(5), 435–441.

Bangor, A., Kortum, P. T., & Miller, J. T. (2008). An empirical evaluation of the System Usability Scale. *International Journal of Human-Computer Interaction, 24*, 574–594.

Beevis, D., & St Denis, G. (1992). Rapid prototyping and the human factors engineering process. *Applied Ergonomics, 23*(3), 155–160.

Beyer, H., & Holtzblatt, K. (2017). *Contextual design: Design for life* (2nd ed.). Cambridge, MA: Morgan Kaufmann.

Boustani, M. A., Frame, A., Munger, S., Healey, P., Westlund, J., Farlow, M., . . . Dexter, P. (2012). Connecting research discovery with care delivery in dementia: The development of the Indianapolis Discovery Network for Dementia. *Clinical Interventions in Aging, 7*, 509–516.

Boustani, M. A., Munger, S., Gulati, R., Vogel, M., Beck, R. A., & Callahan, C. M. (2010). Selecting a change and evaluating its impact on the performance of a complex adaptive health care delivery system. *Clinical Interventions in Aging, 5*, 141–148.

Boustani, M. A., van der Marck, M. A., Adams, N., Azar, J. M., Holden, R. J., Vollmar, H. C., . . . Suarez, S. (2019). Developing the Agile Implementation playbook for integrating evidence-based health care services into clinical practice. *Academic Medicine, 94*, 556–561.

Carayon, P. (2006). Human factors of complex sociotechnical systems. *Applied Ergonomics, 37*, 525–535.

Carayon, P., Cartmill, R., Hoonakker, P., Schoofs Hundt, A., Karsh, B., Krueger, D., . . . Wetterneck, T. B. (2012). Human factors analysis of workflow in health information technology implementation. In P. Carayon (Ed.), *Handbook of human factors and ergonomics in patient safety* (2nd ed., pp. 507–521). Mahwah, NJ: Lawrence Erlbaum.

Charlton, S. G., & O'Brien, T. G. (2010). *Handbook of human factors testing and evaluation.* Boca Raton, FL: CRC Press.

Cornet, V. P., Daley, C., Bolchini, D., Toscos, T., Mirro, M. J., & Holden, R. J. (2019). Patient-centered design grounded in user and clinical realities: Towards valid digital health. *Proceedings of the International Symposium on Human Factors and Ergonomics in Health Care, 8*(1), 100–104.

Cornet, V. P., Daley, C., Cavalcanti, L. H., Parulekar, A., & Holden, R. J. (2020). Design for self-care. In A. Sethumadhavan & F. Sasangohar (Eds.), *Design for health* (pp. 277–302). Amsterdam: Elsevier.

Cornet, V. P., Daley, C. N., Srinivas, P., & Holden, R. J. (2017). User-centered evaluations with older adults: Testing the usability of a mobile health system for heart failure self-management. *Proceedings of the Human Factors and Ergonomics Society Annual Meeting, 61*(1), 6–10.

Cornet, V. P., & Holden, R. J. (2018). Systematic review of smartphone-based passive sensing for health and wellbeing. *Journal of Biomedical Informatics, 77*, 120–132.

Cornet, V. P., Toscos, T., Bolchini, D., Rohani Ghahari, R., Ahmed, R., Daley, C. N., . . . Holden, R. J. (2020, forthcoming). Untold stories in user-centered design of mobile health: Practical challenges and strategies learned from the design and evaluation of an app for older adults with heart failure. Journal of Medical Internet Research mHealth and uHealth. doi:10.2196/17703

Daley, C., Al-Abdulmunem, M., & Holden, R. J. (2019). Knowledge among patients with heart failure: A narrative synthesis of qualitative research. *Heart & Lung, 48*, 477–485.

Daley, C., Bolchini, D., Varrier, A., Rao, K., Joshi, P., Blackburn, J., . . . Holden, R. J. (2018). Naturalistic decision making by older adults with chronic heart failure: An exploratory study using the critical incident technique. *Proceedings of the Human Factors and Ergonomics Society Annual Meeting, 62*, 568–572.

Daley, C., Chen, E. M., Roebuck, A. E., Ghahari, R. R., Sami, A. F., Skaggs, C. G., . . . Toscos, T. R. (2017). Providing patients with implantable cardiac device data through a personal health record: A qualitative study. *Applied Clinical Informatics, 8*(4), 1106–1116.

Dyer, J., Gregersen, H., & Christensen, C. M. (2011). *The innovator's DNA: Mastering the five skills of disruptive innovators.* Boston, MA: Harvard Business Press.

Ghahari, R. R., Holden, R. J., Flanagan, M. E., Wagner, S., Martin, E., Ahmed, R., . . . Allmandinger, T. (2018). Using cardiac implantable electronic device data to facilitate health decision making: A design study. *International Journal of Industrial Ergonomics, 64*, 143–154.

Holden, R. J. (2020). A Simplified System Usability Scale (SUS) for cognitively impaired and older adults. *Proceedings of the International Symposium on Human Factors and Ergonomics in Health Care, 9*(1), 180–182.

Holden, R. J., Asan, O., Wozniak, E. M., Flynn, K. E., & Scanlon, M. C. (2016). Nurses' perceptions, acceptance, and use of a novel in-room pediatric ICU technology: Testing an expanded technology acceptance model. *BioMed Central Medical Informatics and Decision Making, 16*, 145.

Holden, R. J., Binkheder, S., Patel, J., & Viernes, S. H. P. (2018). Best practices for health informatician involvement in interprofessional health care teams. *Applied Clinical Informatics, 9*(1), 141–148.

Holden, R. J., Bodke, K., Tambe, R., Comer, R., Clark, D., & Boustani, M. (2016). Rapid translational field research approach for eHealth R&D. *Proceedings of the International Symposium on Human Factors and Ergonomics in Health Care, 5*(1), 25–27.

Holden, R. J., Cornet, V. P., & Valdez, R. S. (2020). Patient ergonomics: 10-year mapping review of patient-centered human factors. *Applied Ergonomics, 82*, 102972. doi.org/10.1016/j.apergo.2019.102972.

Holden, R. J., Daley, C. N., Mickelson, R. S., Bolchini, D., Toscos, T., Cornet, V. P., . . . Mirro, M. J. (2020). Patient decision-making personas: An application of a patient-centered cognitive task analysis (P-CTA). *Applied Ergonomics, 87*, 103107. doi:https://doi.org/10.1016/j.apergo.2020.103107

Holden, R. J., Joshi, P., Rao, K., Varrier, A., Daley, C. N., Bolchini, D., . . . Martin, E. (2018). Modeling personas for older adults with heart failure. *Proceedings of the Human Factors and Ergonomics Society Annual Meeting, 62*(1), 1072–1076.

Holden, R. J., Kulanthaivel, A., Purkayastha, S., Goggins, K. M., & Kripalani, S. (2017). Know thy eHealth user: Development of biopsychosocial personas from a study of older adults with heart failure. *International Journal of Medical Informatics, 108*, 158–167.

Holden, R. J., McDougald Scott, A. M., Hoonakker, P. L. T., Hundt, A. S., & Carayon, P. (2015). Data collection challenges in community settings: Insights from two field studies of patients with chronic disease. *Quality of Life Research, 24*, 1043–1055.

Holden, R. J., & Mickelson, R. S. (2013). Performance barriers among elderly chronic heart failure patients: An application of patient-engaged human factors and ergonomics. *Proceedings of the Human Factors and Ergonomics Society, 57*(1), 758–762.

Holden, R. J., Rivera-Rodriguez, A. J., Faye, H., Scanlon, M. C., & Karsh, B. (2013). Automation and adaptation: Nurses' problem-solving behavior following the implementation of bar coded medication administration technology. *Cognition, Technology & Work, 15*, 283–296.

Holden, R. J., Schubert, C. C., Eiland, E. C., Storrow, A. B., Miller, K. F., & Collins, S. P. (2015). Self-care barriers reported by emergency department patients with acute heart failure: A sociotechnical systems-based approach. *Annals of Emergency Medicine, 66*, 1–12.

Holden, R. J., Schubert, C. C., & Mickelson, R. S. (2015). The patient work system: An analysis of self-care performance barriers among elderly heart failure patients and their informal caregivers. *Applied Ergonomics, 47*, 133–150.

Holden, R. J., Toscos, T., & Daley, C. N. (2020). Researcher reflections on human factors and health equity. In R. Roscoe, E. K. Chiou, & A. R. Wooldridge (Eds.), *Advancing diversity, inclusion, and social justice through human systems engineering* (pp. 51–62). Boca Raton, FL: CRC Press.

Holden, R. J., Voida, S., Savoy, A., Jones, J. F., & Kulanthaivel, A. (2016). Human factors engineering and human–computer interaction: Supporting user performance and experience. In J. Finnell & B. E. Dixon (Eds.), *Clinical informatics study guide* (pp. 287–307). New York: Springer.

International Organization for Standardization. (2010). Ergonomics of human-system interaction – Part 210: Human-centred design for interactive systems. In (Vol. ISO 9241-210:2010). Geneva, Switzerland.

Jacko, J. A. (Ed.) (2012). *The human-computer interaction handbook: Fundamentals, evolving technologies and emerging applications* (3rd ed.). Mahwah, NJ: Lawrence Erbaum Associates.

Johnson, S. (2010). *Where good ideas come from: The natural history of innovation*. New York: Penguin.

Joseph, A., Henriksen, K., & Malone, E. (2018). The architecture of safety: An emerging priority for improving patient safety. *Health Affairs, 37*(11), 1884–1891.

Kahneman, D. (2003). Maps of bounded rationality: Psychology for behavioral economics. *The American Economic Review, 93*(5), 1449–1475.

Katz, B., & Wagner, J. (2014). *The rise of innovation districts: A new geography of innovation in America*. Washington, DC: Brookings Institution.

Lockwood, T. (2009). *Design thinking: Integrating innovation, customer experience, and brand value*. New York: Allworth Press.

Meister, D., & Enderwick, T. P. (2001). *Human factors in system design, development, and testing*. Boca Raton, FL: CRC Press.

Mickelson, R. S., & Holden, R. J. (2017). Capturing the medication management work system of older adults using a digital diary method. *Proceedings of the Human Factors and Ergonomics Society Annual Meeting, 61*(1), 555–559.

Mickelson, R. S., & Holden, R. J. (2018). Medication management strategies used by older adults with heart failure: A systems-based analysis. *European Journal of Cardiovascular Nursing, 17*(5), 418–428.

Millen, D. R. (2000). Rapid ethnography: Time deepening strategies for HCI field research. *Paper presented at the Proceedings of the 3rd Conference on Designing Interactive Systems: Processes, Practices, Methods, and Techniques*, Brooklyn, New York.

Mirro, M. J., Ghahari, R. R., Ahmed, R., Reining, L., Wagner, S., Lehmann, G., . . . Toscos, T. (2018). A patient-centered approach towards designing a novel CIED remote monitoring report. *Journal of Cardiac Failure, 24*(8), S77.
Nielsen, J. (1993). *Usability engineering.* Boston, MA: Academic Press.
Novak, L. L., Unertl, K. M., & Holden, R. J. (2016). Realizing the potential of patient engagement: Designing IT to support health in everyday life. In E. Ammenwerth & M. Rigby (Eds.), *Evidence-based health informatics* (pp. 237–247). Amsterdam: IOS Press.
Pasmore, W. A. (1988). *Designing effective organizations: The sociotechnical systems perspective.* New York: Wiley.
Rasmussen, J., & Goodstein, L. (1987). Decision support in supervisory control of high-risk industrial systems. *Automatica, 23*(5), 663–671.
Reddy, A., Lester, C. A., Stone, J. A., Holden, R. J., Phelan, C. H., & Chui, M. A. (2019). Applying participatory design to a pharmacy system intervention. *Research in Social and Administrative Pharmacy, 15*(11), 1358–1367.
Riegel, B., Lee, C. S., & Dickson, V. V. (2011). Self care in patients with chronic heart failure. *Nature Reviews Cardiology, 8*(11), 644–654.
Rogers, E. M. (2003). *Diffusion of innovations* (5th ed.). New York: Free Press.
Schall, C. E., & McAlister, C. (2019). Organizational construction and interdisciplinary identity in a new health care organization. *Socius, 5*, doi:10.1177/2378023119861258.
Shah, J. J., Smith, S. M., & Vargas-Hernandez, N. (2003). Metrics for measuring ideation effectiveness. *Design Studies, 24*(2), 111–134.
Simon, H. A., & March, J. G. (1958). *Organizations.* New York: John Wiley.
Srinivas, P., Cornet, V., & Holden, R. J. (2017). Human factors analysis, design, and testing of Engage, a consumer health IT application for geriatric heart failure self-care. *International Journal of Human-Computer Interaction, 33*(4), 298–312.
Valdez, R. S., & Holden, R. J. (2016). Health care human factors/ergonomics fieldwork in home and community settings. *Ergonomics in Design, 24*(4), 44–49.
Weinger, M. B., Wiklund, M., & Gardner-Bonneau, D. (Eds.). (2011). *Handbook of human factors in medical device design.* Boca Raton, FL: CRC Press.
Whitefield, A., Wilson, F., & Dowell, J. (1991). A framework for human factors evaluation. *Behaviour & Information Technology, 10*(1), 65–79.
Young, R. A., Roberts, R. G., & Holden, R. J. (2017). The challenges of measuring, improving, and reporting quality in primary care. *The Annals of Family Medicine, 15*(2), 175–182.

11 Quantitative Methods for Analyzing Experimental Studies in Patient Ergonomics Research

Kapil Chalil Madathil and Joel S. Greenstein
Clemson University

CONTENTS

The purpose of this chapter is to show patient ergonomics researchers how to apply some of the basic quantitative methods used to summarize and analyze the data collected from experimental studies. Patient ergonomics research studies the needs of patients and other nonprofessionals who perform health-related work, followed by the design, development, and evaluation of interventions relevant to specific patient populations, environments, and conditions. An experimental study, typically conducted during the intervention development and evaluation phases, consists of collecting a series of observations from participants in a simulated or real-world setting while deliberately manipulating specific variables to answer a particular research question. Both quantitative and qualitative data result from such patient ergonomics studies. Statistical techniques are used to enable patient ergonomics researchers to summarize and describe the quantitative data, subsequently enabling them to make confidence judgments regarding trends and relationships.

Quantitative experimental studies in patient ergonomics research are used to support a wide range of objectives including exploring variables related to enhancing patient work (Chalil Madathil et al., 2013; Scharett et al., 2017; Valdez et al., 2017), understanding the effect of an intervention (Graham et al., 2011; Turner-McGrievy & Tate, 2011; Valle et al., 2013), verifying cause and effect relationships (Abrutyn & Mueller, 2014; Milner et al., 2014), and conducting functional analysis of a system and comparing multiple systems with the primary goal of enhancing patient well-being (Chalil Madathil et al., 2013; Agnisarman et al., 2017; Narasimha et al., 2018).

This chapter provides an overview of the types of data collected from patient ergonomics studies and describes a selected set of statistical methods that can be used by patient ergonomics researchers to analyze experimental outcomes and draw conclusions. In the case study described in Section 11.1, we explore the efficacy of conversational agents for collecting a family health history (FHx) from a patient perspective. Conversational agents engage the users in a conversational dialogue within the user interface to collect and then display their FHx data. To explore whether such a system should be implemented for collecting FHx, certain questions need to be answered:

1. Does a conversational agent make a practical difference in the time taken and errors made in an FHx collection when compared to a conventional form-based data collection system?
2. Does a conversational agent enhance usability, workload burden, patient preferences, and satisfaction when compared to a conventional form-based data collection system?

With this case study, we illustrate statistical techniques that can be used to summarize data, compute margins of error, and determine statistically significant differences between the two conditions.

11.1 CASE STUDY: CONVERSATIONAL INTERFACES FOR FAMILY HEALTH HISTORY COLLECTION

A patient's FHx helps clinicians to diagnose and manage disease risks at an earlier and more treatable stage. Despite its value, the FHx is often underutilized in the clinical setting because clinicians often lack the time and expertise needed to collect a detailed FHx (Reid et al., 2009). To overcome this limitation, various government, academic, and commercial organizations have developed a wide variety of FHx tools that help patients gather and organize their own FHx information outside the clinic (Welch et al., 2018). Most FHx tools consist of web forms, tables, and/or a series of screens with questions to collect data from users. Even though researchers have found that various aspects of these FHx tools are generally acceptable to patients, the utilization of FHx tools outside of research settings is low. Furthermore, concerns about the usability and appropriateness of these tools for low-literacy, underserved populations have been raised (Wang et al., 2015). To improve the use of FHx in health care, researchers have started exploring the use of artificial conversational entities (i.e., chatbots) and virtual assistants as an alternative approach to collecting FHx. Instead of asking users to work through multiple pages of web forms, tables, drop-down menus, and radio buttons, a dialogue-based interface engages with end users in a natural and intuitive way— through conversational dialogue. These entities engage users in a conversational interaction about their FHx similar to how genetic counselors, the gold standard for FHx collection, collect FHx information from patients (Corti & Gillespie, 2015).

Since the use of conversational entities is so new, the research in this area is limited. Investigation in a controlled setting is needed to compare the conversational interfaces with conventional interfaces for FHx data collection in terms of efficacy, efficiency, usability, workload burden, patient preferences, and satisfaction. This case study presents an experimental study conducted to evaluate a conversational entity to collect FHx (see Figure 11.1) by systematically comparing it with a conventional form-based health history collection system (Ponathil et al., 2018).

A within-subject experimental design was used to evaluate and compare the conversational approach with the conventional approach to FHx collection. That is, the same participants used both interfaces, and one independent variable was investigated—the type of data collection interface. The conventional interface was the current version of the Surgeon General's My Family Health Portrait, which consists of a table of relatives and pop-up dialogue boxes to enter information about them. The conversational approach consists of a conversational dialogue in a chat form along the side, with the family pedigree in the center of the screen. Using convenience sampling, 50 participants with a mean age of 34.3 ($s_D = 6.43$) were recruited via email and word-of-mouth for this study. Participants had to be at least 18 years old and have basic computer skills. They were excluded if they used only tablets and not desktop or laptop computers. None of the participants had previous experience with electronic FHx tools.

Fifty participants were recruited via email and word-of-mouth for this study. To minimize the degree to which the order in which the interfaces were presented to the participants impacted the results, half of the participants were presented the conventional interface first, and the other half were presented the conversational interface first. The researchers greeted the participants and briefed them on the study procedure. After consenting to participate in the study, the participants completed a pre-test questionnaire and were then provided with a printed fictional FHx scenario, which included personal information, family history, and previous cancer history in the family. The researchers asked the participants to complete the following tasks: (1) create a user profile; (2) add their fictional family health history; (3) re-access the platform; (4) edit the information; and (5) share the information with a family member. Once the participants completed the tasks using the first interface, they were given a questionnaire about their satisfaction with it and were then asked to share their experience of working on the interface in a discussion with the researcher. After experiencing both the applications, participants were asked to indicate which interface they preferred. The researchers observed the challenges the participants faced while performing the tasks, recorded the errors committed by them, and timed their progress through the tasks.

The pre-test questionnaire collected demographic data as well as information about how much experience the participant had using the internet and related applications. The number of seconds taken to complete the tasks for each tool was calculated, beginning at the time the participant entered the interface to begin the task

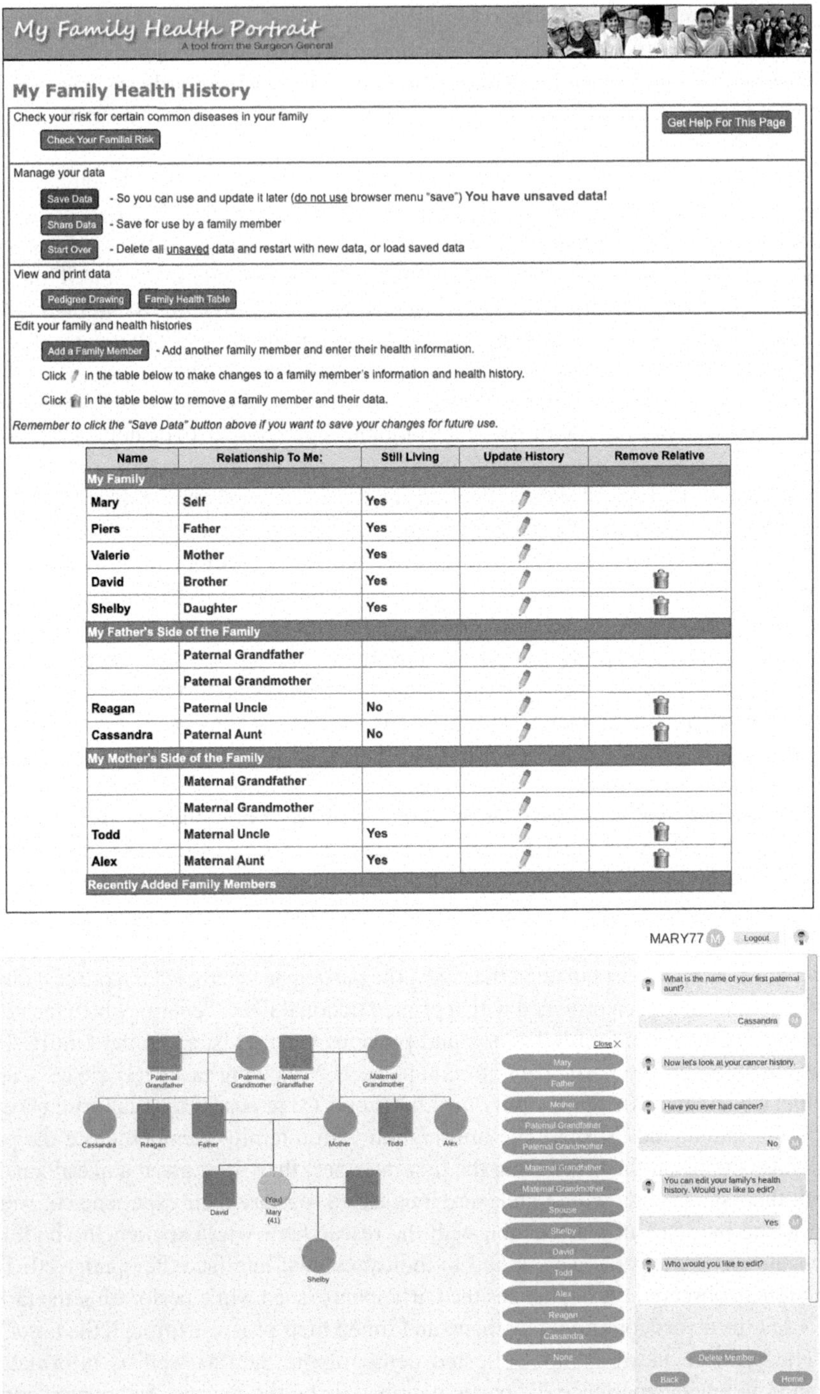

FIGURE 11.1 Conventional (above) and conversational (below) interfaces for FHx collection.

and ending when the participant completed the task. The questionnaires administered each time the participant completed the tasks with an interface addressed the perceived usefulness and ease-of-use of the application. The questions asked were derived from the technology acceptance model (Davis, 1989), NASA task load index (TLX) workload instrument (Hart & Staveland, 1988), and the IBM computer system usability questionnaire (CSUQ) (Lewis, 1995). After using both interfaces and completing the related questionnaires, the participants completed a final post-test questionnaire, ranking the interfaces in terms of preference, with "1" being more preferred and "2" being less preferred.

11.2 DATA COLLECTED FROM EXPERIMENTAL PATIENT ERGONOMICS STUDIES

The experimental design, structure, and measurement type predominantly determine the statistical techniques that can be used to analyze the data collected from patient ergonomics studies. A sound experimental design is required to enable patient ergonomics researchers to gather interpretable comparisons of the effects of manipulated variables. At the very least, a good experimental design consists of identified independent variables and their respective states that will be manipulated or held constant, associated dependent variables that measure the outcomes of the experiment, characteristics and the number of participants to be used, and a scheme for the replication of unique states of the manipulated variables. There are two common methods of collecting data from experimental studies: the between-subjects, or independent, design and the within-subject, or repeated measures, design. The former involves manipulation of the independent variable using different groups of participants, and the latter involves manipulation of the independent variable with the same group. Our case study is an example of a within-subject experimental design. The independent variable in our study is the type of FHx data collection interface. It is tested at two levels—conventional and conversational.

Two types of variability can potentially arise in such experimental studies: (1) systematic variation (i.e., the variability due to the deliberate manipulation of independent variables created by the researcher) and (2) random variation (i.e., the variation due to random factors not controlled by the researcher). Statistical techniques enable us to quantify such systematic and random variations.

Data structure is another factor that determines the analysis technique. It consists of two broad classifications: structured and unstructured. Structured data, also called quantitative data, consist of measures such as rating scales, counts, durations, and category classifications. Examples of data that may be unstructured include observations, comments made in focus groups and on social networks, and video and audio recordings (see Chapters 9 and 12 in this volume).

The measurement type characterizes the amount of information a measure contains with respect to the construct being measured and is broadly classified into four categories ordered from the least to most informative: nominal, ordinal, interval, and ratio levels of measurement. A nominal or categorical measurement refers to the category that an item belongs to (e.g., gender or race). An ordinal measure consists of a rank ordering of items, with no information regarding the magnitude of the difference between consecutive ranks. In our case study, participants were asked to rank-order the two interfaces in terms of preference. Approximately 72% of participants preferred the conversational interface to the conventional one. An interval measurement provides information regarding both the order and magnitude of the difference between values. Interval scales arbitrarily assign a zero score and have equal intervals on the scale representing equal differences in the variable being measured. The 7-point Likert scale used in the IBM CSUQ is an example of an interval scale. The first item of the CSUQ, "Overall, I am satisfied with how easy it is to use this system," invites a response of a number from 1 to 7, where "1" indicates strong agreement and "7" indicates strong disagreement. A ratio measurement, in addition to indicating the order and the magnitude of the difference between values, has a true zero score. The true zero means the complete absence of the attribute being measured. For instance, in the case study, participants were asked to complete the task of adding a family member's health history using two interfaces. The number of errors made by the participant in performing this task is one attribute that was measured to evaluate the interface. In this case, a value of zero has a physical meaning—the participant made no errors while completing the task. Ratio measures permit the determination of averages, ratios, and percentages.

Table 11.1 lists the data collected in our case study along with their respective measurement type. Table 11.2 provides a few examples of quantitative measures that can be collected in patient ergonomics studies.

Statistical inference techniques enable us to draw conclusions about a population from a sample. A sample is a subset of participants from the population about which we are interested in developing conclusions. Analyzing the data collected from patient ergonomics studies, like other experimental studies, consists of determining descriptive statistics (such as the mean, median, and standard deviation) to summarize the data collected for the sample, followed by the determination of inferential statistics (such as confidence intervals and effect sizes) to generalize the findings from the sample to the entire population.

11.3 UNDERSTANDING THE DATA, THEIR DISTRIBUTION, AND MARGINS OF ERROR COMPUTATION

The data generated from patient ergonomics studies are inherently noisy. That is, not every participant responds identically to the same stimulus. Nor does an individual

TABLE 11.1
Data Collected in the Case Study and Their Respective Measurement Type

Measurement Type	Data Collected in the Case Study
Nominal	Gender, race
Ordinal	Educational level, experience with computers, preference rank
Interval	IBM CSUQ, measures derived from the Technology Acceptance Model
Ratio	Task completion time, task completion rate, NASA-TLX index, errors made, age

TABLE 11.2
Examples of Quantitative Data Collected from Patient Ergonomics Studies

Measures	Examples
Efficiency measures	Task completion time, task completion rate, frequency of action, and error rate.
Workload measures	NASA-TLX indices, instantaneous self-assessment scale, mental workload index, primary and secondary task performance metrics.
Situation awareness measures	Situation Awareness Global Assessment Technique (SAGAT), Situation Awareness Rating Technique (SART)
Physiological measures	Heart rate and its variability; eye-gaze data such as blink rate, blink durations, number of fixations, duration of fixations, pupil size variability; integrated scores of physiological data; and electroencephalogram data.
Usability measures	System Usability Scale (SUS), Questionnaire for User Interaction Satisfaction (QUIS), Software Usability Measurement Inventory (SUMI), Computer System Usability Questionnaire (CSUQ), After Scenario Questionnaire (ASQ), Usefulness, Satisfaction, and Ease of Use Questionnaire (USE), Purdue Usability Testing Questionnaire (PUTQ), End-User Computing Satisfaction Questionnaire (EUCS), and metrics derived from the Technology Acceptance Model.
Social network measures	Social relations, interactions, information flow, membership in groups and events, and co-participation in events

participant respond identically to a repeated stimulus. Visual methods and normality tests are frequently used to understand the characteristics of the data's distribution. Frequency distributions, box plots, stem-and-leaf plots, probability–probability plots, and quantile–quantile plots are commonly used visual methods to explore the overall distribution of data and to make an educated decision to include a data point in the statistical analysis. In our case study, the frequency distributions of the perceived workload experienced by the 50 participants as they completed the tasks using the conversational and conventional interfaces are shown in Figure 11.2.

Frequency distributions can be symmetrical or asymmetrical. One often-found symmetrical frequency distribution is the normal distribution. It has a bell-shaped curve with a vast majority of the scores located near the center of the distribution. The frequency distributions of workload scores shown in Figure 11.2 show a symmetric distribution. In contrast, in asymmetrical frequency distributions, the majority of the scores are clustered toward one end of the distribution. Asymmetrical frequency distributions can be positively or negatively skewed, with the former having frequent scores clustered at the lower end of the distribution and the latter having scores clustered at the higher end of the distribution. Commonly used statistical tests to assess the normality of a sample distribution include the Kolmogorov–Smirnov and the Shapiro–Wilk tests. A statistically significant result from these tests suggests the distribution is non-normal. Additional sources provide details on assessing normality and analyzing data with non-normal distributions (Cohen et al., 2014; Cumming & Calin-Jageman, 2016; Field, 2013).

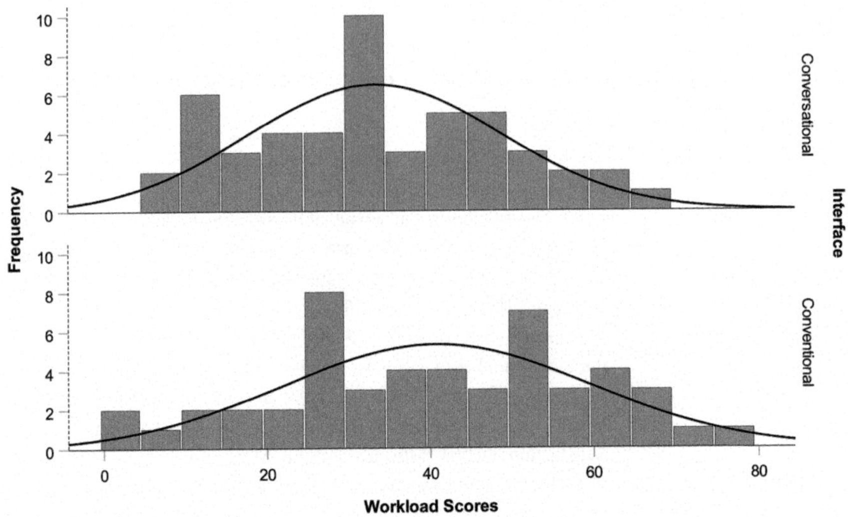

FIGURE 11.2 Frequency histograms displaying the distribution of workload scores for the conversational and conventional interfaces.

11.3.1 Data Summarization

Descriptive statistics include quantitative measures that enable a researcher to summarize the data collected from experimental studies. Some provide information regarding central tendency, including mean, median, and mode; others provide information regarding the dispersion of the data. The mean is the average value of the data set, the median is the value that divides the data set into equal halves, and the mode is the value that occurs most frequently. Though the arithmetic mean, median, and mode are used to summarize the central tendency of the majority of dependent variables, for certain dependent measures, such as the time taken to complete a task, small sample sizes will often have a positively skewed distribution. In this case, the geometric mean is used because it provides a better estimate of the middle value of the population when the distribution is skewed. To quantify the dispersion of data around a measure of central tendency, measures such as the variance, standard deviation, range (difference between the largest and smallest observations), and percentiles are used.

11.3.2 Confidence Intervals (CI)

The lack of access to data from the entire patient population requires patient ergonomics researchers to make estimates of population parameters based on the data collected from a small sample. A confidence interval (CI) is a range of values developed from the data collected from the sample that has a specified chance of containing the true value of the population parameter. Any value provided within the CI is considered to be a plausible one for the population parameter, whereas those outside are considered to be implausible (Smithson, 2002). The three components that affect the length of the CI are (1) confidence level; (2) the variability of the data, estimated using the standard deviation; and (3) the sample size. The confidence level, usually set by convention at 95%, suggests that if a researcher were to sample the population 100 times, the calculated 95% confidence interval would contain the value of the population parameter 95 times. In the next sections, we illustrate the calculation of CIs for the metrics of task completion, task completion time, and subjective ratings in the FHx case study.

11.3.3 Quantifying Successful Task Completion

The case study consists of participants completing an FHx information entry task using a conventional and a conversational interface. This produced the binomial distribution of successes and failures denoted as 0's and 1's, respectively. Forty-six of the 50 participants were successfully able to complete the task for the conversational interface. The adjusted-Wald method (Agresti & Coull, 1998) may be used to calculate the binomial confidence interval. Sauro and Lewis (2010, 2016) discuss the merits of this approach when compared to alternative methods for calculating binomial confidence intervals. Assuming a 95% CI, we can calculate the confidence interval for successful task completion using two steps: (1) calculate the adjusted proportion and adjusted sample size, and (2) use Wald's method to calculate the binomial confidence intervals with the adjusted proportion.

$$\text{Adjusted Wald proporation, } \hat{p} = \left(x + \left(z^2_{\left(1-\frac{\alpha}{2}\right)} \Big/ 2 \right) \right) \Big/ \hat{n} \tag{11.1}$$

where x = number of participants successfully completing the task
n = sample size

$$\hat{n} = \left(n + z^2_{\left(1-\frac{\alpha}{2}\right)} \right) = \text{the adjusted sample size} \tag{11.2}$$

α = likelihood of the true population parameter being outside the confidence interval
$z_{\left(1-\frac{\alpha}{2}\right)}$ = critical value of the normal distribution (For the 95% confidence level, $\alpha = 0.05$ and $z_{(0.975)}$ is 1.96. This value is determined from a statistical table or software.)

The computations are then

$$\hat{p} = \left(46 + \frac{1.96^2}{2} \right) \Big/ \left(50 + 1.96^2 \right) = 0.890$$

$$\begin{aligned} \text{Confidence interval, CI} &= \hat{p} \pm z_{\left(1-\frac{\alpha}{2}\right)} \sqrt{\frac{\hat{p}(1-\hat{p})}{\hat{n}}} \\ &= 0.89 \pm 1.96 \sqrt{\frac{0.89(1-0.89)}{50+1.96^2}} = 0.81, 0.97 \end{aligned} \tag{11.3}$$

Thus, with a 95% confidence level, we estimate that the actual task completion rate on the conversational interface for the patient population from which our sample was taken will be between 81% and 97%.

11.3.4 Quantifying Task Completion Time

For continuous data that follow a normal distribution, we can calculate the mean and standard deviation and use the t-distribution to calculate the CI. In our case study, the mean and standard deviation for the dependent variable of time taken to complete the first task with the conventional interface with 50 participants were 76.02 and 31.35 seconds, respectively. The respective CI can be calculated using the following formula:

$$\mathrm{CI} = \bar{x} \pm t_{\left(\frac{\alpha}{2},\, n-1\right)}\left[\frac{s}{\sqrt{n}}\right] \tag{11.4}$$

where $\bar{x}$ = sample mean
n = sample size
α = likelihood of the true population parameter being outside the confidence interval
s = standard deviation of the sample
$t_{\left(\frac{\alpha}{2},\, n-1\right)}$ = the upper $\alpha/2$ % point of the t-distribution, a function of α and n. This value is determined from a statistical table or software.

For $\alpha = 0.05$ and $n = 50$, $t_{(0.025,\, 49)} = 2.01$

The computations are then

$$\mathrm{CI} = 76.02 \pm 2.01\left[\frac{31.35}{\sqrt{50}}\right] = 67.11,\ 84.93$$

Thus, with a 95% confidence level, we can say that the average time taken to complete the task for the entire population is between 67.11 and 84.93 seconds.

The median value is a better estimate of central tendency for studies with a large sample size and values dispersed as a skewed distribution. Sauro and Lewis (2016) recommend a method akin to calculating CI for percentiles to calculate the confidence intervals around the median value. Sauro and Lewis (2010, 2016) found that the geometric mean, as opposed to the arithmetic mean and mode, is a better estimate of the task time with a sample size of less than 25 participants. They illustrate the steps to compute the geometric mean and associated confidence intervals (Sauro & Lewis, 2016).

11.3.5 Quantifying Responses from Subjective Rating Scales

The responses collected from a subjective rating scale consist of interval data. Patient ergonomics researchers should take care to use reliable and validated scales with acceptable psychometric properties to collect data. The most common analysis procedures include developing a composite score that combines the responses from multiple questions (such as the mean of the responses to multiple questions) or analyzing each item on the scale separately. The use of parametric methods to analyze Likert scale responses has been controversial since they are not ratio data. However, studies using simulated and real data suggest that parametric tests are robust when analyzing Likert scale responses (Norman, 2010). In our case study, we used the IBM CSUQ to analyze overall satisfaction with each interface. Lewis (1995) recommends calculating overall satisfaction by averaging the scores from items 1 to 19 of the questionnaire. Equation (11.4) can then be used to calculate the CIs for satisfaction with the interface.

11.4 COMPARING TWO CONDITIONS

Patient ergonomics studies often involve analyzing the efficacy of an intervention by systematically comparing it with an existing system. In our case study, we compared a conventional and a conversational interface on the time to complete the task and error rate, perceived workload, and user satisfaction. The sample size, variability in the data, and the size of the performance difference between the two interfaces are the key factors that determine statistically significant differences. We will be able to make a confidence judgment provided we have a large enough sample size, low enough data variability, and large enough differences in the values of the dependent variables between the interfaces. To determine statistically significant differences between the means of interval or ratio data, we can use the t-test. The four assumptions that need to be met for a t-test to be appropriate are the following: (1) normality of the sampling distribution of the differences between scores; (2) similar variances for the two data sets being compared; (3) the responses provided by the participants are independent of those of the other participants; and (4) the samples are representative of the patient population. An independent-samples t-test is used for studies using a between-subjects experimental design in which different participants are assigned to different conditions. A dependent-samples t-test is used to analyze studies using a within-subject experimental design as in our case study, in which the participants completed tasks using both interfaces. The evaluation is done by calculating the CI using the following formula:

$$\mathrm{CI} = \hat{D} \pm t_{\left(\frac{\alpha}{2},\ n-1\right)}\left(\frac{s_D}{\sqrt{n}}\right) \tag{11.5}$$

where $\hat{D}$ = mean difference between the scores

s_D = standard deviation of the difference scores

n = sample size

α = likelihood of the true population parameter being outside the confidence interval

$t_{\left(\frac{\alpha}{2},\ n-1\right)}$ = the upper $\alpha/2$ % point of the t-distribution, a function of α and n. This value is determined from a statistical table or software. For $\alpha = 0.05$; $n = 50$, $t_{(0.025,\ 49)} = 2.01$.

In a dependent-samples t-test, we analyze the differences between the scores of two conditions. Using the dependent variable of overall user satisfaction in our case study, we illustrate the steps to compute the confidence interval. The mean overall satisfaction scores for the conventional and conversational interfaces were 5.62 ($s_D = 0.86$) and 4.72 ($s_D = 1.32$) respectively. First, we need to calculate the mean and standard deviation of the differences of the scores between the two conditions; $\hat{D} = 0.897$; $s_D = 1.473$. We can then compute the confidence interval as follows:

$$\text{CI} = \hat{D} \pm t_{\left(\frac{\alpha}{2},\ n-1\right)}\left(\frac{s_D}{\sqrt{n}}\right) = 0.897 \pm 2.01\left(\frac{1.473}{\sqrt{50}}\right) = 1.32, 0.48$$

Thus, at a 95% confidence level, the difference between the overall satisfaction scores of the interfaces lies between 0.48 and 1.32. Since the lower bound on the CI is greater than zero, this difference is statistically significant at the α level of 0.05. It is a good practice to go on to compute the effect size, a standardized measure of the magnitude of the finding, to gauge if it is substantive. Measures such as the Pearson correlation coefficient r and Cohen's d are commonly used measures of effect sizes. The computation of Pearson's correlation coefficient r using the t-statistic and degrees of freedom ($n-1$) is as follows.

$$t = \hat{D} / \left(\frac{s_D}{\sqrt{n}}\right) = \frac{0.897}{\left(\frac{1.473}{\sqrt{50}}\right)} = 4.31$$

$$r = \sqrt{\frac{t^2}{\left(t^2 + df\right)}} = \sqrt{\frac{4.31^2}{4.31^2 + 49}} = 0.5 \tag{11.6}$$

Pearson's correlation coefficients of 0.1, 0.3, and 0.5 represent small, medium, and large effects, respectively. Accordingly, our effect size of 0.5 represents a large effect, representing a substantive finding. We can thus make a confidence judgment that the satisfaction was significantly higher for the conversational interface than the conventional one, $t(49) = 4.31$, $p < 0.05$, $r = 0.5$. A bar chart depicting the results graphically is shown in Figure 11.3.

Analysis of variance (ANOVA) is used to analyze data with more than two conditions. Nonparametric tests are used to analyze data with non-normal distributions. Nonparametric tests often employ a ranking technique, which generates a data set with the higher scores represented by large ranks and the lower scores by small ranks. The most commonly used nonparametric tests include the Mann–Whitney test, the Wilcoxon signed-rank test, Friedman's test, and the Kruskal–Wallis test. Cohen et al. (2002) and Cumming and Calin-Jageman (2017) present the details of the statistical analyses that have been introduced here, as well as those for independent t-tests, ANOVA, and nonparametric tests.

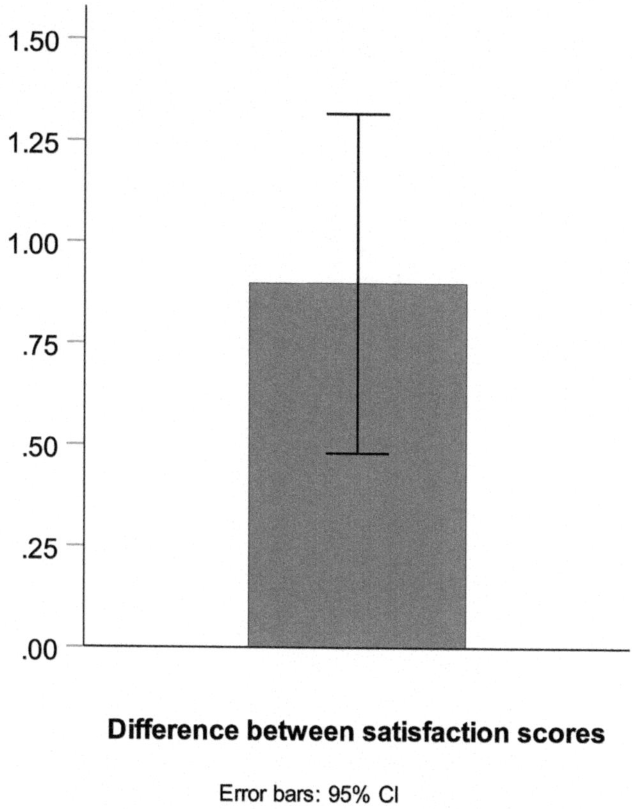

FIGURE 11.3 Representation of the difference between satisfaction scores on a bar graph with the 95% confidence interval.

11.5 RECOMMENDATIONS FOR PATIENT ERGONOMICS RESEARCHERS

Quantitative experimental studies in patient ergonomics play an important role in improving the quality and effectiveness of health care (see examples in Table 11.3). For example, experimental studies have been reported in the literature to evaluate the design and efficacy of technologies used by patients or using patient data, including electronic consenting (Chalil Madathil et al., 2011; Madathil et al., 2013; Sanderson et al., 2013; Koikkara et al., 2015), remote monitoring and telemedicine systems (Agnisarman et al., 2017), electronic health records (Ratwani et al., 2018; Ponathil et al., 2019), decision support systems (Chalil Madathil, 2013; Khasawneh et al., 2018), and online health communities (Chalil Madathil et al., 2013; Narasimha et al., 2019).

Research objectives should govern the data collection and analysis plan for patient ergonomics studies. In addition, a sound experimental design is the key to quantitative patient ergonomics studies as it enables a researcher to use multiple statistical analysis techniques to gain insights. Experimental design should not be a consequence

TABLE 11.3
Examples of Experimental Studies in Patient Ergonomics

Objective	Intervention	Outcome Measures	Analysis Technique	Key Findings
Evaluate the efficacy of e-consenting systems in healthcare settings. (Madathil et al., 2013)	Tablet-based e-consenting system	Task completion time, errors made, NASA-TLX indices, and usability	Descriptive statistics and Analysis of variance (ANOVA)	No significant differences in the time taken to complete the tasks. Participants found the new system to be more usable than the conventional one.
Evaluate the usability of home-based telemedicine systems. (Agnisarman et al., 2017)	Four home-based telemedicine software platforms	Task completion time, errors made, NASA-TLX indices, and usability	Descriptive statistics and ANOVA	Unwieldy session initiation phase, poor interfaces, and low information quality led to lower levels of user satisfaction.
Evaluate the usability of electronic health record prototypes developed for tablet computers. (Karahoca et al., 2010)	Mobile Emergency Department Software and Mobile Emergency Department Software Iconic	Task completion rate, average completion time, and usability.	Descriptive statistics and independent-samples t-test	Perceived usability was higher for Mobile Emergency Department Software Iconic than for the Mobile Emergency Department Software
Evaluate the usability of electronic health records. (Ratwani et al., 2018)	Electronic health record systems (Cerner and Epic)	Task duration, number of clicks needed to complete each task, and accuracy	Descriptive statistics and ANOVA	Wide variability found in task completion time, clicks, and error rates. Highlighted the need for improved evaluation and implementation process
Evaluate the efficacy of concept and mind maps to enhance patient consent comprehension. (Koikkara et al., 2015)	Mind maps and concept maps	Task duration, comprehension, NASA-TLX indices, and usability	Descriptive statistics and ANOVA	Concept and mind maps improve consent comprehension compared to a text-based consent form.
Evaluate the informational needs of ovarian cancer patients and their supporters. (Chalil Madathil et al., 2013)	None	Type of information desired: ovarian cancer specific, treatment-related, or coping	Content analysis and multinomial logistic regression	Treatment-related information was most sought by patients, whereas coping information was most sought by supporters. When forum posts were negative in tone, the information seekers were more likely to be looking for ovarian cancer specific information.

of using a statistical technique; rather, research questions and experimental design should govern the data analysis technique.

A significant challenge with patient ergonomics studies is to obtain high-quality, timely, adequate, and actionable evidence. Some recent patient ergonomic studies have made use of crowdsourcing applications such as Amazon Mechanical Turk and online survey panels such as Qualtrics Research Services to collect data from a representative patient population. For instance, recent patient ergonomics studies (Agnisarman et al., 2018; Chalil Madathil & Greenstein, 2018) collected data using Amazon Mechanical Turk to understand the effectiveness of adding narratives to health-care public reports to support the decision-making process of patients. Research investigating the effectiveness of such crowdsourcing platforms as a behavioral testing platform has demonstrated that studies conducted online can provide results like those obtained from conventional laboratory studies (Mason & Suri, 2012; Paolacci & Chandler, 2014; Peer et al., 2014). Techniques such as *a priori* power analysis can be used to determine an adequate number of participants required for an experimental study. Power analysis techniques are described in detail by Myers et al. (2013).

Before conducting any statistical tests, the nature of the research data should be explored using graphical displays. Statistical tests are based on assumptions regarding the population and the data. The validity of the results of a statistical test is dependent on the satisfaction of these assumptions, with violations leading to unreliable inferences. Patient ergonomics studies may result in data with non-normal distributions with extreme skewness and outliers. In such scenarios, robust statistical techniques, a set of analysis techniques where deviation from a test assumption does not affect the inferences, are appropriate. Cumming and Calin-Jagemen (2016) provide a detailed overview of robust statistical methods.

Null Hypothesis Significance Testing (NHST) has in the past been perhaps the dominant approach used by patient ergonomics researchers to analyze experimental outcomes, where testable statements called experimental and null hypotheses are developed to understand the effect of an intervention on a patient population. The experimental hypothesis suggests the presence of an effect and the null hypothesis, its absence. In the NHST approach, the researcher sets a criterion at a significance level, typically 0.05, and decides whether to reject it ($p<0.05$) or not ($p\geq 0.05$). In contrast, for various reasons including ease of understanding by researchers and their audience, Cumming and Calin-Jagemen (2016) recommend that research questions be posed and that the data be analyzed and inferences generated using interval effect size estimates, the approach that we have chosen to present.

In this chapter, we have illustrated a few basic quantitative techniques for summarizing and generalizing the findings of patient ergonomics studies. Additional parametric techniques such as correlation, regression, and ANOVA as well as non-parametric techniques can also be appropriately used in patient ergonomics studies.

REFERENCES

Abrutyn, S., & Mueller, A. S. (2014). Are suicidal behaviors contagious in adolescence? Using longitudinal data to examine suicide suggestion. *American Sociological Review, 79*(2), 211–227.

Agnisarman, S. O., Chalil Madathil, K., Smith, K., Ashok, A., Welch, B., & McElligott, J. T. (2017). Lessons learned from the usability assessment of home-based telemedicine systems. *Applied Ergonomics, 58*, 424–434.

Agnisarman, S., Ponathil, A., Lopes, S., & Chalil Madathil, K. (2018). An investigation of consumer's choice of a healthcare facility when user-generated anecdotal information is integrated into healthcare public reports. *International Journal of Industrial Ergonomics, 66*, 206–220.

Agresti, A., & Coull, B. A. (1998). Approximate is better than "exact" for interval estimation of binomial proportions. *The American Statistician, 52*(2), 119–126.

Chalil Madathil, K., Koikkara, R., Gramopadhye, A. K., & Greenstein, J. S. (2011). An empirical study of the usability of consenting systems: iPad, touchscreen and paper-based systems. In *Proceedings of the Human Factors and Ergonomics Society's Annual Meeting.* Las Vegas, NV.

Chalil Madathil, K., Greenstein, J., Neyens, D. M., Juang, K., & Gramopadhye, A. K. (2013). An investigation of the informational needs of ovarian cancer patients and their supporters. In *Proceedings of the Human Factors and Ergonomics Society's International Annual Meeting.* San Diego, CA.

Chalil Madathil, K. (2013). *Building the science of healthcare public reporting: Integrating anecdotal information to enhance sensemaking* (Doctoral dissertation). Clemson, SC: Clemson University.

Chalil Madathil, K., & Greenstein, J. S. (2018). An investigation of the effect of anecdotal information on the choice of a healthcare facility. *Applied Ergonomics, 70*, 269–278.

Cohen, P., West, S. G., & Aiken, L. S. (2014). *Applied multiple regression/correlation analysis for the behavioral sciences.* Psychology Press.

Corti, K., & Gillespie, A. (2015). A truly human interface: Interacting face-to-face with someone whose words are determined by a computer program. *Frontiers in Psychology, 6*, 634.

Cumming, G., & Calin-Jageman, R. (2016). *Introduction to the new statistics: Estimation, open science, and beyond.* Routledge, New York, NY 11107.

Davis, F. D. (1989). Perceived usefulness, perceived ease of use, and user acceptance of information technology. *The Mississippi Quarterly, 13*(3), 319–340.

Field, A. (2013). *Discovering statistics using IBM SPSS statistics.* SAGE, London, UK.

Graham, A. L., Cobb, N. K., Papandonatos, G. D., Moreno, J. L., Kang, H., Tinkelman, D. G., . . . Abrams, D. B. (2011). A randomized trial of Internet and telephone treatment for smoking cessation. *Archives of Internal Medicine, 171*(1), 46–53.

Hart, S. G., & Staveland, L. E. (1988). Development of NASA-TLX (Task Load Index): Results of empirical and theoretical research. In P. A. Hancock & N. Meshkati (Eds.), *Advances in psychology* (Vol. 52, pp. 139–183). North Holland Publishing Co. Amsterdam, Netherlands.

Karahoca, A., Bayraktar, E., Tatoglu, E., & Karahoca, D. (2010). Information system design for a hospital emergency department: A usability analysis of software prototypes. *Journal of Biomedical Informatics, 43(2)*, 224–232.

Khasawneh, A., Ponathil, A., Firat Ozkan, N., & Chalil Madathil, K. (2018). How should I choose my dentist? A preliminary study investigating the effectiveness of decision aids on healthcare online review portals. In *Proceedings of the Human Factors and Ergonomics Society's Annual Meeting.* Los Angeles, CA.

Koikkara, R., Greenstein, J. S., & Madathil, K. C. (2015). The effect of graphic organizers on the performance of electronic consenting systems. In *Proceedings of the Human Factors and Ergonomics Society Annual Meeting.* Los Angeles, CA.

Lewis, J. R. (1995). IBM computer usability satisfaction questionnaires: Psychometric evaluation and instructions for use. *International Journal of Human–Computer Interaction, 7*(1), 57–78.

Madathil, K. C., Koikkara, R., Obeid, J., Greenstein, J. S., Sanderson, I. C., Fryar, K., . . . Gramopadhye, A. K. (2013). An investigation of the efficacy of electronic consenting interfaces of research permissions management system in a hospital setting. *International Journal of Medical Informatics, 82*(9), 854–863.

Mason, W., & Suri, S. (2012). Conducting behavioral research on Amazon's Mechanical Turk. *Behavior Research Methods, 44*(1), 1–23.

Milner, A., Page, A., & LaMontagne, A. D. (2014). Cause and effect in studies on unemployment, mental health and suicide: A meta-analytic and conceptual review. *Psychological Medicine, 44*(5), 909–917.

Myers, J. L., Well, A. D., & Lorch, R. F., Jr. (2013). *Research design and statistical analysis.* Routledge, New York, NY 10016.

Narasimha, S., Agnisarman, S., Chalil Madathil, K., Gramopadhye, A., & McElligott, J. T. (2018). Designing home-based telemedicine systems for the geriatric population: An empirical study. *Telemedicine and e-Health, 24*(2), 94–110.

Narasimha, S., Wilson, M., Dixon, E., Davis, N., & Chalil Madathil, K. (2019). An investigation of the interaction patterns of peer patrons on an online peer-support portal for informal caregivers of Alzheimer's patients. *Journal of Consumer Health on the Internet, 23*(4), 313–342.

Norman, G. (2010). Likert scales, levels of measurement and the "laws" of statistics. *Advances in Health Sciences Education: Theory and Practice, 15*(5), 625–632.

Paolacci, G., & Chandler, J. (2014). Inside the Turk: Understanding Mechanical Turk as a participant pool. *Current Directions in Psychological Science, 23*(3), 184–188.

Peer, E., Vosgerau, J., & Acquisti, A. (2014). Reputation as a sufficient condition for data quality on Amazon Mechanical Turk. *Behavior Research Methods, 46*(4), 1023–1031.

Ponathil, A., Firat Ozkan, N., Bertrand, J., Welch, B., & Chalil Madathil, K. (2018). New approaches to collecting family health history--A preliminary study investigating the efficacy of conversational systems to collect family health history. In *Proceedings of the Human Factors and Ergonomics Society's Annual Meeting*, Los Angeles, CA.

Ponathil, A., Firat Ozkan, N., Bertrand, J., Welch, B., & Chalil Madathil, K. (2019). Conversational systems for family health history collection for geriatric population. In *Proceedings of the Human Factors and Ergonomics Society's Annual Meeting.* Los Angeles, CA.

Ratwani, R. M., Savage, E., Will, A., Arnold, R., Khairat, S., Miller, K., . . . Hettinger, A. Z. (2018). A usability and safety analysis of electronic health records: A multi-center study. *Journal of the American Medical Informatics Association, 25*(9), 1197–1201.

Reid, G. T., Walter, F. M., Brisbane, J. M., & Emery, J. D. (2009). Family history questionnaires designed for clinical use: A systematic review. *Public Health Genomics, 12*(2), 73–83.

Sanderson, I. C., Obeid, J. S., Madathil, K. C., Gerken, K., Fryar, K., Rugg, D., . . . Moskowitz, J. (2013). Managing clinical research permissions electronically: A novel approach to enhancing recruitment and managing consents. *Clinical Trials, 10*(4), 604–611.

Sauro, J., & Lewis, J. R. (2010). Average task times in usability tests: What to report? In *Proceedings of the Special Interest Group on Computer-Human Interaction Conference on Human Factors in Computing Systems*, 2347–2350. New York, NY.

Sauro, J., & Lewis, J. R. (2016). *Quantifying the user experience: Practical statistics for user research.* Morgan Kaufmann.

Scharett, E., Madathil, K. C., Lopes, S., Rogers, H., Agnisarman, S., Narasimha, S., . . . Dye, C. (2017). An investigation of the information sought by caregivers of Alzheimer's patients on online peer support groups. *Cyberpsychology, Behavior and Social Networking, 20*(10), 640–657.

Smithson, M. (2002). *Confidence intervals.* SAGE Publications.

Turner-McGrievy, G., & Tate, D. (2011). Tweets, apps, and pods: Results of the 6-month Mobile Pounds Off Digitally (Mobile POD) randomized weight-loss intervention among adults. *Journal of Medical Internet Research*, *13*(4), e120.

Valdez, R. S., Guterbock, T. M., Fitzgibbon, K., Williams, I. C., Wellbeloved-Stone, C. A., Bears, J. E., & Menefee, H. K. (2017). From loquacious to reticent: Understanding patient health information communication to guide consumer health IT design. *Journal of the American Medical Informatics Association*, *24*(4), 680–696.

Valle, C. G., Tate, D. F., Mayer, D. K., Allicock, M., & Cai, J. (2013). A randomized trial of a Facebook-based physical activity intervention for young adult cancer survivors. *Journal of Cancer Survivorship: Research and Practice*, *7*(3), 355–368.

Wang, Bickmore, T., Bowen, D. J., Norkunas, T., Campion, M., Cabral, H., . . . Paasche-Orlow, M. (2015). Acceptability and feasibility of a virtual counselor (VICKY) to collect family health histories. *Genetics in Medicine: Official Journal of the American College of Medical Genetics*, *17*(10), 822–830.

Welch, B. M., Wiley, K., Pflieger, L., Achiangia, R., Baker, K., Hughes-Halbert, C., . . . Doerr, M. (2018). Review and comparison of electronic patient-facing family health history tools. *Journal of Genetic Counseling*, *27*(2), 381–391.

12 Emerging Methods for Patient Ergonomics

Mustafa Ozkaynak
University of Colorado Anschutz Medical Campus

Laurie Lovett Novak
Vanderbilt University Medical Center

Yong K. Choi
Department of Public Health Sciences/
Health Informatics Division
School of Medicine, UC Davis

Rohit Ashok Khot
Royal Melbourne Institute of Technology

CONTENTS

This chapter discusses four emerging methods for data collection and intervention delivery to support patient work: (1) qualitative methods, (2) augmented and virtual reality, (3) Internet of Things (IoT) and sensors, and (4) gamification. Although these can be used for both data collection and intervention delivery, we highlight how (1) and (2) support data collection and how (3) and (4) support intervention delivery. We provide a review of research on the use of these methods in patient work and discuss their strengths and weaknesses. We provide case examples in which these methods were used or, in the absence of examples, discuss potential uses.

12.1 EMERGING QUALITATIVE METHODS

Qualitative methods have long been used to produce rich data on the experience of health and illness in everyday life, including the role and impact of technology. Traditional methods have included ethnographic inquiry, observation of activities, interviews, focus groups, surveys, document and artifact analyses, and spatial analyses (see also Chapter 9 in this volume). Patient work researchers have access to new tools and methods to obtain rich data describing the experience of illness and caregiving in everyday life and interpret the data in ways that lead to insights for the design of new interventions.

12.1.1 Methods of Data Collection

Ethnography sets a high standard for understanding activity in context. Ethnographers engage with research participants in their local settings, documenting detailed accounts of activity in relation to the physical environment, social norms, material artifacts, and other factors. Ethnography takes spatial and temporal relationships into account. Researchers are taking advantage of technology to augment and streamline ethnographic data collection. Video ethnography (Pink et al., 2017) enables the capture of participant's narration of events, enriching the data and contributing to subsequent interpretation. Electronic journaling (Hewitt, 2017) and photovoice (Chew & Lopez, 2018) are approaches for obtaining participant-generated data that allow insight into participant experiences by giving them time to reflect on the text and images they contribute (see Chapter 9 in this volume for a case example). Various apps and tools are available to enable participants to document their experiences through voice, text, and image or video capture. However, it can be difficult to recruit and sustain participation for these studies (Filep et al., 2018; Hayman et al., 2012). To address this, researchers seek to maintain a connection with participants and encourage them to complete the documentation requested. Electronic journaling and photovoice methods work best when deployed for short periods (Hayman et al., 2012). For example, asking a teen to document their experience managing diabetes over six months might yield spotty adherence to the study protocol. However, having them document their experiences for a week during the school year and a week during the summer might produce interesting data on how school structures their activities. Examples might include how the school start time creates a morning schedule that includes eating, monitoring blood glucose, and taking insulin at the same time every weekday, or how carrying a backpack enables the glucose monitor and insulin pen to be carried with the teen all day (or be nearby in the locker) during the school week. Conversely, during the summer, the teen might need to be reminded to monitor glucose or bring supplies on an excursion to the lake.

Given the ubiquity of mobile phones and availability of inexpensive cameras and voice recorders, methods for capturing the role of place in health-related work are also emerging (Richardson et al., 2019). Linking captured data to geographic information system tools (Oyana, 2017) enables researchers to identify spatial patterns in patient work on a scale larger than the household or workplace. Commuting patterns, social interactions, and other routine movements reveal opportunities for intervention and building resilience in self-care.

Social media research has increased dramatically with the rise in social networks such as Facebook and Twitter. These resources are excellent for capturing a large quantity of posts. Quantitative social media research at the population level seeks to identify relationships between the posted information or search terms and health-related outcomes. Given the difficulty in capturing public health data, this is appealing despite issues with accuracy (Lazer et al., 2014). Qualitative social media research can reach further into the data, beyond the mention of a word or the existence of a "like," to explore what people are truly talking about, and how those topics relate to health. Some studies have asked participants for access to their social media posts, as in a study of postpartum depression (Choudhury et al., 2014). Ethical issues related to the use of social media data are especially challenging regarding privacy and informed consent (Rothstein, 2015), and investigators are encouraged to work closely with institutional ethics boards to establish appropriate recruitment and data acquisition practices.

12.1.2 Data Analysis

The fundamental process for qualitative data analysis involves engagement with the data and application of theory to develop insights. This has not changed; however, emerging tools can facilitate the process. Web-based qualitative data analysis (QDA) tools enable teams of qualitative researchers to simultaneously code and analyze qualitative data, increasingly in collaboration with participants (Jennings et al., 2018). Team members may be distributed across space or working different shifts. Online meeting tools can enable teams to discuss and examine data together, including visualizing the findings. An example is the map of "code co-occurrence" available in the Dedoose™ QDA tool that enables a quick assessment of overlap in themes in the data, sometimes resulting in surprising insights that may have eluded the investigators. QDA tools typically include features to test inter-rater reliability among coders.

Patient and caregiver stories are known to be rich with detail and insight (Gubrium, 2009). Methods for the analysis of stories can be gleaned from content and narrative analysis traditions (Carson et al., 2017; Robillard et al., 2017). Content analysis can be triangulated using computer-based natural language processing capabilities (Renz et al., 2018). In multimedia and participatory data collection, investigators may be faced with stories comprising participant-generated video and audio, along with interview transcripts and field notes. Thoughtful approaches to data management help preserve context of these accounts by linking them appropriately. Academic research teams often include students and other temporary personnel. It is helpful to conduct data analysis early in a project to capitalize on the knowledge of the people who collected the data, rather than piecing together the methods post hoc.

12.2 VIRTUAL AND AUGMENTED REALITY

Virtualization or virtual reality (VR) refers to creating virtual (i.e., simulated) representation of physical environments (e.g., homes and neighborhoods) using computer technologies. Augmented reality (AR) is the use of technology to superimpose information—sounds, images, and text—on what is visualized. VR develops computer-generated environments for users to interact with, and be immersed in. AR adds to the reality one would ordinarily see, rather than replacing it. AR includes synthetic objects that are added to the real world in real time, enriching reality with helpful and relevant information (Azuma et al., 2001). AR users see the real world, except that in this real world, the virtual objects are placed or superimposed over it, forming a part of what the user is seeing with the sensation that the virtual and real objects coexist in the same space.

The sense of immersion in a VR environment is largely achieved through visual, auditory, and haptic stimuli that simulate 3D visual, auditory, and haptic cues available in the real world. Visually, this is delivered to the user via desktop computers, a head-mounted display, or Cave Automatic Virtual Environment, which present the computer-generated imagery of the VR scene from the perspective of each of the user's eyes. Immersion has been defined as the extent to which a user feels present in the computer-generated imagery environment, rather than in their actual physical environment. VR can be applied to patient work research by providing an opportunity to evaluate various characteristics of physical environments while conducting various self-management activities (Brennan et al., 2015). Examining the relationship between the physical environment and activities and how this relationship affects other phenomena such as cognition and routine formation can inform various interventions to support individual needs and improve patient outcomes in various contexts including diabetes, pain management, and weight management (Werner et al., 2018).

Although VR and AR applications are more commonly used by clinicians (e.g., for training), they offer potential for novel patient work research to improve patient outcomes. By immersing participants in multisensory, realistic-looking environments, it is possible to simulate a patient's living environment at a different location, such as simulating their home during hospital discharge (Brennan et al., 2015), examining the effects of cue presentation (Baumann & Sayette, 2006; Pericot-Valverde et al., 2016), supporting rehabilitation and cognitive training (e.g., for patients with stroke), and providing cognitive aids to persons living with dementia (Hayhurst, 2018). AR can be helpful to study cue exposure for patients with obesity or substance addiction (Giglioli et al., 2016) or for patient education on food consumption (e.g., counting carbohydrates) (Domhardt et al., 2015). Cue exposure is an effective technique to examine the relationship between the stimulus and response. VR allows researchers to efficiently test links between the various stimulus–context combinations and participants' responses. Understanding these links can inform redesign of the daily living environment to help patients make informed decisions. Moreover, AR can facilitate healthier responses to cues (Metcalf et al., 2018).

Current challenges of using AR and VR in daily living environments include technical problems (e.g., network connectivity, application stability, and battery life), achieving immersion, usability (e.g., cumbersome hand swipes and gestures, and cognitive overload), and VR sickness.

12.3 SENSORS AND IOT

The use of sensor technology has long been an integral part of modern healthcare systems, especially with its use in physiological monitoring of patients in the clinical settings. However, through advances in information technology and low manufacturing costs, sensors are becoming more affordable and pervasive, expanding their impact and use to everyday life outside of clinical settings. The ubiquitous usage of sensors has been strengthened with the introduction of the IoT, a term that refers to the network of devices and sensors equipped with internet connectivity, enabling them to send and receive data (Höller et al., 2014). Today, patients and informal caregivers can choose from an array of smart sensors to objectively collect real-time data to address their unique needs and challenges associated with their medical conditions and to facilitate self-management.

12.3.1 Assessing Patient Work in the Home Environment (Smart Home)

A prominent focus area of using sensor technologies for patient work is collecting and monitoring the health and wellness data of patients in their home environment through embedded sensors and wearables or body-worn sensors. The sensor system combines passive and active monitoring; creates a "smart" home environment with holistic monitoring of the patient's health parameters, daily activities, and behavioral patterns; and provides tailored support (Demiris & Hensel, 2008).

Smart home environments could potentially foster a safe environment for the patient, while making the patient's work easier and more convenient. Artificial intelligence-powered data analytics and smart home systems can identify potential patterns in health, detect anomalous activities, and prompt early intervention to prevent adverse health events. For instance, sensors can be placed throughout the home to detect motion, sound, vital signs, or other environmental situations. The data generated by these sensors could be integrated through IoT networks to be processed and analyzed by a data monitoring service. If there is no movement detected for a predetermined period of time, the system could be programmed to send a notification to a caregiver or emergency medical service.

With data from the smart home sensors, patients can become more aware of their daily routines and behaviors and understand them in a personal context. The self-awareness of health parameters and behaviors and how they impact activities of daily living is especially important to develop appropriate coping strategies for self-management. Recent advances in machine learning algorithms can analyze data from multiple sensors to correctly classify and categorize patients' activities of daily living to assess their functional capacity (Ghayvat et al., 2015; Suryadevara et al., 2013). These algorithms can infer users' activities throughout everyday life and, for example, be used to encourage physical activity for those with sedentary lifestyles (Consolvo et al., 2008).

The use of sensor technology for data collection is appealing due to its objective collection of physiological, behavioral, and environmental data about patients and their home environments. This circumvents the limitations of conventional data collection methods, such as journaling, which rely on subjective experience and recall (see also Chapter 9 in this volume). Results from subjective measures are prone to recall bias, which produces systematic error when study participants cannot remember previous events or experiences or else report them inconsistently (Althubaiti, 2016). Therefore, subjective data may not accurately reflect the patient's actual behaviors, limiting actionable insights.

12.3.2 The Use of IoT Sensors in Sleep Management

Sleep self-management strategies that incorporate sensors have the potential to empower patients to track and improve their sleep quality. With the uptake of smartphone ownership, Depose have been developed that utilize embedded sensors in a smartphone to self-monitor activity levels and visualize sleep patterns. Such apps often instruct a user to connect the phone to the charger and place it on the sleeping surface or under the pillow to passively collect data. Using the data, the sleep tracking apps can provide information on sleep patterns (e.g., bedtime, wakeup time, and average time in bed). Additionally, consumer-grade wearable devices such as wrist-worn activity trackers and smart watches also provide users with estimates of sleep-related parameters using proprietary algorithms, including the amount of time in light, deep, and rapid eye movement stage of sleep (Choi et al., 2018). The wearables can collect biometric parameters such as heart rate and blood pressure and potentially provide more detailed estimates than smartphone sensor-based apps. However, sleep estimates generated by wearable devices are under scrutiny for inaccuracy and cannot be used as a substitute for data collected by polysomnography in a sleep lab (Haghayegh et al., 2019). The limitations of wrist-worn sensors also include limited battery life and discomfort of wearing the device during sleep. Despite the shortcomings, consumer-grade IoT sensors provide simple and economical means to longer-term sleep monitoring.

12.3.3 The Use of IoT Sensors in Fall Detection

Sensors that can automatically detect falls are a critical component of the medical alert system for older adults. Different technologies and methodologies have been studied to detect and even predict fall-related injuries, such as using wearables (Wang et al., 2014), acoustic detection (Salman Khan et al., 2015), or privacy-preserving motion capture images (de Miguel et al., 2017). The fall detection sensor system can detect slips and falls and automatically call for assistance to emergency contacts or local emergency medical services. Therefore, the algorithms must be robust and have a high degree of precision and accuracy to avoid problems caused by false alarms.

12.3.4 The Use of IoT Sensors in Diabetes Management

Integrated digital solutions that incorporate affordable, smart biosensors can accurately capture health parameters such as glucose levels, blood pressure, and heart rate. With these sensors, patients can readily have available information for diabetes self-management in the home setting. More recent advances are in wearable, continuous glucose monitoring systems that can seamlessly track glucose levels throughout the day, rather than the brief snapshot at a single point in time typically obtained with a finger stick. With the continuous data stream, machine learning–based decision aid algorithms are employed to assist patients to determine the correct insulin dose.

12.3.5 Challenges

IoT-based sensor systems offer comprehensive improvements in supporting patient work, maximizing an individual's ability to self-manage, and providing ways to collect patient data in and outside hospital settings. However, despite many potential benefits, using IoT devices to collect and manage patient data creates significant privacy and security challenges for patient ergonomics researchers. For now, consumer devices such as smartphones, wearable devices, and other IoT-based systems that collect health data are not regulated under the Health Insurance Portability and Accountability Act (HIPAA) if the data are collected for personal use and by non-healthcare entities.

Whether HIPAA compliance safeguards are required or not, researchers must incorporate stringent data governance controls to handle patient data collected using these emerging methods to minimize data security risks and ensure patient privacy. First and foremost, researchers must fully inform their research participants the extent of privacy and security risks involved in using these emerging tools and relevant data governance procedures. The informed consent process must be carefully designed to minimize the use of technical jargon and accommodate varying degrees of digital literacy (O'Connor et al., 2017). Researchers should provide participants fine-grained control over the type and access to their health data whenever possible to respect their privacy. Prior to analysis or sharing, the collected data must be rendered anonymous, removing any personally identifiable information (i.e., de-identification) by using sophisticated de-identification techniques (National Committee on Vital and Health Statistics, 2019).

Researchers must also consider various adoption barriers of emerging technology tools for patients. Previous research has shown privacy concerns to be a major barrier in the adoption of sensor technology in a home environment (Chung, 2014; Demiris & Hensel, 2008; Reeder et al., 2016). Patients may be reluctant to use or reject some health monitoring devices because they think others would perceive them as frail and having limited autonomy (Demiris et al., 2004; Steele et al., 2009). In addition, researchers must be aware that the cost of and access to new technology can create or widen health disparities.

12.4 GAMIFICATION

Games are firmly embedded in human culture and greatly influence our social and leisure lives. Games emerge from combinations of rules, structure, voluntariness, uncertain outcomes, conflicts, representations, and resolution in different proportions. Games stimulate intellectual curiosity, sense of achievement, social recognition,

cognition, and self-determination (Blohm & Leimeister, 2013). Gamification is the use of game design elements in nongame contexts such as a patient's self-management of a chronic condition (Deterding et al., 2011). These design elements include documentation of behavior, scoring systems (badges and trophies), ranking (ranks, points, levels, and leaderboards), group tasks (social engagement loops and onboarding), time pressure, challenging tasks, quests, stories, theme avatars, and virtual worlds (Blohm & Leimeister, 2013).

Gamification could be leveraged in developing applications with the potential to better facilitate self-management and improve adherence (Sardi et al., 2017). They can target a wide audience: children, adolescents, and both healthy and ill adults. Previous studies in chronic disease management utilizing gamification accomplished health behavior change by motivating patients (Borghese et al., 2013; Hu et al., 2014; Munson & Consolvo, 2012). Gamification-based interventions have been successful because they (1) take playful aspects of gaming experiences and restructure them so a typically boring activity is something enjoyable, competitive, and engaging; (2) use visual aesthetics and game mechanics to promote play and interaction with other players, while yielding a great level of enjoyment and entertainment; (3) provide assistance to patients with chronic conditions by improving their adherence to medication and treatment plans; and (4) improve communication and bilateral encouragement among users by means of social sharing (e.g., online posts) and instant messaging (Sardi et al., 2017).

Despite their benefits, the implementation of gamification techniques in patient work has significant challenges (Sardi et al., 2017). First, the long-term viability of the gamification effects is still unknown. Some game mechanics (e.g., points and badges) did not provide a tangible health-driven meaning in terms of the user's competence and health skills, and they may be improperly located on the application's display (Zuckerman & Gal-Oz, 2014). Second, some gamification applications offer a valuable reward for an activity that does not require a significant endeavor. Third, fun and motivation are not a "one-size-fits-all" proposition; a game element that seems motivational for one individual may not be for another.

12.5 CASE EXAMPLE: FOOD CONSUMPTION

Food consumption from the lens of Human-Food Interaction (HFI) is a new subdiscipline of Human-Computer Interaction that looks at the role of technology in supporting food-related practices: how we grow, cook, eat, and dispose of the food (Comber et al., 2014). A large part of this emerging field is dedicated to understanding and supporting healthy eating behavior, which covers an important part of patient work, as well as presents methodological challenges.

Eating a healthy, well-balanced diet can improve overall health and may impact chronic illnesses. Most individuals, however, struggle to incorporate a well-balanced diet into their lifestyle. Instead, they consume a diet that is high in calories, saturated fats, and sugars, while being low in fruits, vegetables, and fiber. Such unhealthy eating contributes to adverse health conditions such as obesity (Camilleri et al., 2016).

Technology has been explored for altering such behaviors and encouraging healthier eating habits (Khot & Mueller, 2019). Unlike traditional healthcare delivery, where clinicians assess the patient's dietary intake to develop a plan for behavioral changes, the emphasis here is on empowering individuals, giving the patient responsibility for their own health and dietary behaviors (Kalantarian et al., 2017). An individual in this model is responsible for tracking their dietary behavior using either manual or automated approaches, wherein technology can help individuals make sense of the collected data by offering feedback in a novel and engaging manner. We describe below some of the commonly used techniques for self-monitoring of dietary behavior.

12.5.1 Food Journaling

Food journaling is a commonly used technique that encourages people to document their eating habits to support reflection on their eating behavior. The simplest form of food journaling is to use pen and paper. However, this is time-consuming and is prone to accidental data logging errors. As a remedy, commercial food tracking applications (e.g., MyFoodDiary and MyFitnessPal) are used for food journaling, which can minimize potential data logging inaccuracies while allowing the user to easily search for components of a meal in a food database to retrieve nutrition facts (e.g., calories, protein, carbohydrates, and sugar) relative to their daily goal and consumption. These applications, however, have limitations in tracking diverse meals and accommodating the specific needs of people. For example, people suffering from irritable bowel syndrome (Karkar et al., 2017) and eating disorders (Eikey & Reddy, 2017) need more flexibility in meal logging depending on their lifestyle and stage of their condition. A solution to burdensome text-based food journaling is to use photographs of meals for dietary analysis, using applications such as MealLogger. While photo journaling is a simpler method of data logging than text, its reliability depends on captured food data sets and maturity of the machine learning techniques used to recognize nutritional components of the food (Zhu et al., 2010). Nonetheless, photo-based food journaling of all meals is a habit difficult to adopt, as item-by-item tracking is a tedious task (C. F. Chung et al., 2017).

12.5.2 Sensor-based Dietary Monitoring

With the advancements in sensing and networking technologies, a range of devices have been developed. These range across wearable cameras (Thomaz et al., 2013), proximity sensors (Chun et al., 2018), wearable acoustic sensors (Yatani & Truong, 2012), electromyography (EMG)-measuring eyeglasses (Huang et al., 2017), and in-ear microphones (Gao et al., 2016) to automatically detect, monitor, and record eating activity. Although these technologies can potentially alleviate tedious data entry, the cost of these devices and social awkwardness of using such devices in public tend to hinder their adoption (Kalantarian et al., 2017). Additionally, most automated dietary monitoring technologies work well in controlled environments, but struggle to offer consistent results in real-world environments (Thomaz et al., 2017).

12.5.3 Persuasive Games

Designing persuasive games is another direction that has been explored to promote healthy eating habits. Smartphone-based games such as Monster Appetite (Hwang &

Mamykina, 2017) and National Mindless Eating Challenge (Kaipainen et al., 2012) use avatars to educate players about healthy eating habits. Studies of these games showed an increase in players' nutrition knowledge, besides raising awareness and commitment to initiate and maintain healthy eating behavior. More recently, AR- and VR-based technologies have been explored to promote healthy eating behavior. An example is Feed the Food Monsters, an AR-based game (Arza et al., 2018) that educates people about their chewing behavior in a social dining setting.

To conclude, the field of dietary monitoring has undergone a radical change in the last decade from maintaining a paper-based food diary to automated smartphone-based apps to machine learning–based recognition to AR- and VR-based systems. With advancements in AI and sensing technologies, things will further change in the future, enabling access to much richer and more accurate data about an individual's food and dietary behavior. The pressing question for the research community will change from "how to capture dietary data" to "what we should do with dietary data." Projects such as EdiPulse (Khot et al., 2017), TastyBeats (Khot et al., 2015), and Chorus (Ferdous et al., 2017) put forward a new way of seeing personal data, going beyond rational instrumental goals to viewing data as avenues or enablers of richer food experiences.

12.6 RECOMMENDATION AND FUTURE DIRECTIONS

This chapter focused on methods to support data collection and intervention delivery. However, analysis is also a critical step. Analyzing data after data collection, to design or evaluate an intervention, should be accomplished in a way that findings can be interpreted by patients and clinicians quickly, translated to therapy plans, support adherence, and improve patient–clinician collaboration. Visualization can help with analysis and collaborative interpretation of data (J. Chung et al., 2017; Schroeder et al., 2017).

Novel methods typically emerge with the development of theoretical frameworks. Development of more theoretical frameworks on patient work would highlight the need for these methods. In return, novel methods will inform more theoretical frameworks. For example, studying workflow for daily living settings may be beneficial to better understand adherence, self-management, or frailty (J. Chung et al., 2017). A framework on workflow for daily living settings along with the use of emerging methods can stimulate development and clarify what data to capture and analyze (Ozkaynak et al., 2016; 2018).

REFERENCES

Althubaiti, A. (2016). Information bias in health research: Definition, pitfalls, and adjustment methods. *Journal of Multidisciplinary Healthcare, 9*, 211–217.

Arza, E. S., Kurra, H., Khot, R. A., & Mueller, F. F. (2018). Feed the food monsters! Helping co-diners chew their food better with augmented reality. Paper presented at the *Proceedings of the 2018 Annual Symposium on Computer-Human Interaction in Play Companion Extended Abstracts*, Melbourne, VIC, Australia.

Azuma, R., Behringer, R., Julier, S., Baillor, Y., MacIntyre, B., & Feiner, S. (2001). Recent advances in augmented reality. *Institute of Electrical and Electronics Engineers Computer Graphics and Applications, 21*(6), 34–47.

Baumann, S. B., & Sayette, M. A. (2006). Smoking cues in a virtual world provoke craving in cigarette smokers. *Psychology of Addictive Behaviors, 20*(4), 484–489.

Blohm, I., & Leimeister, J. M. (2013). Gamification - Design of IT-based enhancing services for motivational support and behavioral change. *Business and Information Systems Engineering, 5*(4), 275–278.

Borghese, N. A., Mainetti, R., Pirovano, M., & Lanzi, P. L. (2013). *An intelligent game engine for the at-home rehabilitation of stroke patients.* Paper presented at the *2013* IEEE 2nd International Conference on Serious Games and Applications for Health (SeGAH), Vilamoura, Portugal.

Brennan, P. F., Ponto, K., Casper, G., Tredinnick, R., & Broecker, M. (2015). Virtualizing living and working spaces: Proof of concept for a biomedical space-replication methodology. *Journal of Biomedical Informatics, 57,* 53–61.

Camilleri, G. M., Mejean, C., Bellisle, F., Andreeva, V. A., Kesse-Guyot, E., Hercberg, S., & Peneau, S. (2016). Intuitive eating is inversely associated with body weight status in the general population-based NutriNet-Sante study. *Obesity (Silver Spring), 24*(5), 1154–1161.

Carson, A., Chabot, C., Greyson, D., Shannon, K., Duff, P., & Shoveller, J. (2017). A narrative analysis of the birth stories of early-age mothers. *Sociology of Health and Illness, 39*(6), 816–831.

Chew, H. S. J., & Lopez, V. (2018). Empowered to self-care: A photovoice study in patients with heart failure. *Journal of Transcultural Nursing, 29*(5), 410–419.

Choi, Y. K., Demiris, G., Lin, S. Y., Iribarren, S. J., Landis, C. A., Thompson, H. J., . . . Ward, T. M. (2018). Smartphone applications to support sleep self-management: Review and evaluation. *Journal of Clinical Sleep Medicine, 14*(10), 1783–1790.

Choudhury, M. D., Counts, S., Horvitz, E. J., & Hoff, A. (2014). Characterizing and predicting postpartum depression from shared facebook data. Paper presented at the *Proceedings of the 17th ACM Conference on Computer Supported Cooperative Work; Social Computing*, Baltimore, MD.

Chun, K. S., Bhattacharya, S., & Thomaz, E. (2018). Detecting eating episodes by tracking jawbone movements with a non-contact wearable sensor. *Proceedings of the ACM on Interactive, Mobile, Wearable and Ubiquitous Technologies, 2*(1), 1–21.

Chung, C. F., Agapie, E., Schroeder, J., Mishra, S., Fogarty, J., & Munson, S. A. (2017). When personal tracking becomes social: Examining the use of Instagram for healthy eating. *Proceedings of the Special Interest Group on Computer-Human Interaction Conference on Human Factors in Computing Systems, 2017,* 1674–1687.

Chung, J. (2014). *In-home use of home-based sensor technology for monitoring mobility in community-dwelling Korean American older adults.* Seattle, WA: University of Washington.

Chung, J., Ozkaynak, M., & Demiris, G. (2017). Examining daily activity routines of older adults using workflow. *Journal of Biomedical Informatics, 71,* 82–90.

Comber, R., Choi, J. H.-J., Hoonhout, J., & O'hara, K. (2014). Editorial: Designing for human-food interaction: An introduction to the special issue on "food and interaction design." *International Journal of Human-Computer Studies, 72*(2), 181–184.

Consolvo, S., McDonald, D. W., Toscos, T., Chen, M. Y., Froehlich, J., Harrison, B., . . . Landay, J. A. (2008). Activity sensing in the wild: A field trial of ubifit garden. Paper presented at the *Proceedings of the Special Interest Group on Computer-Human Interaction Conference on Human Factors in Computing Systems*, Florence, Italy.

de Miguel, K., Brunete, A., Hernando, M., & Gambao, E. (2017). Home camera-based fall detection system for the elderly. *Sensors (Basel), 17*(12).

Demiris, G., & Hensel, B. K. (2008). Technologies for an aging society: A systematic review of "smart home" applications. *Yearbook of Medical Informatics*, 33–40.

Demiris, G., Rantz, M., Aud, M., Marek, K., Tyrer, H., Skubic, M., & Hussam, A. (2004). Older adults' attitudes towards and perceptions of "smart home" technologies: A pilot study. *Journal of Medical Internet Research in Medicine, 29*(2), 87–94.

Deterding, S., Dixon, D., Khaled, R., & Nacke, L. (2011). From game design elements to gamefulness: Defining "gamification." Paper presented at the *Proceedings of the 15th International Academic MindTrek Conference: Envisioning Future Media Environments*, Tampere, Finland.

Domhardt, M., Tiefengrabner, M., Dinic, R., Fötschl, U., Oostingh, G. J., Stütz, T., . . . Ginzinger, S. W. (2015). Training of carbohydrate estimation for people with diabetes using mobile augmented reality. *Journal of Diabetes Science and Technology, 9*(3), 516–524.

Eikey, E. V., & Reddy, M. C. (2017). "It's definitely been a journey": A qualitative study on how women with eating disorders use weight loss apps. Paper presented at the *Proceedings of the 2017 CHI Conference on Human Factors in Computing Systems*, Denver, CO.

Ferdous, H. S., Vetere, F., Davis, H., Ploderer, B., O'Hara, K., Comber, R., & Farr-Wharton, G. (2017). Celebratory technology to orchestrate the sharing of devices and stories during family mealtimes. Paper presented at the *Proceedings of the 2017 CHI Conference on Human Factors in Computing Systems*, Denver, CO.

Filep, C. V., Turner, S., Eidse, N., Thompson-Fawcett, M., & Fitzsimons, S. (2018). Advancing rigour in solicited diary research. *Qualitative Research, 18*(4), 451–470.

Gao, Y., Zhang, N., Wang, H., Ding, X., Ye, X., Chen, G., & Cao, Y. (2016, 27–29 June). *iHear food: Eating detection using commodity bluetooth headsets.* Paper presented at the *2016 IEEE First International Conference on Connected Health: Applications, Systems and Engineering Technologies (CHASE)*, Washington, DC, USA.

Ghayvat, H., Mukhopadhyay, S., Gui, X., & Suryadevara, N. (2015). WSN- and IOT-Based smart homes and their extension to smart buildings. *Sensors (Basel), 15*(5), 10350–10379.

Giglioli, I. A. C., Chirico, A., Cipresso, P., Serino, S., Pedroli, E., Pallavicini, F., & Riva, G. (2016). Feeling ghost food as real one: Psychometric assessment of presence engagement exposing to food in augmented reality. In S. Serino, A. Matic, D. Giakoumis, G. Lopez, & P. Cipresso (Eds.), *Pervasive computing paradigms for mental health* (pp. 99–109). Cham: Springer.

Gubrium, A. (2009). Digital storytelling: An emergent method for health promotion research and practice. *Health Promotion Practice, 10*(2), 186–191.

Haghayegh, S., Khoshnevis, S., Smolensky, M. H., Diller, K. R., & Castriotta, R. J. (2019). Accuracy of wristband Fitbit models in assessing sleep: Systematic review and meta-analysis. *Journal of Medical Internet Research, 21*(11), e16273.

Hayhurst, J. (2018). How augmented reality and virtual reality is being used to support people living with dementia—Design challenges and future directions. In T. Jung & M. C. tom Dieck (Eds.), *Augmented reality and virtual reality: Empowering human, place and business* (pp. 295–305). Cham: Springer International Publishing.

Hayman, B., Wilkes, L., & Jackson, D. (2012). Journaling: Identification of challenges and reflection on strategies. *Nursing Research, 19*(3), 27–31.

Hewitt, E. (2017). Building bridges: The use of reflective oral diaries as a qualitative research tool. *International Journal of Research & Method in Education, 40*(4), 345–359.

Höller, J., Tsiatsis, V., Mulligan, C., Karnouskos, S., Avesand, S., & Boyle, D. (2014). Chapter 14- Smart cities. In J. Höller, V. Tsiatsis, C. Mulligan, S. Karnouskos, S. Avesand, & D. Boyle (Eds.), *From machine-to-machine to the internet of things* (pp. 281–294). Oxford: Academic Press.

Hu, R., Fico, G., Cancela, J., & Arredondo, M. T. (2014). Gamification system to support family-based behavioral interventions for childhood obesity. Paper presented at the *Institute of Electrical and Electronics Engineers-Engineering in Medicine and Biology Society International Conference on Biomedical and Health Informatics,* Valencia, Spain.

Huang, Q., Wang, W., & Zhang, Q. (2017). Your glasses know your diet: Dietary monitoring using electromyography sensors. *Institute of Electrical and Electronics Engineers Internet of Things Journal, 4*(3), 705–712.

Hwang, M. L., & Mamykina, L. (2017). Monster appetite: Effects of subversive framing on nutritional choices in a digital game environment. Paper presented at the *Proceedings of the 2017 CHI Conference on Human Factors in Computing Systems*, Denver, CO.

Jennings, H., Slade, M., Bates, P., Munday, E., & Toney, R. (2018). Best practice framework for Patient and Public Involvement (PPI) in collaborative data analysis of qualitative mental health research: Methodology development and refinement. *BioMed Central Psychiatry, 18*(1), 213.

Kaipainen, K., Payne, C. R., & Wansink, B. (2012). Mindless eating challenge: Retention, weight outcomes, and barriers for changes in a public web-based healthy eating and weight loss program. *Journal of Medical Internet Research, 14*(6), e168.

Kalantarian, H., Alshurafa, N., & Sarrafzadeh, M. (2017). A survey of diet monitoring technology. *Institute of Electrical and Electronics Engineers Pervasive Computing, 16*(1), 57–65.

Karkar, R., Schroeder, J., Epstein, D. A., Pina, L. R., Scofield, J., Fogarty, J., . . . Zia, J. (2017). TummyTrials: A feasibility study of using self-experimentation to detect individualized food triggers. Paper presented at the *Proceedings of the 2017 CHI Conference on Human Factors in Computing Systems*, Denver, CO.

Khot, R. A., Aggarwal, D., Pennings, R., Hjorth, L., & Mueller, F. F. (2017). *EdiPulse*: Investigating a playful approach to self-monitoring through 3D printed chocolate treats. Paper presented at the *Proceedings of the 2017 CHI Conference on Human Factors in Computing Systems*, Denver, CO.

Khot, R. A., Lee, J., Aggarwal, D., Hjorth, L., & Mueller, F. F. (2015). TastyBeats: Designing palatable representations of physical activity. Paper presented at the *Proceedings of the 33rd Annual ACM Conference on Human Factors in Computing Systems*, Seoul, Republic of Korea.

Khot, R. A., & Mueller, F. F. (2019). Human-Food interactions and trends. *Human-Computer Interactions, 12*(4), 238–413.

Lazer, D., Kennedy, R., King, G., & Vespignani, A. (2014). Big data. The parable of Google flu: Traps in big data analysis. *Science, 343*(6176), 1203–1205.

Metcalf, M., Rossie, K., Stokes, K., Tallman, C., & Tanner, B. (2018). Virtual reality cue refusal video game for alcohol and cigarette recovery support: Summative study. *Journal of Medical Internet Research Serious Games, 6*(2), e7.

Munson, S. A., & Consolvo, S. (2012). *Exploring goal-setting, rewards, self-monitoring, and sharing to motivate physical activity.* Paper presented at the 2012 6th International Conference on Pervasive Computing Technologies for Healthcare (PervasiveHealth) and Workshops, San Diego, CA, USA.

National Committee on Vital and Health Statistics. (2019). Health information privacy beyond HIPAA: A framework for use and protection Accessed date: Jan 31, 2020, Available at https://ncvhs.hhs.gov/wp-content/uploads/2019/07/Report-Framework-for-Health-Information-Privacy.pdf.

O'Connor, Y., Rowan, W., Lynch, L., & Heavin, C. (2017). Privacy by design: Informed consent and internet of things for smart health. *Procedia Computer Science, 113*, 653–658.

Oyana, T. J. (2017). The use of GIS/GPS and spatial analyses in community-based participatory research. In S. S. Coughlin, S. A. Smith, & M. E. Fernandez (Eds.), *Handbook of community-based participatory research*, Oxford University Press, New York, pp. 39–56.

Ozkaynak, M., Jones, J., Weiss, J., Klem, P., & Reeder, B. (2016). A workflow framework for health management in daily living settings. *Studies in Health Technology and Informatics, 225*, 392–396.

Ozkaynak, M., Valdez, R. S., Holden, R. J., & Weiss, J. (2018). Infinicare framework for an integrated understanding of health-related activities in clinical and daily-living contexts. *Health Systems, 7*(1), 66–78.

Pericot-Valverde, I., Germeroth, L. J., & Tiffany, S. T. (2016). The use of virtual reality in the production of cue-specific craving for cigarettes: A Meta-Analysis. *Nicotine & Tobacco Research, 18*(5), 538–546.

Pink, S., Sumartojo, S., Lupton, D., & Heyes LaBond, C. (2017). Empathetic technologies: Digital materiality and video ethnography. *Visual Studies, 32*(4), 371–381.

Reeder B., Chung J., Joe J., Lazar A., Thompson H.J., Demiris G. (2016). Understanding Older Adults' Perceptions of In-Home Sensors *Using an Obtrusiveness Framework. In: Schmorrow D., Fidopiastis C. (eds) Foundations of Augmented Cognition: Neuroergonomics and Operational Neuroscience.* AC 2016. Lecture Notes in Computer Science, vol 9744. Springer, Cham. https://doi.org/10.1007/978-3-319-39952-2_34.

Renz, S. M., Carrington, J. M., & Badger, T. A. (2018). Two strategies for qualitative content analysis: An intramethod approach to triangulation. *Qualitative Health Research, 28*(5), 824–831.

Richardson, D. M., Pickus, H., & Parks, L. (2019). Pathways to mobility: Engaging Mexican American youth through participatory photo mapping. *Journal of Adolescent Research, 34*(1), 55–84.

Robillard, A. G., Reed, C., Larkey, L., Kohler, C., Ingram, L. A., Lewis, K., & Julious, C. (2017). In their own words: Stories from HIV-positive African American women. *Health Education Journal, 76*(6), 741–752.

Rothstein, M. A. (2015). Ethical issues in big data health research: Currents in contemporary bioethics. *Journal of Law, Medicine, & Ethics, 43*(2), 425–429.

Salman Khan, M., Yu, M., Feng, P., Wang, L., & Chambers, J. (2015). An unsupervised acoustic fall detection system using source separation for sound interference suppression. *Signal Processing, 110*, 199–210.

Sardi, L., Idri, A., & Fernandez-Aleman, J. L. (2017). A systematic review of gamification in e-Health. *Journal of Biomedical Informatics, 71*, 31–48.

Schroeder, J., Hoffswell, J., Chung, C. F., Fogarty, J., Munson, S., & Zia, J. (2017). Supporting patient-provider collaboration to identify individual triggers using food and symptom journals. *CSCW Conference on Computer-Supported Cooperative Work, 2017*, 1726–1739.

Steele, R., Lo, A., Secombe, C., & Wong, Y. K. (2009). Elderly persons' perception and acceptance of using wireless sensor networks to assist healthcare. *International Journal of Medical Informatics, 78*(12), 788–801.

Suryadevara, N. K., Mukhopadhyay, S. C., Wang, R., & Rayudu, R. K. (2013). Forecasting the behavior of an elderly using wireless sensors data in a smart home. *Engineering Applications of Artificial Intelligence, 26*(10), 2641–2652.

Thomaz, E., Essa, I. A., & Abowd, G. D. (2017). Challenges and opportunities in automated detection of eating activity. In J. M. Rehg, S. A. Murphy, & S. Kumar (Eds.), *Mobile health: Sensors, analytic methods, and applications* (pp. 151–174). Cham: Springer International Publishing.

Thomaz, E., Parnami, A., Essa, I., & Abowd, G. D. (2013). Feasibility of identifying eating moments from first-person images leveraging human computation. Paper presented at the *Proceedings of the 4th International SenseCam; Pervasive Imaging Conference*, San Diego, CA.

Wang, J., Zhang, Z., Li, B., Lee, S., & Sherratt, R. S. (2014). An enhanced fall detection system for elderly person monitoring using consumer home networks. *Institute of Electrical and Electronics Engineers Transactions on Consumer Electronics, 60*(1), 23–29.

Werner, N. E., Jolliff, A. F., Casper, G., Martell, T., & Ponto, K. (2018). Home is where the head is: A distributed cognition account of personal health information management in the home among those with chronic illness. *Ergonomics, 61*(8), 1065–1078.

Yatani, K., & Truong, K. N. (2012). BodyScope: A wearable acoustic sensor for activity recognition. Paper presented at the *Proceedings of the 2012 ACM Conference on Ubiquitous Computing*, Pittsburgh, PA.

Zhu, F., Bosch, M., Woo, I., Kim, S., Boushey, C. J., Ebert, D. S., & Delp, E. J. (2010). The use of mobile devices in aiding dietary assessment and evaluation. *Institute of Electrical and Electronics Engineers Journal of Selected Topics in Signal Processing, 4*(4), 756–766.

Zuckerman, O., & Gal-Oz, A. (2014). Deconstructing gamification: Evaluating the effectiveness of continuous measurement, virtual rewards, and social comparison for promoting physical activity. *Personal and Ubiquitous Computing, 18*(7), 1705–1719.

13 Enhancing Patient Ergonomics with Patient and Public Involvement in Research Projects

Dominic Furniss
Human Reliability Associates Ltd.

Alexandra R. Lang
University of Nottingham

Colleen Ewart
NIHR Applied Research Collaborations East Midlands

CONTENTS

Patient and public involvement (PPI) engages patients in the conduct and management of health and social care research projects. This contrasts with traditional approaches to research where the patient role is restricted to being "participants" who contribute their data voluntarily. Interest in PPI is growing internationally, particularly its potential to make research processes and outcomes more patient-centered, harnessing the experiences of service users as they play a larger role in making research more relevant and impactful to them.

Patient ergonomics focuses on the patient, their "work," and related issues as the subject of study from a human factors and ergonomics (HFE) perspective (e.g., Valdez et al., 2014; Holden et al., 2015). PPI for patient ergonomics involves contributions by patients or other service users, such as informal caregivers, in the conduct and management of the project (e.g., by advising on patient information leaflets and recruitment strategy), defining and prioritizing research questions, and conducting data gathering and analysis. PPI aligns well with patient ergonomics because it is already close to patients and their interests due to the user-centered approach inherent to the discipline. Coproducing research with patients and the public can help bridge the gap between academic/industry and layperson perspectives, from conception to dissemination of research projects.

This chapter will define PPI and discuss how PPI can contribute to patient ergonomics, as well as practical challenges and lessons learnt from case studies. It is possible to conceive of PPI as a meta-method, to be used on the research project itself, but really it encapsulates a deeper change in mindset and philosophy in how research is managed.

13.1 INTRODUCTION TO PATIENT AND PUBLIC INVOLVEMENT (PPI)

PPI in research has been defined by INVOLVE (2019) as "research being carried out 'with' or 'by' members of the public rather than 'to,' 'about,' or 'for' them." INVOLVE is part of the UK National Institute for Healthcare Research (NIHR) and has been a leading organization in advocating and actioning public involvement in healthcare research since 1996. It differentiates between the activities of involvement, participation, and engagement:

- **Involvement**—where members of the public are actively involved in the management and conduct of research projects and in research organizations.
- **Participation**—where people take part in research through a variety of methods and activities to contribute their data to the study.
- **Engagement**—where information about research is disseminated through a variety of outlets and media (e.g., public open days and discussions on research), raising awareness of research through traditional and novel digital media.

While offering a potentially useful way to consider PPI, the above categorization is not standardized and as such there is wide variation in how PPI is understood and conducted, varying by discipline and jurisdiction. One example of differing terminology coming from the United States is from the Patient-Centered Outcomes Research Institute (PCORI), who consider PPI under the term of patient engagement, defining it as "patients and other healthcare stakeholders are equitable partners—as opposed to research subjects—who leverage their lived experience and expertise to influence research to be more patient centered, relevant, and useful" (PCORI, 2019). PPI is peppered with a wide range of language to describe the activities of involving and engaging laypeople (e.g., Patient and Family Advisory Councils [PFACs]) (Harrison et al., 2018) and Patient and Service User Engagement (PSUE) (Shippee et al., 2015). In 2014, PCORI commissioned a systematic review to understand the current landscape of PPI practices in health and social care research (Domecq et al., 2014). The report concluded that PPI "increased study enrollment rates and aided researchers in securing funding, designing study protocols and choosing relevant outcomes." However, there were concerns about the extra time and effort needed for PPI and the threat of tokenism, whereby PPI is treated as a shallow tick-box exercise. Ocloo and Matthews (2016) also highlight the problems associated with some current PPI approaches that do not encourage empowerment, diversity, and equality.

Early studies in the field of PPI called for more research on its practice (e.g., Brett et al., 2014; Domecq et al., 2014; Shippee et al., 2015), and the field has matured considerably since then. The now-established journal *Research Involvement and Engagement* contains a wealth of information about the developing practice of PPI. The journal is coproduced with patients and service users and invites research articles, methodologies, protocols, and commentaries, particularly those with patient authors. Many of these articles cover a broad spectrum of work from health services research, but they also address patient experiences and patient work.

Practitioners of ergonomics contribute to the design and evaluation of tasks, jobs, products, environments, and systems in order to make them compatible with the needs, abilities, and limitations of people (IEA, 2019). It can be challenging to design such systems to meet the needs of others who have different experiences to our own, and even more humbling when sensitive subject matter is shared (e.g., bereavement, mental health, and degenerative disease). PPI bridges the gap between research design and patients' knowledge, expertise, preferences, and priorities, to reveal and prepare for known unknowns and unknown unknowns in the research process.

13.1.1 Hierarchy of Involvement

Involvement can be activated at different levels, each level differing in effort, commitment, and potential impact or outcome. This is visualized as a hierarchy of involvement in Figure 13.1.

When considering the hierarchy, one should think about what level of PPI is wanted and needed by the research team, the resources available, and the requirements and preferences of the PPI representatives. For example, how might they be

Ladder of Citizen Engagement		Ladder of Participation		PPI Example
Citizen Control	Citizen Control	Co-Producing	Doing With	Democratic and the joint working groups to set the agenda for research.
Delegation		Co-Designing		Researchers work with PPI reps in converting lived experience into a role play for use in interviews
Partnership		Engaging	Doing For	A design activity sees the co-creation of a participant welcome pack, whereby items within the pack are developed in partnership. PPI reps consider what it important to include in welcome packs with support of researchers.
Placation	Tokenism	Consulting		An iterative process leads to the development of project communications e.g. press release. Initial content was from the research team but has been modified and added to by PPI representatives prior to submission.
Consultation		Informing		PPI reps are invited to review the content of information sheets/ questionnaires from the view of a naïve lay person to ensure understanding.
Informing		Educating	Doing To	PPI reps are provided information about what a research project involves and invited to take on a non-active PPI 'ambassador role' which does not have direct input or impact on the research project.
Therapy	Nonparticipation	Coercing		PPI reps are provided with a finished document which needs submitting in 3 days e.g. ethics application. The timeframe does not enable lay person review and so it encourages acquiescence.
Manipulation		*Not applicable*		PPI reps will be exposed to researcher driven content (potentially biased content) and will be involved in perpetuating that information without any opportunity to consult, review or feedback.

FIGURE 13.1 Hierarchy of involvement with examples, adapting and combining a ladder of citizen engagement (Arnstein, 1969) and ladder of participation (Hart, 1992).

able to contribute to the research process and how might this involvement be facilitated to optimize the experience for them and the research project? How do professional researchers support and train their nonprofessional research partners to enable them to contribute effectively and confidently? The hierarchy stretches from tokenism to being fully embedded. Projects may move up and down the ladder at different stages, and researchers may move up and down the ladder depending on their comfort and project circumstances. Comfort is mentioned here because the higher levels involve sharing power and giving control, which not everyone might be comfortable with, particularly those whose experience is limited to traditional patient "participant roles" in research. These higher levels are distinctive due to the different mindset and philosophy alluded to earlier, where patients are in control and have tangible ownership in the direction and design of enquiries. The use of this hierarchy is advocated as a tool for reflection on practice and to provide researchers a way to consider how they are approaching and resourcing PPI and where and how they might achieve higher levels of involvement described in the hierarchy. It is not designed to be prescriptive, but pragmatic examples of how PPI can be implemented are described in the case studies and the recommendations at the end of the chapter. As this is a new and growing cross-disciplinary area, future work may be able to reflect on best practice and populate this table further.

13.1.2 Timeline of Involvement

Whereas the hierarchy focuses more on the level or depth of PPI within a research process, the timeline of PPI looks more at which project activities may be supported with involvement. This can include setting the research agenda for activities such as developing commission briefs, involvement in funding decisions, designing study protocols, recruitment of participants, data collection and analysis, dissemination, uptake, and evaluating research (see Marjanovic et al., 2019 for a more detailed breakdown). These different activities need to be properly planned into the research project with appropriate resourcing.

Due to the way in which research is funded, it can be difficult to predict timelines and extent of PPI involvement. Specifically affected is the involvement prior to a project award being funded and time allocation for PPI within a research project, the latter of which will depend entirely on the scope and scale of a project. Previous experience has seen extreme variations in PPI time and resource within funded studies, where nominal allocations of time vary and need to be responsive to peaks and troughs of requirement over different stages of a project timeline. The case studies below provide further guidance and examples of PPI in practice.

Over the course of a research study, from conception of the idea to impact and dissemination, it is important to undertake reflective practices of PPI activities to understand what went well and why, and what could be better done for future projects, why, and how. This is one way to inform and enable moving away from a tokenistic PPI approach, ensuring it has more significance to the research process and maximizes its value to all stakeholders. Researcher perspectives of PPI might also develop between projects as new relationships offer new perspectives and support development of new ways of working.

13.2 CASE STUDIES OF PPI FOR PATIENT ERGONOMICS

We present four case studies of conducting PPI from our experience. They illustrate a range of different configurations, benefits, and challenges of PPI.

13.2.1 Case Study 1: Child and Adolescent Mental Health

The STAndardised DIagnostic Assessment for children and adolescents with emotional difficulties (STADIA) trial seeks to test a standard diagnostic questionnaire when young people first present with emotional difficulties and are referred to Child and Adolescent Mental Health Services (CAMHS), to support clinical decision-making and personalize treatment. STADIA embeds PPI with a range of involvement levels: two PPI coinvestigators (a parent with lived experience and HFE expert) on the trial management team; two PPI representatives on the trial steering committee; an adult PPI advisory panel; and a young person's PPI advisory panel. Colleen describes her experience as a parent PPI coinvestigator involved in several STADIA activities (Box 13.1).

The integration of PPI input through the formal role of a person with lived experience on the trial management team led to the development of a proposal in which

BOX 13.1 COLLEEN'S EXPERIENCES AS A STADIA PROJECT PARENT PPI COINVESTIGATOR

I was involved during the proposal development and drafted the lay summary with colleagues, sought feedback from PPI colleagues, and then assisted in refining the final submission. It became clear that my previous co-applicant roles had been "tokenistic"; this was the real deal!

To recruit lay representatives for the trial steering committee, I helped interview parents/caregivers nominated through project networks and selected two based on their lived experience, community links, and transferable skills in business and education. I recruited two advisory groups from my voluntary sector connections: one for parents/caregivers and another for young people aged 11+ years who had experience of being referred to CAMHS within the last 3 years.

I was asked to consult a group of young people about a STADIA logo! This contributed to creating a visual identity for the project that would speak to a range of stakeholders. I led this activity while learning at the same time. Fourteen logos were developed by email consultation with young people, parents/caregivers, and STADIA researchers. Using my community links, I gave a presentation to twenty young people, eight of whom agreed to support the task: five males and three females (aged 12–18). The 14 options were placed on walls around the room and the young people were given "rounds" to vote on those images, placing a tick or cross on each image. The least popular was removed each time and the process repeated until two remained. The group were happy for the project management team to make the final decision as they could not decide between themselves. We had no budget for this activity, so I gave out personalized certificates, thanking them for their valuable contribution, which were well received.

I also led a parent/caregiver consultation to feedback on study questionnaires and information sheets with the parents/caregivers of the young people I had consulted about the logo. I presented the resources and what was required of the group during the activity. This group provided insightful and clear feedback for amendments to the documents in readiness for our ethics submission. I paid all 17 parents/caregivers in line with project guidance. This activity enabled me to build rapport with a group steeped in "lived experience." I recruited our "core" parent/caregiver advisory panel from this group. The evaluation from our first PPI meeting reported 100% of participants found attendance to be invaluable.

the service users' voice was always present. This strategy ensured that the research agenda was always cognizant of the potential beneficiaries of the intervention, not only clinicians but ultimately recipients of CAMHS services. This role on the team was identified 18 months prior to proposal submission and in these formative stages, PPI input was offered with no reimbursement. Although funders are increasingly looking for PPI inclusion in funding awards, mechanisms rarely exist to facilitate this in a way that recognizes and reimburses PPI representatives. Researchers need to be creative in how they engage service users with little or no budget prior to funding awards to try and truly coproduce research agendas (e.g., working lunches), invitation to free development opportunities, and offering of the skills and expertise of the researcher to the PPI representatives.

In this study, the PPI parent coinvestigator is treated in the same way as any other coinvestigator, with active invitation of their views being prevalent within meetings. The personal skills and expertise of the PPI expert are understood and encouraged so they are empowered to suggest their own ideas and be proactive in line with their personal experience and proficiencies. They are paid (in accordance with guidance from INVOLVE [2020]) for approximately 2 hours of work per week over the full duration of the 5-year project. Despite this provision, there has been growth in the request and need for PPI service user input and activities. The evolving role and opportunities for PPI may not always be evident from the outset, so research teams face challenges in reconciling and prioritizing which PPI activities to implement.

13.2.2 Case Study 2: Caregivers' Experiences of Home Enteral Feeding

Home enteral feeding involves giving artificial nutrition through a tube inserted in the patient's nose or straight into their stomach, generally because they have problems ingesting food through their mouth and throat. Patients may be prescribed medications administered through these tubes (e.g., they may need to crush them and dissolve them manually) which could lead to blockages. We wanted to explore caregivers' experiences, errors, and strategies to cope with home enteral feeding.

A focus group was organized, but we struggled to recruit participants and ended up with just two people who had experience of enteral feeding, but many years ago. Our recruitment and data collection had not gone well. On reflection, we should have realized that a midday meeting in central London was not convenient to many caregivers, who would be busy caring, and we did not have the connections to reach out to enough people successfully. The model "if we build it, they will come" had not worked this time.

We developed a Plan B, which was to try to gather the experiences and data remotely. However, we involved people in the community to check that our plans were practical and achievable before committing to them. We fortuitously heard about someone who was seconded to help build community connections at the university and enlisted their help. With her involvement, we established a remote PPI panel consisting of three caregivers with experience of home enteral feeding and a parenteral nutrition patient who also had an administrative role within the charity PINNT (Patients on Intravenous and Naso-gastric Nutrition Treatment). Our panel proved invaluable for the success of the remainder of the project through a variety of activities: piloting the survey; reviewing the survey for readability and contributing additional questions considered relevant; providing advice and acting as gatekeepers to networks for recruitment of participants; and reviewing summary findings and output from the project (Alsaeed et al., 2018), which was republished in PINNT's quarterly magazine to disseminate this work to stakeholders. All communication with the panel was done through email, so they had time and flexibility to choose when to respond. We offered £100 in gift vouchers for their contribution.

Establishing and working with a PPI panel transformed the project from its naïve beginnings with little or no traction with the community to one guided and shaped by people in the community (e.g., the panel contributed new questions to the survey and helped with recruitment and dissemination). We had reflected on what was wrong with our initial approach and adapted the project with PPI input. We initially thought we had just a recruitment problem and needed ways to reach our target community. However, as we learned more about PPI, we realized PPI input could contribute more. It had the potential to make our survey more relevant, accessible, and impactful for the community. From our initial experience, we thought getting people into meetings would be difficult and did everything remotely; email provided a flexible tool for caregivers. We devised a fairly straightforward plan: to do a survey for data gathering, but to have PPI input into the creation of it, advice on how to conduct and disseminate it, and PPI reflections to help make sense of the results. The person who helped us to make community connections outside of the university found out about PINNT and recruited someone who was well connected to them and the PPI panel. This was particularly fortuitous for recruitment and dissemination for the project.

13.2.3 Case Study 3: Patient Perinatal Mental Health Journeys

Healthcare professionals and researchers in the mental health field have identified that adult mental health services were not adequately meeting the needs of patients experiencing perinatal mental health difficulties. A proposal was developed to recruit service users and capture their patient journeys to analyze where and how service provision could be improved for better patient experience. As the scoping research would delve into the sensitive topic of perinatal mental health, PPI was considered to be a vital element in ensuring that the focus of the enquiry, methods used, and outputs were appropriate to service users and the investigation focused on real-world issues.

To formally embed PPI, the yearlong project recruited a part-time lived experience researcher already employed by the academic institution and a Lived Experience Advisory Panel (LEAP). It was felt this would provide a balance of perspectives and collaboration in developing a research agenda which was fit for purpose and targeted efforts where most needed. The realities of effectively activating the LEAP and supporting the lived experience researcher during the funded period posed significant challenges. The PPI proposal perhaps raised ambitions beyond the point of what was feasible during the small-scale and minimally resourced project. To combat these issues, smaller projects such as this one can take a more ad hoc, informal approach, utilizing tokens of appreciation and goodwill gestures. Use of existing PPI advisory groups and remote/online activities rather than face-to-face methods is advocated to make the most of limited resources. Larger projects should formally calculate time and budget to enable a consistent PPI presence in all aspects of a research study, as one would calculate the time and effort required of any other research team member or participant. Flexibility and accessibility are key to determining how resources can be optimized to enable time for and extent of PPI in any given study. A rubric from PCORI provides practical examples of PPI activities that can support different stages of a research project (PCORI, 2015). The administration and logistics of effective PPI are not to be underestimated; for example, difficulties were experienced in the timeliness and arranging of routine LEAP meetings to inform various stages of the project. As the project was exploratory in nature, there was also the question of whether or not PPI representatives could also be active participants in the research. Ordinarily, PPI representatives would hold a completely separate role profile to that of a research participant. However, the nature of this study meant that PPI volunteers were also keen to contribute their experiences to the data gathered. This could be considered a conflict of interest, but those involved believed this joint role made their contributions doubly valuable. The project involved several researchers with no previous PPI experience and as such this project (while small-scale) required rapid learning and development of new skills to meet the PPI project plan. Advice to researchers embarking on their first projects including PPI would be:

- Utilize existing guidance from organizations, available through funding bodies and/or healthcare service providers (e.g., INVOLVE and PCORI).
- Talk to other researchers experienced in PPI in research. Seek informal mentorship to support and sense-check decision-making with more experienced persons.
- When there is little or no resource for PPI during scoping work or research ideation, be creative in ways to reimburse individuals. This could be providing access to reading material which might not be available to them, invitations to events (development or knowledge transfer), and simple gestures to demonstrate their contribution is valued and appreciated (e.g., meet at a café for a coffee and feedback status of research regardless of outcome).
- Approach PPI as a partnership.
- Work to understand the skills and expertise available from the research team and the PPI representatives. People involved in PPI have lives outside of that role, which can be useful to a project.

There are lessons to be learned in managing the scale of PPI so that it is proportionate to the scale of the research project. While good intentions might drive an ambitious PPI strategy which aims for the higher rungs on the involvement ladder (Figure 13.1), plans need to be feasible so that expectations can be met. When resources (time, manpower, and funding) are scarce, a pragmatic approach to PPI is required, examples of which are included in this chapter.

13.3 HOW PPI CAN POSITIVELY RESONATE WITH PATIENT ERGONOMICS PROJECTS

Data gathering and applying methods are often seen as the core of a research study. However, conducting any research project involves the interplay between many different project functions that occur upstream and downstream to these core activities (see Chapters 9–12 in this volume). For example, problems defining the right research question will have downstream implications, and disseminating the findings for impact might be hindered by upstream functions (e.g., Furniss et al., 2016a). We next describe how PPI can positively resonate with patient ergonomics projects.

13.3.1 Defining and Prioritizing Research Questions

In an ideal world, PPI should be an integral component of formative work to identify and develop research concepts. This is often difficult to achieve due to funding mechanisms which ultimately rely on the goodwill of PPI volunteers to support the development of research proposals. One notable exception is the approach taken and funds provided by a UK-based charity—The James Lind Alliance (JLA, 2020). The JLA brings patients, caregivers, and clinicians together in Priority Setting Partnership to identify and prioritize unanswered research questions or evidence uncertainties that they agree are the most important. The aim of this is to make sure health research funders are aware of the issues that matter most to the people who need to use the research in their everyday lives. This scheme is currently limited to healthcare-specific research domains.

PPI involvement can refine and change questions that drive research studies. For example, case study 3 exploring perinatal mental health journeys was derived as a concept from a research "sandpit" workshop in which the research questions were produced by an interdisciplinary team involving health service researchers, designers, and a clinician. It focused on how a design method (design card deck) could be used to examine problems within adult mental health services. The original questions posed by the research community were not wholly incorrect but the perspectives from which they were derived had their own bias. The focus of the enquiry changed following the inclusion of PPI input, whereby experiential accounts from service users encouraged researchers to "step back" from focusing on a specific design methodology as an intervention and move to holistically consider service provision as a temporal occurrence and thus shift the focus to examine these services in respect to interactions over time (see also Chapter 4 in this volume). This enabled the researchers to connect the project to existing work on "trajectories of user experience" (Benford & Giannachi, 2008) and modify the approach to reflect how service users perceived the research could evolve. As such, PPI at early prefunded stages can be important to provide a balance to perspective and ensure the proposed research addresses real-world problems experienced and gives a voice to those with lived experience.

At the mid-levels of the hierarchy of involvement, PPI can include defining and prioritizing questions within the bounds of a project that already has funding, direction, and focus. For example, on one of our projects, we needed to interview patients about their experiences of a service but we did not have predefined questions. Instead, the plan was to hold a PPI workshop to develop the questions we should be asking (Furniss et al., 2016b). This felt more meaningful and relevant.

13.3.2 Examining Ethical Issues

PPI input can contribute to ethical approval through protocol development. Practically, this can involve writing and reviewing project documentation for submission, writing a lay summary, and working through issues that come up in the design of the protocol or in the approval process. It is remarkable how writing in plain English can be challenging for academics. Talking to and writing with people outside of scientific circles can help clarify ideas and simplify their presentation. More concretely, for observational HFE studies, PPI representatives can highlight practical matters and how to tackle ethical issues. For example, a PhD student wanted approval to shadow pharmacists in a hospital to observe their communication and working patterns. The ethics committee said they were concerned the researcher might be inadvertently exposed to personal information and suggested obtaining consent from every patient prior to visiting a ward. This would have made her research unworkable. The issue was brought to a PPI panel established for another project for their opinion. The panel concurred the concerns were misplaced and the advice disproportionate to the risks. This was fed back to the committee who were happy with the panel's view.

13.3.3 Sensitizing Researchers to Issues

Furniss et al. (2016b) describe a PPI workshop primarily used to inform research questions on a patient safety project and to review patient documentation. However, an additional advantage was that this early exposure to patients and their stories made us more sensitive to their experiences and concerns, long before we conducted research in the field. Particularly important was understanding the sensitivities around discussing error and what could go wrong with patients as it could make them more concerned about their safety and feel more vulnerable. As a result, we tried to avoid terms such as "error" and "harm." The lesson learned was to be more emotionally attuned to patients, not just developing rules of how to act and what to say.

Similar experiences occurred throughout the STADIA project, a specific example being consideration of the topic of data linkage between current and future data sets by adult and teenage PPI groups. This complex issue for which a decision is required of participants in real time about potential, unknown future linking of personal data sets is particularly sensitive as it potentially involves not only trial data but also data from health and education institutes in the future. Although these data linkage approaches are increasingly common in research, PPI contributions have an important role to support and enhance communication about and dissemination of why these opportunities are important and how they might occur with minimum impact to participants.

13.3.4 Producing Project Documentation for Patients

Reviewing patient-facing documentation, such as patient information leaflets and consent forms, is standard in contemporary PPI. In addition, welcome packs for PPI collaborators (STADIA) and participants (Perinatal Mental Health Journeys) have been developed in partnership with PPI collaborators to support patient engagement regardless of role (participation or PPI). These packs provide a range of documents to support nonacademic research contributors in their roles and can include role profile; maps and general information; support and signposting to advice for PPI representatives; code of conduct; financial reimbursement information where appropriate;

research glossary (MindTech, 2018); and project lay summary. Development and use of these packs proved useful to ensure a baseline level of knowledge and support for those involved in PPI activities, with the benefit of standardizing PPI best practice within and between research projects.

13.3.5 Recruitment

PPI representatives can be well connected and embedded within communities of interest to the project, which can enable recruitment strategies. However, there are more novel ways in which PPI can support recruitment. STADIA PPI representatives and researchers codesigned role-plays and scenarios to assess potential applicants for researcher team positions. Role-plays were designed to assess candidates' ability to problem-solve, communicate, and empathize with potential study participants. This novel application of PPI enabled recruitment of the right person for a project investigating potentially sensitive issues. The authors are unaware of any similar approach to recruitment enhanced by PPI and believe this innovative application should be further explored.

13.3.6 Collecting Data

Garfield et al. (2015) report a study that explored the benefits and challenges of having laypeople collect data. They found that laypeople brought different perspectives to the research, which added value to the data collection and findings. However, there were pragmatic challenges to including laypeople, such as getting appropriate permissions to be on wards in this research capacity and training them in research procedures (e.g., taking consent). One might also imagine scenarios where a patient researcher with lived experience of relevance to the research enquiry might be able to build better rapport and common ground with a patient than a researcher without that experience.

13.3.7 Analyzing Data

Previous work (Jennings et al., 2018) has identified and explored best practices for collaborative data analysis (CDA) between researchers and PPI representatives—notably in mental health research. The methodology suggests how to integrate PPI into CDA and lists the four main characteristics of successful CDA: (1) coproduction; (2) realistic time and resource allocation; (3) manageable demands on CDA contributors; and (4) management of expectations and team dynamics. Our own experience also shows value in including PPI representatives in the analysis process as they can bring in new ideas, force clarity, and challenge and validate assumptions. This corroborates previous reports that lay members can enrich an analysis when they are involved (Garfield et al., 2016; Locock et al., 2019).

13.3.8 Planning Dissemination and Pathways to Impact

PPI representatives who are active in professional and more informal networks may be able to inform the research team about what messages might be receptive to whom, help design key messages at the end of a study, and provide a sounding board for what forms of dissemination may and may not work. One of our projects held a PPI workshop at the end of the project to work out different strategies for dissemination.

In today's technology-enabled society, there are many ways in which PPI work and approaches can be communicated and also how PPI can contribute to the dissemination of research more widely and publicly. Accessible outlets such as social media, layperson articles in the print, and online media provide a wealth of means for PPI to directly produce, disseminate, and engage with research outputs. Additionally, there are many more opportunities to produce creative content representing research practices and outputs. These include options such as video content, animations, and artistic visualizations. These can be used as modes of communication and enable engagement by people for whom traditional forms of communication might be more challenging or stressful. A key point here is for researchers to understand their PPI partners, their capabilities, and interests so that these can be tapped into and utilized to support communication of research.

13.4 CONCLUSION AND RECOMMENDATIONS

PPI advice is hard to capture in simple rules. Just as research projects need to be designed to suit the context and adapted when plans do not work out, so does PPI. PPI can aim to be an integral part of all stages of the research process, applied from concept development through impact and dissemination. Alternatively, it can perform a discrete role. However, more deeply it involves a change in mindset and philosophy about how research is conducted to ensure that projects are done with and for service users. In this chapter, we have defined and explained what PPI is, highlighted some of our experiences with PPI through case studies, and discussed how PPI can positively resonate to enhance patient ergonomics projects. We end with four recommendations for others who wish to start exploring PPI.

13.4.1 Plan: Design Appropriate PPI

Each research project has its own focus, budgets, timescales, and nuances, including attitudes to PPI in the project and the role that PPI should play (these attitudes will depend on the funding routes for the research and the experiences of researchers within those teams). These need to be considered. We saw in case study 2 that PPI was not originally included, leading to problems with the effective conduct of the study. At the other end of the spectrum, we saw in case study 3 that PPI plans were overly ambitious and disproportionate to the resources that were available for the project. PPI plans should be appropriate for the project. Researchers should think about appropriate compensation for the time people spend reading, reviewing, and attending meetings, using the hierarchy of involvement and timelines available to plan the depth, activities, and extent of PPI within different projects. For those just starting out, merely having a chat with PPI representatives over coffee could provide a low-cost but valuable opportunity for exchanging information. Reflexivity should also be part of the researcher's repertoire, to think about their own skills, networks, and experiences to support PPI. PPI does not have to happen all at once; it can develop between projects, as experiences are gained and relationships are built.

13.4.2 Prepare: Think about How to Facilitate PPI Once a Plan Has Formed

Patients and the public may be unfamiliar with the activities they are asked to do, so supporting them through the process can facilitate successful PPI. For example, we saw in the STADIA project that PPI welcome packs were created to give people a baseline understanding about expectations and how PPI activities and engagement would be organized within the project. For simpler PPI activities, an induction could be more appropriate, but even simpler things such as taking the time to listen to people's concerns and stories, being respectful, and building relationships are important. People with lived experience should not be perceived as a resource or commodity to further research endeavors; however, a partnership can lead to important insight, practical guidance, and a different way of perceiving research enquiries, offering significant benefits.

13.4.3 Adapt: Be Responsive and Flexible

Research and PPI plans do not always run as intended. Indeed, part of the rationale for PPI involvement is to challenge assumptions that might affect the course of the research. Adaptations can be called upon at a small-scale (e.g., when a PPI workshop takes a different route to that which was planned) or a large-scale (e.g., when the research questions that were driving the project are not the ones that are important to patients).

13.4.4 Learn: Reflect and Learn with People

PPI involvement is a different type of activity to regular recruitment of study participants. It takes time to become familiar with the different ways of working and partnering with PPI representatives. This opens up space for learning with people. PPI efforts and practices will not all go right, but if it is conducted in an open way, with integrity, then everyone can benefit. There will therefore be higher levels of learning at the project level to determine what worked well and why, as well as what could be improved and how, moving PPI strategies from the lower levels of the hierarchy of involvement toward the higher levels. This reflection can occur throughout the project and importantly should always include a "closing of the loop" at the end of a project with all PPI contributors so that they understand where, how, and why their contribution has had value and impact. There is an increasing trend for academic journals to now accept publications on PPI to articulate and spread this learning wider. We would recommend people familiarize themselves with this material so they can prepare for PPI.

ACKNOWLEDGMENTS

The authors acknowledge the contributions of all PPI representatives across all case studies. The STADIA trial (award ID 16/96/09) is funded by the National Institute for Health Research (NIHR). Patient Perinatal Mental Health Journeys was funded by the Nottingham Biomedical Research Centre.

REFERENCES

Alsaeed, D., Furniss, D., Blandford, A., Smith, F., & Orlu, M. (2018). Carers' experiences of home enteral feeding: A survey exploring medicines administration challenges and strategies. *Journal of Clinical Pharmacy and Therapeutics*, *43*(3), 359–365.

Arnstein, S. R. (1969). Eight rungs on the ladder of citizen participation. *Journal of the American Institute of Planners*.

Benford, S., & Giannachi, G. (2008). Temporal trajectories in shared interactive narratives. In Proceedings of the SIGCHI Conference on Human Factors in Computing Systems (CHI '08). Association for Computing Machinery, New York, NY, 73–82.

Brett, J., Staniszewska, S., Mockford, C., Herron-Marx, S., Hughes, J., Tysall, C., & Suleman, R. (2014). Mapping the impact of patient and public involvement on health and social care research: A systematic review. *Health Expectations*, *17*(5), 637–650.

Domecq, J. P., Prutsky, G., Elraiyah, T., Wang, Z., Nabhan, M., Shippee, N., & Erwin, P. (2014). Patient engagement in research: A systematic review. *BioMed Central Health Services Research*, *14*(1), 89.

Furniss, D., Curzon, P., & Blandford, A. (2016a). Using FRAM beyond safety: A case study to explore how sociotechnical systems can flourish or stall. *Theoretical Issues in Ergonomics Science*, *17*(5–6), 507–532.

Furniss, D., Iacovides, I., Lyons, I., Blandford, A., & Franklin, B. D. (2016b). Patient and public involvement in patient safety research: A workshop to review patient information, minimise psychological risk and inform research. *Research Involvement and Engagement*, *2*(1), 19.

Garfield, S., Jheeta, S., Husson, F., Jacklin, A., Bischler, A., Norton, C., & Franklin, B. D. (2016). Lay involvement in the analysis of qualitative data in health services research: A descriptive study. *Research Involvement and Engagement*, *2*(1), 29.

Garfield, S., Jheeta, S., Jacklin, A., Bischler, A., Norton, C., & Franklin, B. D. (2015). Patient and public involvement in data collection for health services research: A descriptive study. *Research Involvement and Engagement*, *1*(1), 8.

Harrison, J. D., Anderson, W. G., Fagan, M., Robinson, E., Schnipper, J., Symczak, G., . . . Duong, J. (2018). Patient and family advisory councils (PFACs): Identifying challenges and solutions to support engagement in research. *The Patient-Patient-Centered Outcomes Research*, *11*(4), 413–423.

Hart, R. A. 1992. "Children's participation: From tokenism to citizenship," Papers inness92/6, Innocenti Essay.

Holden, R. J., Schubert, C. C., & Mickelson, R. S. (2015). The patient work system: An analysis of self-care performance barriers among elderly heart failure patients and their informal caregivers. *Applied Ergonomics*, *47*, 133–150.

IEA. (2019). Retrieved from https://www.iea.cc/whats/index.html

JLA. (2020). Retrieved from http://www.jla.nihr.ac.uk/

Jennings, H., Slade, M., Bates, P., Munday, E., & Toney, R. (2018). Best practice framework for Patient and Public Involvement (PPI) in collaborative data analysis of qualitative mental health research: Methodology development and refinement. *BioMed Central Psychiatry, 18*, 213.

INVOLVE. (2019). Retrieved from https://www.invo.org.uk/find-out-more/what-is-public-involvement-in-research-2/

INOLVE. (2020). Retrieved from https://www.invo.org.uk/resource-centre/payment-and-recognition-for-public-involvement/involvement-cost-calculator/

Locock, L., Kirkpatrick, S., Brading, L., Sturmey, G., Cornwell, J., Churchill, N., & Robert, G. (2019). Involving service users in the qualitative analysis of patient narratives to support healthcare quality improvement. *Research Involvement and Engagement*, *5*(1), 1.

Marjanovic, S., Harshfield, A., Carpenter, A., Bertscher, A., Punch, D., & Ball, S. (2019). Involving patients and the public in research. *This.Institute*, 13.

Mindtech. (2018). Nottingham BRC mental health and technology theme jargon buster and glossary of terms. Retrieved from https://www.mindtech.org.uk/images/Jargon_Buster_sept18JW.pdf

Ocloo, J., & Matthews, R. (2016). From tokenism to empowerment: Progressing patient and public involvement in healthcare improvement. *The BMJ Quality & Safety*, *25*(8), 626–632.

PCORI. (2015). PCORI (Patient-Centred Outcomes Research Institute) engagement rubric. 15th October 2015. Retrieved from https://www.pcori.org/document/engagement-rubric

PCORI. (2019). Retrieved from https://www.pcori.org/about-us/our-programs/engagement/public-and-patient-engagement/value-engagement

Shippee, N. D., Domecq Garces, J. P., Prutsky Lopez, G. J., Wang, Z., Elraiyah, T. A., Nabhan, M.,. . . . Erwin, P. J. (2015). Patient and service user engagement in research: A systematic review and synthesized framework. *Health Expectations*, *18*(5), 1151–1166.

Valdez, R. S., Holden, R. J., Novak, L. L., & Veinot, T. C. (2014). Transforming consumer health informatics through a patient work framework: Connecting patients to context. *Journal of the American Medical Informatics Association*, *22*(1), 2–10.

Section V

Conclusion

14 Applying Human Factors and Ergonomics to Study and Improve Patient Work

Key Takeaways and Next Steps

Richard J. Holden
Indiana University School of Medicine

Rupa S. Valdez
University of Virginia

Alexandra R. Lang
University of Nottingham

CONTENTS

Historically, the formal discipline of human factors and ergonomics (HFE) was born in communities of practice concerned with improving the performance of military personnel (Meister, 1999). HFE subsequently expanded to support and protect workers operating machinery in industries such as manufacturing, aviation, power, telecommunications, and office work (Stuster, 2006). As the discipline evolved, it took root in the service industries and consumer products, directly benefiting the lay public through safer and more usable keyboards, theme parks, cars, computing devices, children's backpacks, websites, and more. The multitude and flexibility of HFE tools and methods made them ideal for broad use (Stanton & Young, 1999). Moreover, as a discipline suited to studying or solving problems involving human performance in any sociotechnical system, leaders in HFE have argued for its applicability to the full spectrum of social issues and settings (Roscoe et al., 2020; Thatcher et al., 2018).

14.1 HFE IN HEALTH CARE

The 1960s witnessed the first published examples of HFE in health care (Chapanis & Safrin, 1960; Hindle, 1968; Rappaport, 1970). Thereafter, growth spurts in certain areas such as anesthesiology (Gaba et al., 1995) and related inpatient settings (e.g., surgery) (Stone & McCloy, 2004) continued until the explosion of HFE in health care at the turn of the century due to increased public attention, funding, and regulation regarding patient safety and health information technology around the world (Carayon, 2007). Today, health care is perhaps the most prevalent application domain for HFE in economically developed countries. No doubt HFE has made an important contribution to promoting patient safety, efficient healthcare processes, more usable technologies, and better work environments for healthcare professionals. However, the majority (>75%) of published applications of HFE in the healthcare arena are focused on the work of professionals—physicians, nurses, pharmacists, etc.—and not on the work done by patients, families, and other nonprofessionals (Holden, Cornet, et al., 2020). Fortunately, growing recognition that "patient work" performed by nonprofessionals is pervasive and consequential is accompanied by increased application of what we call patient ergonomics, or *the science and engineering of patient work.*

14.2 TEN TAKEAWAY THEMES ABOUT PATIENT ERGONOMICS

Indeed, the patient ergonomics community of practice has been so productive that the present two-volume handbook on the subject evolved from an idea to a necessity. Volume I of *The Patient Factor* laid the groundwork for further growth in patient ergonomics by defining patient ergonomics and arguing the premise that HFE can improve patient work (Chapter 1, this volume); discussing and illustrating how cognitive, physical, and organizational ergonomics apply to patient work (Chapters 2–4); reviewing HFE for the patient work domains of technology use, communication, self-care, and patient safety (Chapters 5–8); and presenting a variety of methods for patient ergonomics (Chapters 9–13). These foundational chapters not only support future work but reveal key themes to take away about patient ergonomics.

14.2.1 Patient Work Is Jointly Cognitive, Physical, and Organizational

Patient work can be examined from the lens of cognitive ergonomics, physical ergonomics, and macroergonomics, but in reality, all three lenses are jointly required for a realistic view of patient work (see discussion in Chapter 3). For example, the work of self-administering oral medications has cognitive components such as deciding when to take a medication, remembering to take it, and attending to, perceiving, and processing visual information at the time of self-administration. However, medication taking is clearly also physical, requiring the use of physical containers, manipulating the pill (e.g., crushing, splitting and lubricating), and conveying it from container to mouth. Moreover, home medication use requires organizational processes for tracking, purchasing, storing, communicating about, and otherwise managing supply. These cognitive, physical, and organizational processes intersect, overlap, influence each other, and combine—sometimes indistinguishably—in all manners of patient work.

14.2.2 Patient Work Is a Journey

The above example of medication use depicts segments of activity, upon which one may easily expand to add activities such as adjustments to the medication regimen due to disease onset or progression, care transitions, symptoms or side effects, and changes in insurance plans or medication suppliers. Medication use can also involve processes for using tools such as pill organizers, reminder systems, informational websites, or online patient portals (see Chapter 5). Other examples and discussions (e.g., in Chapter 4) depict patient work as a journey, or a combination of interrelated processes comprised of interrelated tasks and interactions over time and place. This is true for acute illness journeys, in which people transition from good health to poor and seek to recover their health, and for chronic illness journeys. Either of these can be prolonged experiences characterized by either fluctuations or deterioration in health. In many cases, an HFE approach to patient work must adopt the journey as its unit of analysis, as rarely will the appropriate analysis be a single task or a single time period (e.g., detecting a piece of information on a medication label, swallowing the pill, or becoming infected with a virus). HFE expertise in the mapping of journeys and processes as a result offers unique value to parties seeking to understand or improve patient work journeys (Carayon et al., 2020).

14.2.3 Patient Work Occurs in Diverse Systems Contexts

The processes and journeys of patient work occur in what one might call real-world contexts, which is to say in sociotechnical systems of dynamically interacting components: task, person, tool, and environmental factors. These systems can be characterized and their relevant components measured to depict the reality of patient work and to set up interventions that will work in practice, not just under ideal circumstances (Werner et al., in press) (see e.g., Chapters 4, 5, and 7). To continue the medication example, understanding medication self-administration out of context might result in creating a smartphone app that issues a daily auditory alarm to remind a patient to take their morning pills. The reminder may theoretically solve the problem of forgetting, but does it take into account the patient's hearing or noise in their environment? Will it work if the patient shares their phone with a family member, has unreliable internet, or leaves the phone in their bedroom during morning bathroom use? Does the alarm help or harm in the case when the patient has already taken their pill on schedule? Does the alarm create embarrassment, privacy threats, or annoyance when the patient is in a social setting? Does the alarm diminish the patient's autonomy and cognitive skill? These and other questions reveal some of the many system factors that shape patient work performance and must be considered in the creation of interventions that support patient "work as done" rather than patient "work as imagined" (Braithwaite et al., 2016; Catchpole & Alfred, 2018).

It is also clear that the work systems in which patient work occurs are diverse (Yin et al., 2020) (see Volume II of this handbook). Not only are there a variety of people, tools, tasks, and environments that have been examined in the patient ergonomics literature, but the combination or configuration of these system components is considerably varied, as well. For example, the conditions shaping patient work—i.e., the patient work system (Holden, Schubert, & Mickelson, 2015; Holden et al., 2017)—can vary drastically for a high- versus low-resource environment, older versus younger patients, or pre- versus posthospitalization. In reality, each individual may have a varied configuration of their patient work system (Holden, Schubert, Eiland, et al., 2015), requiring interventions that are precisely responsive to their system: a sort of precision engineering in parallel to the precision medicine movement.

14.2.4 Patient Work Is Distributed

Across the literature on patient work and patient ergonomics, one of the strongest recurring themes is the distribution of patient work across people, artifacts, and settings. Almost every chapter in Volume I of this handbook addresses the distributed nature of patient work. Work distribution in part refers to the joint contribution to patient work by many actors, including patients, families, clinicians, and members of online or local communities. Although these actors do not always meet the classic requirements of teams (e.g., having a common goal and complementary expertise), they nevertheless perform activities together, in parallel, or in sequence that contribute to the patient's health-related outcomes. Patients and informal (or family) caregivers, in particular, are frequently found to share the "patient work," and in some cases, the unit performing work is the dyad, family, or network, not just the individual patient (Menefee et al., 2016; Valdez & Brennan, 2015). Patient–clinician collaboration is also fundamental to patient work, so much so that we define patient ergonomics as the application of HFE to study and improve work performed by nonprofessionals *either independent of or in concert with healthcare professionals* (see Chapter 1). An apparent misconception about patient work and patient ergonomics is that they ignore the healthcare professional; far from it, patient ergonomics seeks to support the work of both professionals and nonprofessionals, as long as patients or other nonprofessionals are recognized to contribute to it.

Other types of distribution are well described, including the use of artifacts such as tools and physical space to support patient work. An example of the former is patients' ubiquitous use of pen and ink to mark up printed information (e.g., annotating medical instructions or medication lists), write notes and reminders, and document events (e.g., weight or blood pressure logs and personal diaries). An example of the latter is organizing all of one's health-related objects in one "health workspace" (Holden et al., 2017, p. 31) or placing medications in a prominent place as a reminder to take them. The intelligent use of cognitive artifacts and space is well documented in HFE and related writings on professional work (Hollan et al., 2000; Norman, 1986; Xiao, 2005) and is without doubt present in patient work (Mickelson & Holden, 2017, 2018; Mickelson et al., 2015).

Work distribution also occurs across environments and, relatedly, clinical disciplines. HFE in health care has generally followed a path from inpatient (e.g., hospital) or similar (e.g., emergency room) settings to outpatient (e.g., primary care) settings and only recently to other places where health-related activities occur such as retail pharmacies, social care services, the home, and community. HFE has also recently been applied to care transitions, which by definition require attention to multiple settings and the unfolding of work over time. Patient ergonomics studies help expand the settings of interest because they typically focus on the patient, who traverses (and transitions between) many contexts and receives care from many formal and informal sources, from social work to nursing to community and spiritual services. However, even in patient ergonomics, some settings have been relatively neglected, compared to the patient home setting (see Volume II of this handbook).

14.2.5 Technologies to Aid Patient Work Must Be Designed and Tested for Usability

As in other domains of work, technology is a typical facet of patient work. If designed properly, technology can improve such work, but if not, it can contribute to additional burden, inefficiencies, and errors (see Chapters 5–7). For example, research on patient–professional communication shows technology can either facilitate or worsen communication and engagement, depending on how it is designed and used (Chapter 6). Thus, user-centered design and testing for usability and acceptance in context are important HFE approaches to avoid users rejecting the technology or experiencing worse performance. HFE methods are also needed to support successful and sustained implementation of technology for patient work, for example, helping to integrate technologies such as telehealth, online information sites, and mobile apps into the broader process of medical diagnosis (Chapters 5 and 7).

14.2.6 Patients Are Experts Capable of Contributing in Multiskilled Teams

Patients are central to many health- and healthcare-related processes, including activities such as possessing, acquiring, seeking, applying, and sharing information (Chapters 7 and 8). In many cases, patients are not only involved in these processes but also the sole, lead, or the most experienced actor. Take for example a patient with cancer, who may have unique information about how her body and mind respond to treatment or experience symptoms. This patient may also be the one constant across multiple care teams and settings, serving as the coordinator and conveyor of information and treatment plans across interactions with her clinicians. As a recipient of various therapies, she may have knowledge of them on par with a medical professional; as a member of online cancer patient and survivor communities, she may indeed know more than some professionals regarding the variety of experiences others have had. Expertise among patients, their families, and other nonprofessionals speaks to the need for respect and value placed on what they know and can contribute but should not diminish the expertise of clinical professionals. Instead, the presence of expertise among multiple parties, especially when each has unique or specialized expertise, strongly argues for the inclusion of both professionals and nonprofessionals in multiskilled teams. However, how best to design teams composed of professionals and nonprofessionals requires further work.

14.2.7 Traditional HFE Methods Can (and Must!) Be Adapted to Patient Work

Many methods and tools are catalogued in HFE publications and toolkits (Gawron, 2019; Nemeth, 2004; Stanton et al., 2013; Wilson & Corlett, 2005). One can apply standard HFE methods to perform field research (e.g., Chapter 9), develop and evaluate innovative interventions (e.g., Chapter 10), or conduct experimental studies (e.g., Chapter 11). Novel methods have also emerged for using portable sensors, the Internet of Things, virtual reality, and gamification for patient work (e.g., Chapter 12). These methods can all be applied, and there are accruing instructions and examples for doing so. Just as importantly, methods that were once designed for and tested with professionals (e.g., nurses, pilots) or with healthier nonprofessionals (e.g., college students) are being adapted to accommodate the unique characteristics of patient work, the people who perform it, and their environments. For example, privacy and logistic issues related to the home-based and round-the-clock nature of diabetes self-management work require several adjustments for an observational field study (Yin et al., 2018) (cf. Chapter 9). Likewise, cognitive or visual disability among certain populations necessitates simplifying procedures, for example, eliciting data for cognitive task analysis (Holden, Daley, et al., 2020) or usability assessment (Holden, 2020) (cf. Chapter 10).

The adapted methods must also accommodate the range of people and populations, including and especially those who are hard to reach or marginalized. These groups are ironically both at higher risk and less likely to be the target of research and interventions (Cheraghi-Sohi et al., 2020). Patient ergonomics professionals therefore need to be aware of the representativeness of their populations and to be more inclusive, even at the expense of ease and expedience. This means including individuals who may be otherwise excluded due to clinical conditions such as diagnosis of mental illness, social or economic conditions such as education and literacy, and logistic issues such as email and internet use. Excluding people because of inattention to these factors during methods design, recruitment, or data analysis can produce an inaccurate picture of health and interventions that benefit the "haves" and may even harm the "have-nots" (Veinot et al., 2018).

14.2.8 Applying Patient Ergonomics Methods Has Ethical Implications

Just as important as selecting and adapting methods for patient work is being able to apply them in practice. Application involves navigating ethical considerations. One issue is whether studies of patient work can be performed with nonpatients, for example, healthy college students performing a simulated self-care task or using a health website. Methods that are adapted or imported from other fields or from work with other populations raise additional questions. A method for video recording professionals doing physical tasks (e.g., for physical ergonomics risk assessment) takes on different implications when the participant is in their home, self-conscious about an overt physical disability, or not fully able to provide consent, assent, or dissent to be recorded. Patient ergonomics professionals must also grapple with the level and nature of stakeholder involvement (e.g., Chapter 13), especially when the purpose of the project is creating an intervention for these stakeholders. Participatory methods may be warranted, wherein patients and other stakeholders serve as study investigators, codesigners, advisors, citizen scientists, or in other roles (Petersen et al., 2020). The notion of participatory patient ergonomics is not new to HFE (Haines et al., 2002; Noro & Imada, 1991) but is attended by several ethical, practical, and scientific challenges (see Chapter 13). For example, factory workers participating on a project to redesign factory workflow might be able to do so during paid work hours at their work site, and their input could bear timely, direct benefits. A patient with late-stage cancer participating in the redesign of outpatient therapy workflow might do so on their own time, incur inconveniences due to transportation or childcare costs, and die before the new workflow is in place.

14.2.9 Applying Patient Ergonomics Methods Has Practical Implications

Several publications have described the practical challenges of applying patient ergonomics, especially with vulnerable populations and in home and community settings (Blandford et al., 2016; Holden, McDougald Scott, et al., 2015; Valdez & Holden, 2016). Other work provides advice for applying user-centered and participatory design in these contexts (Cornet et al., 2019, 2020; Holden, Toscos, et al., 2020) (see Chapters 10 and 13). Frequently cited practical issues include accessing patients and other nonprofessionals, especially when seeking a diverse and representative sample; building trust and long-term relationships; accommodating the schedules, needs, preferences, and abilities of participants; and managing privacy and safety concerns for both participants and the project team (Valdez & Edmunds, 2020). Fortunately, the practical challenges described are receiving more attention from scholars and practitioners, who increasingly document their own joys and struggles.

14.2.10 Patient Ergonomics Is about Both the Basic and Applied Science of Patient Work

The final theme emerging from the first volume on patient ergonomics is the dual focus on basic discovery of how things are versus the engineering of systems to support patient work. We previously defined patient ergonomics as the science and engineering of patient work (Chapter 1). This was intentionally done to highlight the ultimate goal of patient ergonomics, namely to support or improve patient work in a way that improves outcomes for patients and others. We have reviewed elsewhere the tendency of patient ergonomics research to describe and study patient work, using HFE methods to discover how the work is performed, by whom, with what tools, and so on (Holden, Cornet, et al., 2020). At the same time, many case exemplars in this volume, and a growing number in general, describe practical applications of HFE to create tools, train or educate patients, or otherwise redesign the work system to facilitate the performance of patient work. We therefore remain hopeful that the patient ergonomics community of practice will retain its orientation on both the basic and applied science of patient work.

14.3 NEXT STEPS IN PATIENT ERGONOMICS

What comes next for the community of practice we call patient ergonomics? It would appear patient ergonomics is growing in volume, expanding in scope, and otherwise maturing. Perhaps, it is even gaining a foothold within related scientific disciplines or practice-oriented communities concerned with policy, advocacy, or technology to support patients. For now, patient ergonomics is not a household term and the concept it represents—the science and engineering of patient work—might seem foreign or difficult to reconcile with related concepts. Patient ergonomics is not alone in seeking to establish itself in a crowded market. Rather than wait for patient ergonomics to be discovered, which will depend on extraordinary good fortune, it may be time to promote patient ergonomics as an innovation poised to deliver on a credible promise (cf. Carayon, 2010). Experts in product development and sales would observe that patient ergonomics needs a strong brand, a convincing value proposition, and a cadre of marketing and salespeople working to raise awareness and convert interest into action. If successful in this way, we shall soon see scientists and, better still, nonscientists demanding consulting services, methods, and case studies on patient ergonomics—or *PxErgo*, as the kids will call it. The vision for a world made better by patient ergonomics must surely include the voice of the patient and other stakeholders, calling on health systems, technology vendors, and regulators to make patient work easier, technologies more usable, and processes more human-centered. When the demand is there, the patient ergonomics community must be prepared to show why it is perfectly suited to fulfill it.

As to the more proximate next steps, we turn to Volume II of this handbook. Whereas the first volume provided a theoretical and methodological foundation for patient ergonomics, the second is about context and application. Volume II addresses the diversity of places and populations in the literature on patient ergonomics, consistent with the above observation that patient work occurs in diverse contexts. These application contexts include the emergency room, care transitions, home and community settings, online communities, veterans, pediatrics, older adults, underserved populations, and health seekers. The last of these, health seekers, represent a domain of health relatively untouched by HFE (Holden, Cornet, et al., 2020), namely health "beyond disease" (Holden & Valdez, 2019). Arguably, patient ergonomics can have its biggest impact by addressing not only the work and care of individuals with acute or chronic conditions, but also individuals and groups engaged in activities promoting health, primary or secondary prevention, and well-being.

The expanding scope of patient ergonomics and the many opportunities in this space suggest one other future direction: multidisciplinarity. Patient ergonomics already boasts multidisciplinary roots, as a community of practice overlapping with human factors engineering, psychology, health informatics, user-centered design, nursing, public health, various social sciences, and other disciplines. This multidisciplinarity is apparent in the backgrounds of contributors to patient ergonomics, the methods and theories they use, and the topics they address (e.g., helping the underserved, aging, cognitive and physical performance, and organizational design). Even so, there is room for growth, to incorporate the values, theories, methods, and priorities of other scientific and practical disciplines into patient ergonomics. Patient ergonomics needs contributions from across the clinical specialties, from medicine, nursing, and pharmacy to social work, occupational therapy, and public health. Patient ergonomics also needs partners who can help address levels of analysis at the level of society, government, and community, including experts in public policy, cultural studies, political science, global and population health, business development, commissioning and procurement, and more.

Lastly, patient ergonomics must enhance its real and perceived practical value to others. It must become—and be seen as—the "right tool for the job" anytime a need arises to study or improve the health-related activities of patients or other nonprofessionals. If so, the best days are ahead for the maturing patient ergonomics community of practice!

REFERENCES

Blandford, A., Furniss, D., & Makri, S. (2016). Qualitative HCI research: Going behind the scenes. *Synthesis Lectures on Human-Centered Informatics, 9*(1), 1–115.

Braithwaite, J., Wears, R. L., & Hollnagel, E. (Eds.). (2016). *Resilient health care, volume 3: Reconciling work-as-imagined and work-as-done*. Boca Raton, FL: CRC Press.

Carayon, P. (2010). Human factors in patient safety as an innovation. *Applied Ergonomics, 41*(5), 657–665.

Carayon, P. (Ed.) (2007). *Handbook of human factors and ergonomics in patient safety*. Mahwah, NJ: Lawrence Erlbaum.

Carayon, P., Wooldridge, A., Hoonakker, P., Hundt, A. S., & Kelly, M. M. J. A. E. (2020). SEIPS 3.0: Human-centered design of the patient journey for patient safety. *Applied Ergonomics, 84*, 103033.

Catchpole, K., & Alfred, M. (2018). Industrial conceptualization of health care versus the naturalistic decision-making paradigm: Work as imagined versus work as done. *Journal of Cognitive Engineering and Decision Making, 12*(3), 222–226.

Chapanis, A., & Safrin, M. A. (1960). Of misses and medicines. *Journal of Chronic Diseases, 12*, 403–408.

Cheraghi-Sohi, S., Panagioti, M., Daker-White, G., Giles, S., Riste, L., Kirk, S., . . . Sanders, C. (2020). Patient safety in marginalised groups: A narrative scoping review. *International Journal for Equity in Health, 19*, 26.

Cornet, V. P., Daley, C., Bolchini, D., Toscos, T., Mirro, M. J., & Holden, R. J. (2019). Patient-centered design grounded in user and clinical realities: Towards valid digital health. In the *Proceedings of the International Symposium on Human Factors and Ergonomics in Health Care, 8*(1), 100–104.

Cornet, V. P., Toscos, T., Bolchini, D., Ghahari, R. R., Ahmed, R., Daley, C., . . . Holden, R. J. (2020). Untold stories in user-centered design of mobile health: Practical challenges and strategies learned from the design and evaluation of an app for older adults with heart failure. *Journal of Medical Internet Research mHealth and uHealth, 8*(7), e17703.

Gaba, D. M., Howard, S. K., & Small, S. D. (1995). Situation awareness in anesthesiology. *Human Factors, 37*(1), 20–31.

Gawron, V. J. (2019). *Human performance, workload, and situational awareness measures handbook* (3rd ed.). Boca Raton, FL: CRC Press.

Haines, H., Wilson, J. R., Vink, P., & Koningsveld, E. (2002). Validating a framework for participatory ergonomics (the PEF). *Ergonomics, 45*, 309–327.

Hindle, A. (1968). *A systems approach to hospital management.* Paper presented at the Ergonomics Research Society Annual Conference.

Holden, R. J. (2020). A simplified system usability scale (SUS) for cognitively impaired and older adults. In the *Proceedings of the International Symposium on Human Factors and Ergonomics in Health Care, 9*(1), 180–182.

Holden, R. J., Cornet, V. P., & Valdez, R. S. (2020). Patient ergonomics: 10-year mapping review of patient-centered human factors. *Applied Ergonomics, 82*, 102972.

Holden, R. J., Daley, C. N., Mickelson, R. S., Bolchini, D., Toscos, T., Cornet, V. P., . . . Mirro, M. J. (2020). Patient decision-making personas: An application of a patient-centered cognitive task analysis (P-CTA). *Applied Ergonomics, 87*, 103107.

Holden, R. J., McDougald Scott, A. M., Hoonakker, P. L. T., Hundt, A. S., & Carayon, P. (2015). Data collection challenges in community settings: Insights from two field studies of patients with chronic disease. *Quality of Life Research, 24*, 1043–1055.

Holden, R. J., Schubert, C. C., Eiland, E. C., Storrow, A. B., Miller, K. F., & Collins, S. P. (2015). Self-care barriers reported by emergency department patients with acute heart failure: A sociotechnical systems-based approach. *Annals of Emergency Medicine, 66*, 1–12.

Holden, R. J., Schubert, C. C., & Mickelson, R. S. (2015). The patient work system: An analysis of self-care performance barriers among elderly heart failure patients and their informal caregivers. *Applied Ergonomics, 47*, 133–150.

Holden, R. J., Toscos, T., & Daley, C. N. (2020). Researcher reflections on human factors and health equity. In R. Roscoe, E. K. Chiou, & A. R. Wooldridge (Eds.), *Advancing diversity, inclusion, and social justice through human systems engineering* (pp. 51–62). Boca Raton, FL: CRC Press.

Holden, R. J., & Valdez, R. S. (2019). Beyond disease: Technologies for health promotion. In the *Proceedings of the International Symposium on Human Factors and Ergonomics in Health Care, 8*(1), 62–66.

Holden, R. J., Valdez, R. S., Schubert, C. C., Thompson, M. J., & Hundt, A. S. (2017). Macroergonomic factors in the patient work system: Examining the context of patients with chronic illness. *Ergonomics, 60*(1), 26–43.

Hollan, J., Hutchins, E., & Kirsh, D. (2000). Distributed cognition: Toward a new foundation for human-computer interaction research. *Association for Computing Machinery Transactions on Computer-Human Interaction, 7*, 174–196.

Hutchins, E. (1995). *Cognition in the wild.* Cambridge, MA: MIT Press.

Meister, D. (1999). *The history of human factors and ergonomics*. Boca Raton, FL: CRC Press.

Menefee, H. K., Thompson, M. J., Guterbock, T. M., Williams, I. C., & Valdez, R. S. (2016). Mechanisms of communicating health information through Facebook: Implications for consumer health information technology design. *Journal of Medical Internet Research, 18*(8), e218.

Mickelson, R. S., & Holden, R. J. (2017). Capturing the medication management work system of older adults using a digital diary method. In the *Proceedings of the Human Factors and Ergonomics Society Annual Meeting, 61*(1), 555–559.

Mickelson, R. S., & Holden, R. J. (2018). Medication management strategies used by older adults with heart failure: A systems-based analysis. *European Journal of Cardiovascular Nursing, 17*, 418–428.

Mickelson, R. S., Willis, M., & Holden, R. J. (2015). Medication-related cognitive artifacts used by older adults with heart failure. *Health Policy & Technology, 4*, 387–398.

Nemeth, C. P. (2004). *Human factors methods for design: Making systems human-centered*. Boca Raton, FL: CRC Press.

Norman, D. A. (1986). Cognitive engineering. In D. A. Norman & S. W. Draper (Eds.), *User centered system design: New perspectives on human-computer interaction* (pp. 31–61). Hillsdale, NJ: Lawrence Erlbaum.

Noro, K., & Imada, A. S. (Eds.). (1991). *Participatory ergonomics*. London: Taylor & Francis.

Petersen, C., Austin, R. R., Backonja, U., Campos, H., Chung, A. E., Hekler, E. B., . . . Salmi, L. (2020). Citizen science to further precision medicine: From vision to implementation. *Journal of the American Medical Informatics Association Open, 3*(1), 2–8.

Rappaport, M. (1970). Human factors applications in medicine. *Human Factors, 12*(1), 25–35.

Roscoe, R. D., Chiou, E. K., & Wooldridge, A. R. (2020). *Advancing diversity, inclusion, and social justice through human systems engineering*. Boca Raton, FL: CRC Press.

Stanton, N. A., Salmon, P. M., Rafferty, L. A., Walker, G. H., Baber, C., & Jenkins, D. P. (2013). *Human factors methods: A practical guide for engineering and design*. Surrey, UK: Ashgate.

Stanton, N. A., & Young, M. S. (1999). What price ergonomics? *Nature, 399*(6733), 197–198.

Stone, R., & McCloy, R. (2004). Ergonomics in medicine and surgery. *The BMJ, 328*(7448), 1115–1118.

Stuster, J. (Ed.) (2006). *The human factors and ergonomics society: Stories from the first 50 years*. Santa Monica, CA: Human Factors and Ergonomics Society.

Thatcher, A., Waterson, P., Todd, A., & Moray, N. (2018). State of science: Ergonomics and global issues. *Ergonomics, 61*(2), 197–213.

Valdez, R. S., & Brennan, P. F. (2015). Exploring patients' health information communication practices with social network members as a foundation for consumer health IT design. *International Journal of Medical Informatics, 84*, 363–374.

Valdez, R. S., & Edmunds, D. S. (2020). Reimagining community-based research and action in human factors: A dialogues across disciplines. In R. Roscoe, E. K. Chiou, & A. R. Wooldridge (Eds.), *Advancing diversity, inclusion, and social justice through human systems engineering* (pp. 267–276). Boca Raton, FL: CRC Press.

Valdez, R. S., & Holden, R. J. (2016). Health care human factors/ergonomics fieldwork in home and community settings. *Ergonomics in Design, 24*, 44–49.

Veinot, T. C., Mitchell, H., & Ancker, J. S. (2018). Good intentions are not enough: How informatics interventions can worsen inequality. *Journal of the American Medical Informatics Association, 25*(8), 1080–1088.

Werner, N. E., Ponnala, S., Doutcheva, N., & Holden, R. J. (in press). Human factors/ergonomics work system analysis of patient work: State of the science and future directions. *International Journal for Quality in Health Care*. doi:10.1093/intqhc/mzaa099

Wilson, J. R., & Corlett, N. (Eds.). (2005). *Evaluation of human work* (3rd ed.). Boca Raton, FL: CRC Press.

Xiao, Y. (2005). Artifacts and collaborative work in healthcare: Methodological, theoretical, and technological implications of the tangible. *Journal of Biomedical Informatics, 38*, 26–33.

Yin, K., Harms, T., Ho, K., Rapport, F., Vagholkar, S., Laranjo, L., . . . Lau, A. Y. (2018). Patient work from a context and time use perspective: a mixed-methods study protocol. *The BMJ Open, 8*(12), e022163.

Yin, K., Jung, J., Coiera, E., Laranjo, L., Blandford, A., Khoja, A., . . . Lau, A. Y. (2020). Patient work and their contexts: Scoping review. *Journal of Medical Internet Research, 22*(6), e16656.

Index

Note: **Bold** page numbers refer to tables and *italic* page numbers refer to figures.